Lecture Notes in Physics
Monographs

Springer
Berlin
Heidelberg
New York
Barcelona
Hong Kong
London
Milan
Paris
Singapore
Tokyo

Physics and Astronomy

ONLINE LIBRARY

http://www.springer.de/phys/

The Editorial Policy for Monographs

The series Lecture Notes in Physics reports new developments in physical research and teaching - quickly, informally, and at a high level. The type of material considered for publication in the monograph Series includes monographs presenting original research or new angles in a classical field. The timeliness of a manuscript is more important than its form, which may be preliminary or tentative. Manuscripts should be reasonably self-contained. They will often present not only results of the author(s) but also related work by other people and will provide sufficient motivation, examples, and applications.

The manuscripts or a detailed description thereof should be submitted either to one of the series editors or to the managing editor. The proposal is then carefully refereed. A final decision concerning publication can often only be made on the basis of the complete manuscript, but otherwise the editors will try to make a preliminary decision as definite as they can on the basis of the available information.

Manuscripts should be no less than 100 and preferably no more than 400 pages in length. Final manuscripts should be in English. They should include a table of contents and an informative introduction accessible also to readers not particularly familiar with the topic treated. Authors are free to use the material in other publications. However, if extensive use is made elsewhere, the publisher should be informed. Authors receive jointly 30 complimentary copies of their book. They are entitled to purchase further copies of their book at a reduced rate. No reprints of individual contributions can be supplied. No royalty is paid on Lecture Notes in Physics volumes. Commitment to publish is made by letter of interest rather than by signing a formal contract. Springer-Verlag secures the copyright for each volume.

The Production Process

The books are hardbound, and quality paper appropriate to the needs of the author(s) is used. Publication time is about ten weeks. More than twenty years of experience guarantee authors the best possible service. To reach the goal of rapid publication at a low price the technique of photographic reproduction from a camera-ready manuscript was chosen. This process shifts the main responsibility for the technical quality considerably from the publisher to the author. We therefore urge all authors to observe very carefully our guide-lines for the preparation of camera-ready manuscripts, which we will supply on request. This applies especially to the quality of figures and halftones submitted for publication. Figures should be submitted as originals or glossy prints, as very often Xerox copies are not suitable for reproduction. For the same reason, any writing within figures should not be smaller than 2.5 mm. It might be useful to look at some of the volumes already published or, especially if some atypical text is planned, to write to the Physics Editorial Department of Springer-Verlag direct. This avoids mistakes and time-consuming correspondence during the production period.

As a special service, we offer free of charge LATEX and TEX macro packages to format the text according to Springer-Verlag's quality requirements. We strongly recommend authors to make use of this offer, as the result will be a book of considerably improved technical quality.

For further information please contact Springer-Verlag, Physics Editorial Department II, Tiergartenstrasse 17, D-69121 Heidelberg, Germany.

Series homepage – http://www.springer.de/phys/books/lnpm

Volker Perlick

Ray Optics,
Fermat's Principle,
and Applications
to General Relativity

 Springer

Author

Volker Perlick
Technische Universität Berlin
Sekr. PN 7-1
Hardenbergstrasse 36
10623 Berlin, Germany

Library of Congress Cataloging-in-Publication Data

Perlick, Volker, 1956-
   Ray optics, Fermat's principle, and applications to general relativity / c Volker Perlick.
     p. cm. -- (Lecture notes in physics. Monographs, ISSN 0940-7677 ; v. m61)
   Includes bibliographical references and index.
   ISBN 3540668985 (alk. paper)
     1. Light--Transmission--Mathematical models. 2. Maxwell equations. 3. General
   relativity (Physics) I. Title. II. Lecture notes in physics. New series m, Monographs ;
   m61.

   QC389 .P37 2000
   523.01'53--dc21

                                                                        99-089303

ISSN 0940-7677 (Lecture Notes in Physics. Monographs)
ISBN 3-540-66898-5 Springer-Verlag Berlin Heidelberg New York

© Springer-Verlag Berlin Heidelberg 2000
Printed in Germany

The use of general descriptive names, registered names, trademarks, etc. in this publication does not imply, even in the absence of a specific statement, that such names are exempt from the relevant protective laws and regulations and therefore free for general use.

Typesetting: Camera-ready by the author
Cover design: *design & production*, Heidelberg

Printed on acid-free paper
SPIN: 10644482   55/3144/du - 5 4 3 2 1 0

Preface

All kind of information from distant celestial bodies comes to us in the form of electromagnetic radiation. In most cases the propagation of this radiation can be described, as a reasonable approximation, in terms of rays. This is true not only in the optical range but also in the radio range of the electromagnetic spectrum. For this reason the laws of ray optics are of fundamental importance for astronomy, astrophysics, and cosmology.

According to general relativity, light rays are the light-like geodesics of a Lorentzian metric by which the spacetime geometry is described. This, however, is true only as long as the light rays are freely propagating under the only influence of the gravitational field which is coded in the spacetime geometry. If a light ray is influenced, in addition, by an optical medium (e.g., by a plasma), then it will not follow a light-like geodesic of the spacetime metric. It is true that for electromagnetic radiation traveling through the universe usually the influence of a medium on the path of the ray and on the frequency is small. However, there are several cases in which this influence is very well measurable, in particular in the radio range. For example, the deflection of radio rays in the gravitational field of the Sun is considerably influenced by the Solar corona. Moreover, current and planned Doppler experiments with microwaves in the Solar system reach an accuracy in the frequency of $\Delta \omega / \omega \cong 10^{-15}$ which makes it necessary to take the influence of the interplanetary medium into account. Finally, even in cases where the quantitative influence of the medium is negligibly small it is interesting to ask in which way the qualitative aspects of the theory are influenced by the medium. The latter remark applies, in particular, to the intriguing theory of gravitational lensing.

Unfortunately, general-relativistic light propagation in media is not usually treated in standard textbooks, and the more specialized literature is concentrated on particular types of media and on particular applications rather than on general methodology. In this sense a comprehensive review of general-relativistic ray optics in media would fill a gap in the literature. It is the purpose of this monograph to provide such a review.

Actually, this monograph grew out of a more special idea. It was my original plan to write a review on variational principles for light rays in general relativistic media, and to give some applications to astronomy and astrophysics, in particular to the theory of gravitational lensing. However, I soon realized the necessity of precisely formulating the mathematical theory of light rays in general before I could tackle the question of whether these light rays are characterized by a variational principle. The sections on variational principles and on applications are now at the end of Part II, in which a general mathematical framework for ray optics is set up. This is written in the language of symplectic geometry, thereby elucidating the well-known analogy between ray optics and the phase-space formulation of classical mechanics.

Moreover, I found it desirable to also treat the question of how to derive ray optics as an approximation scheme from Maxwell's equations. This is the topic of Part I which serves the purpose of physically motivating the fundamental definitions of Part II. In vacuo, the passage from Maxwell's equations to ray optics is, of course, an elementary textbook matter and the generalization to isotropic and non-dispersive media is quite straightforward. However, for anisotropic and/or dispersive media this passage is more subtle. In Part I two types of media are discussed in detail, viz., an anisotropic one and a dispersive one, and the emphasis is on general methodology.

I have organized the material in such a way that it should be possible to read Part II without having read Part I. This is not recommended, of course, but the reader might wish to do so. Both parts begin with an introductory section containing a brief guide to the literature and a statement of assumptions and notations used throughout. Whenever the reader feels that a symbol needs explanation or that the underlying assumptions are not clearly stated, he or she should consult the introductory section of the respective part. Also, the index might be of help if problems of that kind occur.

Large parts of this monograph present material which, in essence, is not new. However, I hope that the formulation chosen here might give some new insight. As to Part I, our discussion of the passage from Maxwell's equations to ray optics includes several mathematical details which are difficult to find in the literature, although the general features are certainly known to experts. To mention just one example, it is certainly known to experts that in a linear but anisotropic medium on a general-relativistic spacetime the light rays are determined by two "optical Finsler metrics"; to the best of my knowledge, however, a full proof of this fact is given here for the first time. As to Part II, the basic formalism is just the 170-year-old Hamiltonian optics, rewritten in modern mathematical terminology and adapted to the framework of general relativity. However, the presentation is based on some general mathematical definitions which have not been used before. This remark applies, in particular, to Definition 5.1.1, which is the definition of what I call "ray-optical structures". This definition formalizes the widely accepted idea that all of ray optics can be derived from a "dispersion relation". (The term "ray system" is sometimes used by Vladimir Arnold and his collaborators in a similar though not quite identical sense.) It also applies, e.g., to Definition 5.4.1, on "dilation-invariant" ray optical structures, which characterizes dispersion-free media in a geometric way.

On the other hand, I want to direct the reader's attention to the fact that this monograph contains some particular results which, as far as I know, have not been known before. These include, e.g.:

- the general redshift formula for light rays in media on a general-relativistic spacetime in Sect. 6.2;
- the results on light bundles in isotropic non-dispersive media on a general-relativistic spacetime in Sect. 6.4, in particular the generalized "reciprocity theorem" (Theorem 6.4.3);
- Theorem 7.3.1, which can be viewed as a version of Fermat's principle for light rays in (possibly anisotropic and dispersive) media on general-relativistic spacetimes;
- Theorem 7.5.4, which generalizes the "Morse index theorem" of Riemannian geometry to the case of light rays in stationary media on stationary general-relativistic spacetimes.

Some of the questions raised in this monograph remain unanswered, i.e., to some extent this is an interim report on work in progress. In particular, this remark applies to the following two special issues. (a) In Part I we are able to prove that for the linear medium treated in Chap. 2 ray optics is associated with approximate solutions of Maxwell's equations, i.e., that ray optics gives a viable approximation scheme for electromagnetic radiation. Unfortunately, we are not able to prove a similar result for the plasma model of Chap. 3. This is a gap which should be filled in the future. (b) In Part II we are able to establish a Morse index theorem for light rays in stationary media. However, it is still an open question whether these results can be generalized to the non-stationary case in which, up to now, a Morse theory exists only for vacuum rays. With Fermat's principle in the form of Theorem 7.3.1 we have a starting point for setting up a Morse theory for light rays in arbitrary (non-stationary) media. This is an interesting problem to be tackled in future work.

This monograph in its present form is a slightly revised version of my *Habilitation* thesis. I would like to use this opportunity to thank the members of the Habilitation Committee, Karl-Eberhard Hellwig, Erwin Sedlmayr, Bernd Wegner, John Beem, Friedrich Wilhelm Hehl, and Gernot Neugebauer, for their interest in this work and for several useful comments. In particular, I would like to thank Bernd Wegner for paving the way to having this text published with Springer Verlag.

While working at this monograph I have profited from many discussions, in particular with my academic teacher Karl-Eberhard Hellwig and his collaborators at the Technical University in Berlin, but also with other colleagues. Special thanks are due to Wolfgang Hasse and Marcus Kriele for collaboration on various aspects of light propagation in general relativity; to Wolfgang

Rindler for hospitality at the University of Texas at Dallas and for discussions on the fundamentals of general relativity; to John Beem for hospitality at the University of Missouri at Columbia and for discussions on Lorentzian geometry; to Gernot Neugebauer and his collaborators for hospitality at the University of Jena and for discussions on various aspects of general relativity; to Paolo Piccione, Fabio Giannoni, and Antonio Masiello for hospitality during several visits to Italy and to Brazil and for collaboration on Morse theory; and to Jürgen Ehlers and Arlie Petters for fruitful discussions on Fermat's principle and gravitational lensing. Also, I have enjoyed discussions on this subject with students during seminars and classes in Berlin, Osnabrück, and São Paulo.

Finally, I am grateful to the Deutsche Forschungsgemeinschaft for sponsoring this work with a *Habilitation* stipend, and to the Wigner Foundation, to the Deutscher Akademischer Austauschdienst, and to the Fundação de Amparo á Pesquisa do Estado de São Paulo for financially supporting my visits to Dallas, Columbia, and São Paulo.

Berlin, August 1999   *Volker Perlick*

Contents

Part I. From Maxwell's equations to ray optics

1. Introduction to Part I 3
 1.1 A brief guide to the literature 3
 1.2 Assumptions and notations 5

2. Light propagation in linear dielectric and permeable media 7
 2.1 Maxwell's equations in linear dielectric and permeable media . 7
 2.2 Approximate-plane-wave families 14
 2.3 Asymptotic solutions of Maxwell's equations 17
 2.4 Derivation of the eikonal equation and transport equations .. 19
 2.5 Discussion of the eikonal equation 24
 2.6 Discussion of transport equations and the introduction of rays 31
 2.7 Ray optics as an approximation scheme 36

3. Light propagation in other kinds of media 43
 3.1 Methodological remarks on dispersive media 44
 3.2 Light propagation in a non-magnetized plasma 46

Part II. A mathematical framework for ray optics

4. Introduction to Part II 61
 4.1 A brief guide to the literature 61
 4.2 Assumptions and notations 63

5. Ray-optical structures on arbitrary manifolds 67
 5.1 Definition and basic properties of ray-optical structures 67
 5.2 Regularity notions for ray-optical structures 76
 5.3 Symmetries of ray-optical structures 82
 5.4 Dilation-invariant ray-optical structures 87
 5.5 Eikonal equation 92
 5.6 Caustics... 100

X   Contents

6. **Ray-optical structures on Lorentzian manifolds** 111
    6.1 The vacuum ray-optical structure 111
    6.2 Observer fields, frequency, and redshift 113
    6.3 Isotropic ray-optical structures 120
    6.4 Light bundles in isotropic media 123
    6.5 Stationary ray-optical structures 131
    6.6 Stationary ray optics in vacuum and in simple media 141

7. **Variational principles for rays** 149
    7.1 The principle of stationary action: The general case 149
    7.2 The principle of stationary action: The strongly regular case . 154
    7.3 Fermat's principle 156
    7.4 A Hilbert manifold setting for variational problems 165
    7.5 A Morse theory
         for strongly hyperregular ray-optical structures 168

8. **Applications** 183
    8.1 Doppler effect, aberration, and drag effect in isotropic media . 183
    8.2 Light rays in a uniformly accelerated medium
         on Minkowski space 190
    8.3 Light propagation in a plasma on Kerr spacetime 193
    8.4 Gravitational lensing 199

References 211

Part I

From Maxwell's equations to ray optics

1. Introduction to Part I

In Part I we recapitulate the general ideas of how to derive the laws of ray optics from Maxwell's equations. We presuppose a general-relativistic spacetime as background, and we consider media which are general enough to elucidate all relevant features of the method. Chapter 2 treats the case of a linear (not necessarily isotropic) dielectric and permeable medium in full detail. Chapter 3 discusses dispersive media in general and a simple plasma model in particular. In this way the material presented in Part I serves two purposes. First, it motivates our mathematical frame-work for ray optics, to be set up in Part II below. Second, it provides us with physically important examples of ray optical structures to which we shall recur frequently.

1.1 A brief guide to the literature

In Part I we have to assume some familiarity on the reader's side with Maxwell's equations in matter on a general-relativistic spacetime. Whereas vacuum Maxwell's equations are detailed in any textbook on general relativity, the matter case is not usually treated in extenso. For general aspects of the phenomenology of electromagnetic media in general relativity we refer to Bressan [17] who gives many earlier references. The case of a linear (not necessarily isotropic) dielectric

and permeable medium which is at the basis of Chap. 2 is briefly treated by Schmutzer [127], Chap. IV, following an original article by Marx [91]. The general-relativistic plasma model which is at the basis of Chap. 3 is systematically treated in two articles by Breuer and Ehlers [18] [19]; for earlier work on the same subject we refer to Madore [90], to Bičák and Hadrava [14], and to Anile and Pantano [5] [6].

As an aside it should be mentioned that the phenomenological theory of electromagnetic media can be derived from electron theory by statistical methods, i.e., that the macroscopic (phenomenological) Maxwell equations can be derived from a sort of microscopic Maxwell equations. For linear isotropic media in inertial motion on flat spacetime this is a standard textbook matter; the generalization to accelerated media is due to Kaufmann [67]. For a general-relativistic plasma, the derivation of phenomenological properties from the kinetic theory of photons is discussed in the above-mentioned article by Bičák and Hadrava [14].

The main topic of Part I is the derivation of the laws of ray optics from Maxwell's equations. The basic idea is to make an *approximate-plane-wave ansatz* for the electromagnetic field and to assume that this ansatz satisfies Maxwell's equations in an asymptotic sense for high frequencies. This results in a dynamical law for wave surfaces which can be rewritten equivalently as a dynamical law for rays. In optics the dynamical law for wave surfaces is usually called the *eikonal equation*. It is formally analogous to the *Hamilton-Jacobi equation* of classical mechanics, whereas the dynamical law for rays is formally analogous to *Hamilton's equations*. Mathematically, this so-called *ray method* is, of course, not restricted to Maxwell's equations but applies equally well to other partial differential equations with or without relevance to physics. In this sense, the ray method has applications not only to optics but also to acoustics and to wave mechanics. In the latter context, the ray method is known as *JWKB method*, refering to the pioneering work of Jeffreys, Wentzel, Kramers and Brioullin, and is detailed in virtually any textbook on quantum mechanics.

In this brief guide to the literature we shall concentrate on the ray method in optics. As to other applications we refer to the comprehensive list of references given in monographs such as Keller, Lewis and Seckler [70] or Jeffrey and Kawahara [66]. Purely mathematical aspects of the ray method can be found in textbooks on partial differential equations. Particularly useful for our purposes are, e.g., the books by Chazarain and Piriou [26] and by Egorov and Shubin [36].

Whereas rudiments of the ray method can be traced back to work of Liouville and Green around 1830, it was first carried through in the context of optics by Sommerfeld and Runge [132] in the year 1911, following a suggestion by Debye. The work of Sommerfeld and Runge was restricted to the vacuum Maxwell equations in an inertial system, and the only goal was to derive the corresponding eikonal equation. Their treatment was generalized and systematized by Luneburg [88] who considered infinite asymptotic series solutions rather than just asymptotic solutions of lowest order as Sommerfeld and Runge did. Later, the method was extended from the vacuum case to the case of light propagation in matter. This was a very active field of reasearch in the 1960s, see, e.g., Lewis [84], Chen [27] and Kravtsov [75]. All these papers are restricted to special relativity in the sense that they are presupposing a flat spacetime. Nonetheless, the techniques used are of interest also in view of general relativity. The reason is that vacuum Maxwell's equations on a general-relativistic spacetime are very similar to Maxwell's equations in an inhomogeneous medium on flat spacetime, at least locally. This was first observed by Plebański [120]. Note, however, that global aspects which do not carry over to general relativity are brought into play whenever temporal Fourier expansions (as e.g. by Lewis [84]) and/or spatial

Fourier expansions (as e.g. by Chen [27]) are used. A global treatment of the ray method that does carry over to general relativity is possible in terms of the *Lagrangian manifold* techniques introduced in the 1960s by Maslov and Arnold, see Arnold [7], Maslov [94], Duistermaat [31] or Guillemin and Sternberg [55]. In Part I we are concerned with local questions only. However, we shall touch upon Lagrangian manifold techniques and their relevance for the investigation of *caustics* in Part II below.

In *general relativity*, the passage from Maxwell's equations to ray optics was carried through for the first time by Laue [77] in the year 1920. In this paper, which is the written version of a talk given by Laue at the 86. Naturforscherversammlung, the author demonstrated how to derive from vacuum Maxwell's equations on a curved spacetime the light-like geodesic equation for the rays. Laue's treatment followed closely the seminal paper by Sommerfeld and Runge [132]. A more systematic general-relativistic treatment of the ray method in optics, including asymptotic solutions of arbitrarily high order, was brought forward much later by Ehlers [38]. He considered linear isotropic non-dispersive media on an arbitrary general-relativistic spacetime and derived not only the eikonal equation for the rays but also *transport equations* of arbitrary order for the polarization plane along the rays. In particular, his results put earlier findings about light propagation in such media by Gordon [50] and Pham Mau Quan [117] on a mathematically firm basis. At least for the vacuum case, the main results can now be found in many textbooks on general relativity, see, e.g., Misner, Thorne and Wheeler [98], Straumann [136], or Stephani [133]. The general-relativistic relevance of higher order terms in the asymptotic series expansion was discussed by Dwivedi and Kantowski [32] and by Anile [4]. A general-relativistic treatment of the ray method for dispersive media, exemplified with a special plasma model, is due to Breuer and Ehlers [18] [19] who modified and enhanced earlier work by Madore [90], by Bičák and Hadrava [14], and by Anile and Pantano [5] [6].

1.2 Assumptions and notations

We assume a general-relativistic spacetime, i.e., a four-dimensional C^∞ manifold with a metric of Lorentzian signature $(+, +, +, -)$. On this spacetime background we consider Maxwell's equations, using units making the dielectricity and permeability constants of vacuum equal to one, $\varepsilon_o = \mu_o = 1$. Thereby, in particular, the vacuum velocity of light is set equal to one. We restrict ourselves to the C^∞ category in the sense that throughout Part I all maps and tensor fields are tacitly assumed to be infinitely often differentiable. We work in local coordinates using standard index notation. Throughout, Einstein's summation convention is in force with latin indices running from 1 to 4 and with greek indices running from 1 to 3. The (covariant) components of the spacetime metric will be denoted by g_{ab}. As usual, we define g^{bc} by $g_{ab} g^{bc} = \delta_a^c$, where δ_a^c denotes the Kronecker delta, and we use g_{ab} (and g^{bc}, respectively) to lower (and raise, respectively) indices. With respect to a co-ordinate system $x = (x^1, x^2, x^3, x^4)$, partial derivatives $\frac{\partial}{\partial x^a}$ will be denoted by ∂_a for short, whereas ∇_a means covariant derivative with respect to the Levi-Civita connection of our metric. For the sake of brevity, we shall speak of "a tensor field Q^{abc}" if we mean "a tensor field whose contravariant components in a coordinate system are Q^{abc}" etc. Our treatment will be purely local throughout Part I. Therefore, the use of local coordinates and index notation is no restriction whatsoever.

2. Light propagation in linear dielectric and permeable media

On our spacetime manifold we consider Maxwell's equations in a linear but not necessarily isotropic medium, i. e., in a medium phenomenologically characterized by a dielectricity tensor field and a permeability tensor field. It is our goal to derive and to discuss the laws of ray optics in such a medium. The standard textbook problem of light propagation in vacuo is, of course, included as a special case.

The results of this chapter cover a wide range of applications including light propagation in gases (isotropic case) and crystals (anisotropic case) as long as dispersion is ignored. For dispersive media we refer to Chap. 3 below. In view of applications to astrophysics, the isotropic case is more interesting than the (much more complicated) anisotropic case. On the other hand, a thorough treatment of the anisotropic case is highly instructive from a methodological point of view. In particular, it gives us the opportunity to discuss the phenomenon of birefringence.

2.1 Maxwell's equations in linear dielectric and permeable media

On our spacetime manifold, the source-free Maxwell equations for (macroscopic) electromagnetic fields in matter can be written in local coordinates as

$$\eta^{abcd} \nabla_b F_{cd} = 0 \quad \text{and} \quad \nabla^b G_{bc} = 0 \,, \tag{2.1}$$

or, using partial rather than covariant derivatives, as

$$\eta^{abcd} \partial_b F_{cd} = 0 \quad \text{and} \quad \eta^{abcd} \partial_b \big(\eta_{cd}{}^{ef} G_{ef} \big) = 0 \,. \tag{2.2}$$

In (2.1) and (2.2), η_{abcd} denotes the totally antisymmetric *Levi-Civita tensor* field (volume form) of our metric which is defined by the equation

$$\eta_{1234} = \pm\sqrt{|\det(g_{cd})|} \,. \tag{2.3}$$

Here the plus sign is valid if the coordinate system is right-handed and the minus sign is valid if it is left-handed. In other words, we have to choose an orientation on the domain of our coordinate system to fix the sign ambiguity of the Levi-Civita tensor field. However, this is irrelevant since Maxwell's equations are invariant under $\eta_{abcd} \longmapsto -\eta_{abcd}$, and so are all the relevant results in Part I. If the reader is not familiar with volume elements he or she may consult, e.g., Wald [146], p. 432.

F_{ab} and G_{ab} denote the *electromagnetic field strength* and the *electromagnetic excitation*, respectively, both of which are antisymmetric second rank tensor fields. With respect to a reference system, given in terms of a time-like vector field U^a with $U^a U_a = -1$, we can introduce the *electric field strength*

$$E_a = F_{ab} U^b \tag{2.4}$$

and the *magnetic field strength*

$$B_a = -\tfrac{1}{2} \eta_{abcd} U^b F^{cd} \tag{2.5}$$

such that

$$F_{ab} = -\eta^{cd}{}_{ab} B_c U_d + E_b U_a - E_a U_b \, . \tag{2.6}$$

Here we have used the familiar property

$$\eta_{abcd} \, \eta^{aefk} = -\delta^e_b \, \delta^f_c \, \delta^k_d - \delta^e_c \, \delta^f_d \, \delta^k_b - \delta^e_d \, \delta^f_b \, \delta^k_c +$$
$$\delta^e_d \, \delta^f_c \, \delta^k_b + \delta^e_c \, \delta^f_b \, \delta^k_d + \delta^e_b \, \delta^f_d \, \delta^k_c \tag{2.7}$$

of the Levi-Civita tensor field, cf., e.g., Wald [146], equation (B.2.12).

Similarly, we introduce the *electric excitation*

$$D_a = G_{ab} U^b \tag{2.8}$$

and the *magnetic excitation*

$$H_a = -\tfrac{1}{2} \eta_{abcd} U^b G^{cd} \tag{2.9}$$

such that

$$G_{ab} = -\eta^{cd}{}_{ab} H_c U_d + D_b U_a - D_a U_b \, . \tag{2.10}$$

With respect to the reference system used for their definitions, the electric and magnetic field strengths are purely spatial one-forms, and so are the electric and magnetic excitations,

$$E_a U^a = B_a U^a = 0 \quad \text{and} \quad D_a U^a = H_a U^a = 0 \, . \tag{2.11}$$

Our terminology of calling E_a and B_a the "field strengths" (in german: *Feldstärken*) and D_a and H_a the "excitations" (in german: *Erregungen*) follows Gustav Mie and Arnold Sommerfeld. This terminology is reasonable

since E_a and B_a determine the Lorentz force exerted on a charged test particle whereas, in the presence of field-producing charges and currents, D_a and
H_a are the fields "excited" by those sources via Maxwell's equations. The
traditional terminology of calling H_a the "magnetic field strength" is misleading. Moreover, it is highly inconvenient from a relativistic point of view
where E_a and B_a, rather than E_a and H_a, are united into an antisymmetric
second rank tensor field on spacetime.

In what follows we consider Maxwell's equations in the form (2.2). As
long as only the metric is known, (2.2) gives us eight component equations for
twelve unknown functions. (The unknown functions are the six independent
components of the electromagnetic field strength plus the six independent
components of the electromagnetic excitation.) Hence, (2.2) is an *underde-
termined* system of partial differential equations. It must be supplemented
by *constitutive equations* relating the electromagnetic field strength with the
electromagnetic excitation. Thereby the medium is characterized in a phenomenological way. In this chapter we consider linear dielectric and permeable
media according to the following definition.

Definition 2.1.1. *A linear dielectric and permeable medium is, by definition, a medium characterized by constitutive equations that take the form*

$$D_a = \varepsilon_a{}^b E_b \quad and \quad B_a = \mu_a{}^b H_b , \qquad (2.12)$$

*in some reference system U^b, with second rank tensor fields $\varepsilon_a{}^b$ and $\mu_a{}^b$ sat-
isfying the following conditions:*

(a) $U^a \varepsilon_a{}^b = 0$ *and* $U^a \mu_a{}^b = 0$.
(b) $\varepsilon^{ab} = \varepsilon^{ba}$ *and* $\mu^{ab} = \mu^{ba}$.
(c) $\varepsilon^{ab} Z_a Z_b > 0$ *and* $\mu^{ab} Z_a Z_b > 0$ *for all* $(Z_1, Z_2, Z_3, Z_4) \neq (0, 0, 0, 0)$ *with*
$U^a Z_a = 0$.

*We refer to the distinguished reference system U^b as to the rest system, to
$\varepsilon_a{}^b$ as to the dielectricity tensor field and to $\mu_a{}^b$ as to the permeability tensor
field of the medium.*

Condition (a) of Definition 2.1.1 guarantees that the constitutive equations (2.12) are in agreement with $D_a U^a = 0$ and $B_a U^a = 0$. Conditions (b)
and (c) imply that in the rest system of the medium the *energy density*

$$w = \tfrac{1}{2} \left(D_a E^a + B_a H^a \right) , \qquad (2.13)$$

of the electromagnetic field is positive definite. Altogether, conditions (a),
(b) and (c) guarantee that the dielectricity and permeability tensor fields are
"spatially invertible". We can, thus, define $(\mu^{-1})_a{}^b$ by the properties

$$U^a (\mu^{-1})_a{}^b = 0 ,$$
$$(\mu^{-1})^{ab} = (\mu^{-1})^{ba} , \qquad (2.14)$$
$$(\mu^{-1})_a{}^b \mu_b{}^c = \delta_a^c + U_a U^c .$$

The constitutive equations (2.12) can then be united in a single equation,

$$G_{ab} = \left(\tfrac{1}{2} \eta^{cd}{}_{ab} (\mu^{-1})_c{}^e \eta_{er}{}^{pq} U^r U_d + \varepsilon_b{}^p U^q U_a - \varepsilon_a{}^p U^q U_b \right) F_{pq} . \quad (2.15)$$

The following special case deserves particular interest.

Definition 2.1.2. *A linear dielectric and permeable medium is called isotropic if the dielectricity and permeability tensor fields are of the special form*

$$\varepsilon_a{}^b = \varepsilon \left(\delta_a^b + U_a U^b \right) \quad \text{and} \quad \mu_a{}^b = \mu \left(\delta_a^b + U_a U^b \right) , \quad (2.16)$$

with some scalar functions ε and μ. (Condition (c) of Definition 2.1.1 then requires ε and μ to be strictly positive.)

In the isotropic case, (2.15) reduces to

$$G_{ab} = \frac{1}{\mu} \left(F_{ab} + (1 - \varepsilon\mu) \left(F_{ad} U_b - F_{bd} U_a \right) U^d \right) . \quad (2.17)$$

In particular, vacuum can be characterized as a linear isotropic medium with $\varepsilon = \mu = 1$. In this case (and, more generally, in any isotropic medium with $\varepsilon\mu = 1$) U^a drops out from (2.17) and the constitutive equations take the form (2.12) in *any* reference system. This is in agreement with the obvious fact that for vacuum any reference system can be viewed as the rest system of the medium.

We emphasize that our phenomenological constitutive equations are physically reasonable in the rotational as well as in the irrotational case, i.e., U^a need not be hypersurface-orthogonal. Although this should be clear from the general rules of relativity, there is still a debate on this issue, even in the case of an isotropic medium on flat spacetime, see, e.g., Pellegrini and Swift [106].

We are now going to analyze the dynamics of electromagnetic fields in a linear dielectric and permeable medium. We have already mentioned that Maxwell's equations (2.2) alone give us eight equations for twelve unknown functions. With (2.12) at hand, and assuming that U^a, $\varepsilon_a{}^b$ and $\mu_a{}^b$ are known, we can eliminate six of the unknown functions. Now (2.2) gives us eight equations for six functions, i.e., the system looks overdetermined. However, only six of those eight equations are evolution equations, governing the dynamics of electromagnetic fields, whereas the other two equations are constraints.This is most easily verified in a local coordinate system (x^1, x^2, x^3, x^4) in which the hypersurfaces $x^4 = \text{const.}$ are space-like such that x^4 can be viewed as a local time function. Owing to the antisymmetry of the Levi-Civita tensor, the $a = 4$ components of equations (2.2) do not involve any ∂_4 derivative. Hence, these two equations are to be viewed as *constraints* whereas the remaining six equations, i.e., the $a = 1, 2, 3$ components of equations (2.2), are the *evolution equations* governing the dynamics. Again owing to the antisymmetry of the Levi-Civita tensor field, the evolution equations preserve the constraints in the following sense. If the constraints are written in the form

$C_1 = 0$ and $C_2 = 0$, then the evolution equations imply that $\partial_4 C_1 = f_1 C_1$ and $\partial_4 C_2 = f_2 C_2$ with some spacetime functions f_1 and f_2. Hence, if a solution of the evolution equations satisfies the constraints on some initial hypersurface $x^4 = $ const., then it satisfies the constraints everywhere (on some neighborhood of any point of the initial hypersurface, that is). In other words, locally around any one point all solutions of Maxwell's equations can be found in the following way.

Step 1. Choose a space-like hypersurface through that point.

Step 2. Choose a local coordinate system such that the chosen hypersurface is given by the equation $x^4 = $ const.

Step 3. Solve the evolution equations with all initial data that satisfy the constraints.

In the rest of this section we shall prove that the initial value problem considered in Step 3 is well-posed in the sense that it is characterized by a local existence and uniqueness theorem, provided that the initial hypersurface has been chosen appropriately. Conditions (a), (b) and (c) of Definition 2.1.1 will prove essential for this result.

First we introduce special coordinates according to the following definition.

Definition 2.1.3. *Let U^a denote the rest system of a linear dielectric and permeable medium and fix a spacetime point x_0. Then a local coordinate system (x^1, x^2, x^3, x^4), defined on a neighborhood of x_0, is called adapted to U^a near x_0 if*

(a) *U^a is given by the equation $U^a = \dfrac{1}{\sqrt{-g_{44}}} \delta_4^a$,*

(b) *$g_{4\mu} = 0$ for $\mu = 1, 2, 3$ at the point x_0 .*

For any linear dielectric and permeable medium, it is obvious that adapted coordinates are characterized by the following existence and uniqueness property. If we choose a spacetime point x_0 and a hypersurface S that is orthogonal to U^a at x_0, then there is a coordinate system adapted to U^a near x_0 such that S is represented by the equation $x^4 = $ const. Another coordinate system (x'^1, x'^2, x'^3, x'^4) is, again, adapted to U^a near x_0 if and only if it is related to (x^1, x^2, x^3, x^4) by a coordinate transformation of the special form

$$x^\rho \longmapsto x'^\rho(x^1, x^2, x^3)$$
$$x^4 \longmapsto x'^4(x^1, x^2, x^3, x^4)$$

(2.18)

with $\left. \dfrac{\partial x'^4}{\partial x^\nu} \right|_{x_0} = 0$.

Condition (b) of Definition 2.1.3 makes sure that at the point x_0 the hypersurface $x^4 = $ const. intersects the respective integral curve of U^a orthogonally. This implies that, on a sufficiently small neighborhood of x_0, all hypersurfaces $x^4 = $ const. are space-like. Of course, they cannot be orthogonal to U^a on a whole neighborhood unless the medium is non-rotating. Hence, in an adapted coordinate system the mixed components $g_{\mu 4}$ and $g^{\mu 4}$ of the metric need not vanish except at the central point x_0. The spatial components give positive definite 3×3 matrices $(g_{\mu\nu})$ and $(g^{\mu\nu})$ on some neighborhood of x_0; at the point x_0 these matrices are inverse to each other. The temporal components g_{44} and g^{44} are strictly negative functions on some neighborhood of x_0; at the point x_0 they are inverse to each other.

Now we consider a linear dielectric and permeable medium in a coordinate system adapted to its rest system U^a near some point x_0. Then (2.11) reduces to

$$E_4 = B_4 = 0 \quad \text{and} \quad D_4 = H_4 = 0 \tag{2.19}$$

owing to condition (a) of Definition 2.1.3. Hence, (2.12) simplifies to

$$D_\sigma = \varepsilon_\sigma{}^\rho E_\rho \quad \text{and} \quad B_\sigma = \mu_\sigma{}^\rho H_\rho . \tag{2.20}$$

Conditions (b) and (c) of Definition 2.1.1 guarantee that $\varepsilon_\sigma{}^\rho$ and $\mu_\sigma{}^\rho$ are positive definite and symmetric with respect to $g^{\rho\tau}$. We can, thus, define $v_\sigma{}^\rho$ and $w_\sigma{}^\rho$, which are again positive definite and symmetric with respect to $g^{\rho\tau}$, by

$$\mu_\nu{}^\sigma v_\sigma{}^\tau v_\tau{}^\rho = \delta_\nu^\rho \quad \text{and} \quad \varepsilon_\nu{}^\sigma w_\sigma{}^\tau w_\tau{}^\rho = \delta_\nu^\rho . \tag{2.21}$$

For the following it will be convenient to introduce the quantities

$$Z_\rho = v_\rho{}^\sigma B_\sigma \quad \text{and} \quad Y_\rho = w_\rho{}^\sigma D_\sigma , \tag{2.22}$$

and to use $Z_1, Z_2, Z_3, Y_1, Y_2, Y_3$ for the six independent components of the electromagnetic field. That is to say, we start from Maxwell's equations (2.2) with (2.6) and (2.10); we use part (a) of Definition 2.1.3 and equation (2.19); we eliminate E_τ and H_τ with the help of (2.20); finally, we express D_σ and B_σ in terms of Z_ρ and Y_ρ by means of (2.22). After a little bit of algebra, the $a = 1, 2, 3$ components of Maxwell's equations (2.2) give us evolution equations of the form

$$L^a \partial_a \begin{pmatrix} Z \\ Y \end{pmatrix} + M \begin{pmatrix} Z \\ Y \end{pmatrix} = \begin{pmatrix} 0 \\ 0 \end{pmatrix} \tag{2.23}$$

for the dynamical variables

$$Z = \begin{pmatrix} Z_1 \\ Z_2 \\ Z_3 \end{pmatrix} \quad \text{and} \quad Y = \begin{pmatrix} Y_1 \\ Y_2 \\ Y_3 \end{pmatrix} . \tag{2.24}$$

Here L^1, L^2, L^3 and L^4 are x-dependent 6×6 matrices of the form

$$L^4 = \begin{pmatrix} 1 & Q^T \\ Q & 1 \end{pmatrix} \quad \text{and} \quad L^\rho = \begin{pmatrix} 0 & (A^\rho)^T \\ A^\rho & 0 \end{pmatrix} , \tag{2.25}$$

where Q is a 3×3 matrix with components

$$Q_\lambda{}^\tau = w_\lambda{}^\nu \, \eta_{\nu 4}{}^{4\gamma} \, v_\gamma{}^\tau = w_\lambda{}^\nu \, \eta_\nu{}^{\beta 4 \gamma} \, v_\gamma{}^\tau \, g_{\beta 4} \; ; \tag{2.26}$$

A^ρ is a 3×3 matrix with components

$$A^\rho{}_\lambda{}^\tau = w_\lambda{}^\nu \, \eta_{\nu 4}{}^{\rho \gamma} \, v_\gamma{}^\tau \; ; \tag{2.27}$$

$(\cdot)^T$ means transposition with respect to $g^{\rho \tau}$ such that, e.g.,

$$(Q^T)_\lambda{}^\tau \, g^{\rho \lambda} = Q_\lambda{}^\rho \, g^{\tau \lambda} \; ; \tag{2.28}$$

M is a 6×6 matrix whose components involve the spacetime metric along with $v_\lambda{}^\sigma$ and $w_\rho{}^\tau$.

For the investigation of the evolution equations (2.23) the following two observations are crucial.

(a) At the central point x_0 of our adapted coordinate system we have $g_{\beta 4} = 0$ and, thus, $Q = 0$. By continuity, L^4 is invertible on some neighborhood of x_0 to which we can restrict our considerations. Hence, (2.23) can be solved for the ∂_4 derivative.

(b) L^1, L^2, L^3 and L^4 are symmetric ($=$ self-adjoint) with respect to the positive definite scalar product

$$\begin{pmatrix} Z_1 \\ Y_1 \end{pmatrix} \cdot \begin{pmatrix} Z_2 \\ Y_2 \end{pmatrix} = \tfrac{1}{2} \left(Z_1 \cdot Z_2 + Y_1 \cdot Y_2 \right) \tag{2.29}$$

Here the dots on the right-hand side refer to the scalar product defined by

$$a \cdot b = g^{\mu \nu} \, \overline{a_\nu} \, b_\mu \tag{2.30}$$

for any two $\mathbb{C}^3$-valued functions a and b on the neighborhood considered, with the overbar denoting complex conjugation. (To be sure, in (2.23) all quantities are real. For later purposes, however, we need the complex version of this scalar product.)

These two observations imply that (2.23) satisfies the defining properties of a *symmetric hyperbolic* system of partial differential equations. By a well-known theorem (see, e.g., Theorem 4.5 in Chazarain and Piriou [26] or Sect. 4.12 in Egorov and Shubin [36]) this guarantees local existence and uniqueness of a solution Z, Y for any initial data Z_0, Y_0 given on our hypersurface $x^4 = \text{const}$. (Please recall our stipulation of tacitly working in the C^∞ category throughout Part I. Had we restricted ourselves to the analytic category instead, property (a) alone would guarantee local existence and uniqueness of a solution to any initial data, owing to the well-known Cauchy-Kovalevsky theorem.) Moreover, the fact that (2.23) is symmetric hyperbolic implies that solutions Z, Y are bounded in terms of so-called *energy inequalities*, see, e.g.,

Theorem 4.3 in Chazarain and Piriou [26] or Theorem 2.63 in Egorov and Shubin [36]. We are going to employ these facts later.

The explicit form of the matrix M in (2.23) will be of no interest for us in the following. What really matters is the structure of the L^a, i.e., the information contained in the 6×6 matrix

$$L(x, p) = p_a L^a(x). \tag{2.31}$$

Here the first argument $x = (x^1, x^2, x^3, x^4)$ ranges over the coordinate neighborhood considered and the second argument $p = (p_1, p_2, p_3, p_4)$ ranges over $\mathbb{R}^4$. The matrix $L(x, p)$ defined by (2.31) is called the *principal matrix* or the *characteristic matrix* of the system of differential equations (2.23). Its determinant, which gives a homogeneous polynomial of degree six in the p_a, is called the *principal determinant* or the *characteristic determinant* of (2.23). We shall see later that the laws of ray optics in our medium are coded in the characteristic determinant.

The notions of characteristic matrix and characteristic determinant can be introduced for any system of kth order partial differential equations, linear in the highest order derivatives, that gives n equations for n dynamical variables. The characteristic matrix is then formed in a fashion similar to (2.31) from the coefficients of the highest order derivatives. If these coefficients are independent of the unknown functions, i.e., if the system of differential equations is semi-linear, the characteristic matrix is of the form

$$L(x, p) = p_{a_1} \cdots p_{a_k} L^{a_1 \cdots a_k}(x). \tag{2.32}$$

Hence, its determinant gives a homogeneous polynomial of degree nk with respect to the p_a.

2.2 Approximate-plane-wave families

In the preceding section we have discussed Maxwell's equations in a linear dielectric and permeable medium. The laws of light propagation in such a medium are determined by the dynamics of wavelike solutions of those equations. In this section we clarify what is meant by the attribute "wavelike". The following definition is basic.

Definition 2.2.1. *An* approximate-plane-wave family *is a one-parameter family of antisymmetric second rank tensor fields of the form*

$$F_{ab}(\alpha, x) = \text{Re}\{e^{iS(x)/\alpha} f_{ab}(\alpha, x)\} \tag{2.33}$$

with the following properties.

(a) *The coordinates $x = (x^1, x^2, x^3, x^4)$ range over some open subset of the spacetime manifold and the parameter α ranges over the strictly positive real numbers, $\alpha \in \mathbb{R}^+$.*

(b) *S is a real-valued function whose gradient has no zeros, i.e.,*

$$\partial S(x) = \Big(\partial_1 S(x), \partial_2 S(x), \partial_3 S(x), \partial_4 S(x)\Big) \neq (0, 0, 0, 0) \qquad (2.34)$$

for all x in the neighborhood considered. We refer to S as to the eikonal function *of the approximate-plane-wave family.*

(c) *For each* $\alpha \in \mathbb{R}^+$, $f_{ab}(\alpha, \cdot)$ *is a complex-valued antisymmetric second rank tensor field. Moreover,* f_{ab} *admits a Taylor expansion of the form*

$$f_{ab}(\alpha, x) = \sum_{N=0}^{N_0+1} \alpha^N f_{ab}^N(x) + O(\alpha^{N_0+2}) \qquad (2.35)$$

for all integers $N_0 \geq -1$, *where*

$$f_{ab}^N(x) = N! \lim_{\alpha \to 0} \frac{\partial^N}{\partial \alpha^N} f_{ab}(\alpha, x) . \qquad (2.36)$$

We refer to f_{ab}^N *as to the* N^{th} *order amplitude of the approximate-plane-wave family.*

(d) *For all x in the neighborhood considered,*

$$\big(f_{ab}^0(x)\big) \neq \mathbf{0} . \qquad (2.37)$$

In (2.33), i denotes, of course, the imaginary unit, $i^2 = -1$, and Re denotes the real part of a complex number.

We call S the "eikonal function" because an approximate-plane-wave family satisfies Maxwell's equations in an asymptotic sense to be discussed later only if S satisfies a partial differential equation which is known as the *eikonal equation*. The term "eikonal", which was introduced in 1895 by Bruns [23] in a more special context, is derived from the greek word *eikon* which means "image". This terminology is, indeed, justified since the eikonal equation is the fundamental equation of ray optics; so it governs, in particular, the ray optical laws of image formation.

According to our general stipulation that all maps and tensor fields are tacitly assumed to be infinitely often differentiable it goes without saying that $f_{ab}(\alpha, x)$ is a C^∞ function of $\alpha \in \mathbb{R}^+$. A Taylor expansion of the form (2.35) is valid if and only if this function admits a C^∞ extension into the point $\alpha = 0$. Note that we do not assume that the $O(\alpha^{N+1})$ term in (2.35) goes to zero for $N \to \infty$, i.e., we do not assume analyticity with respect to α.

It is important to realize that an approximate-plane-wave family cannot converge for $\alpha \to 0$. This is an immediate consequence of the following lemma which will often be used in the following.

Lemma 2.2.1. *Let S be the eikonal function of an approximate-plane-wave family, according to Definition 2.2.1 (b). Let u be a complex valued function defined on the same open subset of spacetime as the approximate-plane-wave family. (As always in Part I, we tacitly assume that u is of class C^∞ and, thus, continuous). Then $\lim_{\alpha \to 0} \text{Re}\{e^{iS/\alpha}u\}$ exists pointwise only if $u = 0$.*

Proof. If u is different from zero at some point, it is different from zero, by continuity, on a whole neighborhood. For almost all points x of this neighborhood, (2.34) implies that $S(x) \neq 0$ and the limit does not exist. $\square$

We are now going to justify the name "approximate-plane-wave family". (More fully, (2.33) should be called a "locally-approximate-plane-and-monochromatic-wave family". This terminology, however, seems a little bit too cumbersome.) The physical idea behind Definition 2.2.1 becomes clear if we consider the special case that the tensor fields $\partial_a S$ and $f_{ab}(\alpha, \cdot)$ are covariantly constant (and non-zero, as assured by (2.34) and (2.37)), i.e., that the equations $\nabla_b \partial_a S = 0$ and $\nabla_c f_{ab}(\alpha, \cdot) = 0$ are satisfied. Then (2.33) gives a one-parameter family of monochromatic plane waves. With respect to an inertial system (i.e., a covariantly constant time-like vector field V^a with $g_{ab} V^a V^b = -1$), the frequency of such a wave is given by $\omega = \frac{1}{\alpha} V^a \partial_a S$ and the spatial wave covector is given by $k_a = \frac{1}{\alpha} \partial_a S - \omega g_{ab} V^b$. Hence, the limit $\alpha \to 0$ corresponds to infinitely high frequency with respect to all inertial systems V^a with $V^a \partial_a S \neq 0$.

Now this is a very special case since on a spacetime without symmetry there are no non-zero covariantly constant vector fields. Therefore, as we want to work with ansatz (2.33) on an arbitrary spacetime, we cannot assume that $\partial_a S$ and $f_{ab}(\alpha, \cdot)$ are covariantly constant. However, if we restrict our consideration to a sufficiently small neighborhood, $\partial_a S$ and $f_{ab}(\alpha, \cdot)$ deviate arbitrarily little from being covariantly constant. Similarly, on a sufficiently small neighborhood, any time-like vector field V^a with $g_{ab} V^a V^b = -1$ deviates arbitrarily little from an inertial system. However small this neighborhood may be, by choosing α sufficiently small we can have arbitrarily many wave periods in this small spacetime region.

This reasoning justifies the terminology introduced in Definition 2.2.1. Please note that (2.34) and (2.37) are essential to guarantee that (2.33) gives an approximately plane and monochromatic wave near each point for α sufficiently close to zero.

In correspondence with this interpretation we shall refer to the hypersurfaces $S = $ const. as to the *wave surfaces* of our approximate-plane-wave family. The alternative terms *eikonal surfaces* and *phase surfaces* are also common. Moreover, we call

$$\omega(\alpha, x) = -\frac{1}{\alpha} \partial_a S(x) V^a(x) \tag{2.38}$$

its *frequency function* and we call

$$k_a(\alpha, x) = \frac{1}{\alpha} \partial_a S(x) - \omega(\alpha, x) g_{ab}(x) V^b(x) \tag{2.39}$$

its *spatial wave covector field* with respect to the observer field V^a; here all those time-like vector fields with $g_{ab} V^a V^b = -1$ are admitted for which $V^a \partial_a S$ has no zeros.

It is worthwhile to note that from an approximate-plane-wave family (2.33) we can produce *non-monochromatic* waves by integrating over α with an appropriate *density function* w,

$$\tilde{F}_{ab}(x) = \int_{\alpha_1}^{\alpha_2} F_{ab}(\alpha, x)\, w(\alpha)\, d\alpha \,. \tag{2.40}$$

This can be viewed as a generalized *Fourier synthesis*. Here we have to assume that $\alpha_1 < \alpha_2$, with α_2 sufficiently small to justify the approximate-plane-wave interpretation. Moreover, it is also possible to form superpositions of approximate-plane-wave families with different eikonal functions S.

2.3 Asymptotic solutions of Maxwell's equations

To study the dynamics of wavelike electromagnetic fields in our medium we have to plug our approximate-plane-wave ansatz (2.33) into Maxwell's equations, i.e., into (2.2) supplemented with our constitutive equations. Unfortunately, only in very special cases is it possible to determine the eikonal function S and the amplitudes f_{ab}^N in such a way that the resulting equations are exactly satisfied for some $\alpha \in \mathbb{R}^+$. It is the characteristic feature of the *ray method* to determine S and f_{ab}^N in such a way that Maxwell's equations are satisfied, rather than for some finite value of α, asymptotically for $\alpha \to 0$. In this way the ray method gives us the dynamics of wave surfaces and wave amplitudes in the high frequency limit. To put this rigorously we introduce the following notation.

Definition 2.3.1. *For $N \in \mathbb{Z}$, an approximate-plane-wave family $F_{ab}(\alpha, \cdot)$ in the sense of Definition 2.2.1 is called an Nth order asymptotic solution of Maxwell's equations if*

$$\lim_{\alpha \to 0} \left(\tfrac{1}{\alpha^N}\, \eta^{abcd}\, \partial_b F_{cd}(\alpha, \cdot) \right) = 0 \,,$$
$$\lim_{\alpha \to 0} \left(\tfrac{1}{\alpha^N}\, \eta^{abcd}\, \partial_b \big(\eta_{cd}{}^{ef} G_{ef}(\alpha, \cdot) \big) \right) = 0 \,. \tag{2.41}$$

Here $G_{ef}(\alpha, \cdot)$ is related to $F_{ab}(\alpha, \cdot)$ by the constitutive equations of the medium.

In (2.41), the limits are meant to be performed pointwise with respect to the spacetime coordinates; we shall restrict ourselves to neighborhoods on which the convergence is uniform. For the evaluation of (2.41), the following two observations are crucial.

(a) The metric is independent of α and so are the other tensor fields that enter into the constitutive equations, i.e., U^a, $\varepsilon_a{}^b$ and $\mu_a{}^b$. Hence, the special form in which α enters into the approximate-plane-wave ansatz (2.33) together with the linearity of the constitutive equations implies that (2.41) is trivially satisfied for $N < -1$.

(b) If (2.41) holds for $N = N_0$, then it holds all the more for $N \leq N_0$.

These two observations together suggest that N_0^{th} order asymptotic solutions can be found for arbitrarily large $N_0 \geq 0$ by first evaluating (2.41) for $N = -1$ and then proceeding step by step up to $N = N_0$. We shall see in the following that this inductive procedure gives us dynamical laws for the eikonal function S and, step by step, for the amplitudes f_{ab}^N up to arbitrarily large order $N = N_0$.

If we want to get dynamical laws for f_{ab}^N for all $N \in \mathbb{N}$, we have to assume that our approximate-plane-wave family (2.33) satisfies (2.41) for all $N \in \mathbb{N}$ or, what is the same, for all $N \in \mathbb{Z}$. In this case $F_{ab}(\alpha, \cdot)$ is called an *infinite asymptotic series solution* of Maxwell's equations.

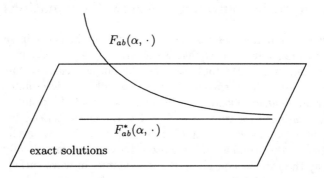

Fig. 2.1. For $N \geq 0$, an Nth order asymptotic solution $F_{ab}(\alpha, \cdot)$ of Maxwell's equations approaches the space of exact solutions of Maxwell's equations asymptotically for $\alpha \to 0$, as will be proven in Sect. 2.7.

(2.41) does, of course, not imply that the one-parameter family $F_{ab}(\alpha, \cdot)$ converges pointwise (or in any other sense) towards an exact solution of Maxwell's equations for $\alpha \to 0$. We have already emphasized that for an approximate-plane-wave family the limit $\lim_{\alpha \to 0} F_{ab}(\alpha, \cdot)$ cannot exist. This raises the question of whether asymptotic solutions can be viewed as approximate solutions. This question will be answered in Sect. 2.7 below by proving the following result. Let $F_{ab}(\alpha, \cdot)$ be an approximate-plane-wave family that is an Nth order asymptotic solution of Maxwell's equations in a linear dielectric and permeable medium for some $N \geq 0$. Then there exists, locally around any one point, a one-parameter family $F_{ab}^*(\alpha, \cdot)$ of exact solutions of Maxwell's equations such that $F_{ab}^*(\alpha, \cdot) - F_{ab}(\alpha, \cdot)$ goes to zero in the pointwise sense (and even with respect to some finer norms involving arbitrarily high derivatives) as α^{N+1} for $\alpha \to 0$. In other words, for α sufficiently small the members of our approximate-plane-wave family can be viewed as arbitrarily good approximations to exact solutions of Maxwell's equations. Figure 2.1 illustrates this situation in the infinite-dimensional space of (C^∞)

antisymmetric second-rank tensor fields defined on some open spacetime domain.

2.4 Derivation of the eikonal equation and transport equations

In this section we derive, in a linear dielectric and permeable medium, the dynamical equations for wave surfaces and for wave amplitudes in the high frequency limit. We do that locally around any spacetime point x_0. As a preparation, we prove the following fact.

Proposition 2.4.1. *Consider an approximate-plane-wave family* $F_{ab}(\alpha, \cdot)$ *that is an* N^{th} *order asymptotic solution of Maxwell's equations in a linear dielectric and permeable medium for some* $N \geq -1$. *Then the frequency function* (2.38) *of* $F_{ab}(\alpha, \cdot)$ *with respect to the rest system of the medium* $(V^a = U^a)$ *has no zeros.*

Proof. We introduce, around any spacetime point x_0, a coordinate system adapted to U^a in the sense of Definition 2.1.3. We are done if we can show that $\partial_4 S$ is different from zero at x_0. By assumption, our approximate-plane-wave family satisfies (2.41) for $N = -1$, i.e.

$$\eta^{abcd} \, \partial_b S \, f^0_{cd} = 0 \,, \tag{2.42}$$

$$\eta^{abcd} \, \partial_b S \, \eta_{cd}{}^{ef} \, g^0_{ef} = 0 \,, \tag{2.43}$$

where g^0_{ef} is related to f^0_{cd} by the constitutive equations. (Here we made use of Lemma 2.2.1.) Now let us assume that $\partial_4 S = 0$ at x_0. At this point, the $a = 4$ component of (2.42) implies

$$g^{\mu\tau} \, \partial_\mu S \, b^0_\tau = 0 \tag{2.44}$$

for the magnetic part b^0_τ of f^0_{ab}, whereas the $a = \rho$ components of (2.42) imply

$$\partial_\nu S \, e^0_\mu - \partial_\mu S \, e^0_\nu = 0 \tag{2.45}$$

for the electric part e^0_ν of f^0_{ab}. Similarly, (2.43) results in

$$g^{\mu\tau} \, \partial_\mu S \, d^0_\tau = 0 \,, \tag{2.46}$$

and

$$\partial_\nu S \, h^0_\mu - \partial_\mu S \, h^0_\nu = 0 \tag{2.47}$$

for the electric part d^0_τ and for the magnetic part h^0_τ of g^0_{ab}. Note that S is real whereas the amplitudes are complex. (2.45) and (2.46) imply

$$g^{\mu\tau} \, e^0_\mu \, \overline{d^0_\tau} \, \partial_\nu S = 0 \,. \tag{2.48}$$

Similarly, (2.44) and (2.47) imply

$$g^{\mu\tau} h_\mu^0 \overline{b_\tau^0} \partial_\nu S = 0 . \tag{2.49}$$

Recall that we are at a point where $\partial_4 S = 0$. Thus, condition (2.34) requires $(\partial_1 S, \partial_2 S, \partial_3 S) \neq (0,0,0)$. Hence, by condition (c) of Definition 2.1.1, (2.48) implies that $(e_1^0, e_2^0, e_3^0) = (0,0,0)$ and (2.49) implies that $(b_1^0, b_2^0, b_3^0) = (0,0,0)$. This shows that our hypothesis of $\partial_4 S$ having a zero gives a contradiction to (2.37). $\qquad\square$

To analyze the dynamics of wave surfaces and amplitudes in the high frequency limit near an arbitrary spacetime point x_0, we introduce near x_0 a coordinate system which is adapted to the rest system U^a of the medium in the sense of Definition 2.1.3. We can then express electromagnetic fields in terms of the dynamical variables $Z_1, Z_2, Z_3, Y_1, Y_2, Y_3$ introduced in (2.22). Then any approximate-plane-wave family takes the form

$$\begin{pmatrix} Z(\alpha, x) \\ Y(\alpha, x) \end{pmatrix} = \mathrm{Re} \left\{ e^{iS(x)/\alpha} \sum_{N=0}^{N_0+1} \alpha^N \begin{pmatrix} z^N(x) \\ y^N(x) \end{pmatrix} + O(\alpha^{N_0+2}) \right\} , \tag{2.50}$$

for any integer $N_0 \geq -1$. Here the complex amplitudes f_{ab}^N from (2.35) are expressed in terms of $\mathbb{C}^3$-valued functions z^N and y^N. The following proposition gives necessary and sufficient conditions on the eikonal function S and on the amplitudes z^N, y^N such that (2.50) is an asymptotic solution of Maxwell's equations.

Proposition 2.4.2. *Consider, locally around any spacetime point x_0, a coordinate system (x^1, x^2, x^3, x^4) adapted to the rest system U^a of a linear dielectric and permeable medium. Then an approximate-plane-wave family, represented in this coordinate system in the form (2.50), is an asymptotic solution of Maxwell's equations in lowest non-trivial order $N = -1$ if and only if $\partial_4 S$ has no zeros and*

$$\partial_a S \, L^a \begin{pmatrix} z^0 \\ y^0 \end{pmatrix} = \begin{pmatrix} 0 \\ 0 \end{pmatrix} . \tag{2.51}$$

For $N_0 \geq 0$, such an approximate-plane-wave family is an N_0^{th} order asymptotic solution of Maxwell's equations if and only if, in addition,

$$\left(L^a \partial_a + M \right) \begin{pmatrix} z^N \\ y^N \end{pmatrix} = -i \, \partial_a S \, L^a \begin{pmatrix} z^{N+1} \\ y^{N+1} \end{pmatrix} \tag{2.52}$$

for $0 \leq N \leq N_0$. Here L^a and M denote the same matrices as in the evolution equation (2.23).

Proof. In our adapted coordinate system, we decompose the asymptotic Maxwell's equations (2.41) into constraint part ($a = 4$) and evolution part

$(a = \rho)$. If these equations are satisfied by an approximate-plane-wave family for some $N \geq -1$, Proposition 2.4.1 implies that $\partial_4 S$ has no zeros. Under this condition the evolution part of (2.41) alone already implies the constraint part of (2.41). This is easy to verify using the fact that, as outlined in Sect. 2.1, the evolution equations preserve the constraints. In other words, we can forget about the constraints and concentrate on evaluating the evolution part of (2.41). According to (2.23), this takes the form

$$\lim_{\alpha \to 0} \left(\frac{1}{\alpha^N} \left(L^a \partial_a + M \right) \begin{pmatrix} Z(\alpha, \cdot) \\ Y(\alpha, \cdot) \end{pmatrix} \right) = \begin{pmatrix} 0 \\ 0 \end{pmatrix} \tag{2.53}$$

in terms of the variables $Z_1, Z_2, Z_3, Y_1, Y_2, Y_3$. Hence, our approximate-plane-wave family is an asymptotic solution of Maxwell's equations to lowest non-trivial order $N = -1$ if and only if $\partial_4 S$ has no zeros and (2.53) is satisfied for $N = -1$. By feeding (2.50) into (2.53) for $N = -1$ we see that the latter condition is equivalent to (2.51), owing to Lemma 2.2.1.

For $N_0 \geq 0$, our approximate-plane-wave family is an N_0^{th} order solution if and only if in addition (2.53) is satisfied for all $0 \leq N \leq N_0$. Upon feeding (2.50) into (2.53), it is easy to prove by induction over N that this is true if and only if (2.52) is satisfied for $0 \leq N \leq N_0$.   □

Condition (d) of Definition 2.2.1 requires that, if (2.50) represents an approximate-plane-wave family, z^0 and y^0 do not vanish simultaneously. Clearly, such a solution z^0, y^0 of (2.51) exists if and only if

$$\det\left(\partial_a S\, L^a\right) = 0 . \tag{2.54}$$

This is a first order partial differential equation for S, homogeneous of degree six with respect to the components of the gradient of S. If S satisfies (2.54) and if $\partial_4 S$ has no zeros, S is called a *solution of the eikonal equation* of the linear dielectric and permeable medium considered. By Proposition 2.4.2, this is a necessary and sufficient condition for S to be the eikonal function of an approximate-plane-wave family that satisfies Maxwell's equations asymptotically to order $N = -1$ at least. In the theory of partial differential equations (2.54) is called the *characteristic equation* of the system of evolution equations (2.23).

In the next section we discuss the eikonal equation in our medium in more detail. In particular, we free ourselves from the special coordinates used so far.

If we have a solution S of the eikonal equation, Proposition 2.4.2 can be used to construct an asymptotic solution of arbitrarily high order. To that end the amplitudes z^N and y^N have to be determined inductively with the help of (2.51) and (2.52). Clearly, z^{N+1} and y^{N+1} are not uniquely determined through z^N and y^N since, for a solution of the eikonal equation, $\partial_a S\, L^a$ has a non-trivial kernel. Let $P_S(x)$ denote the 6×6 matrix that projects orthogonally onto the kernel of $\partial_a S(x)\, L^a(x)$, where "orthogonally" refers to the scalar product (2.29). For any solution S of (2.54) the rank of $P_S(x)$ is bigger than or equal to one. We shall prove later that, owing to the special form of the matrices $L^a(x)$, the rank of $P_S(x)$ cannot be bigger than two. In general, the rank depends, of course, on x.

Let us write

$$\begin{pmatrix} z_\perp^N \\ y_\perp^N \end{pmatrix} = P_S \begin{pmatrix} z^N \\ y^N \end{pmatrix} \quad \text{and} \quad \begin{pmatrix} z_\parallel^N \\ y_\parallel^N \end{pmatrix} = \begin{pmatrix} z^N \\ y^N \end{pmatrix} - \begin{pmatrix} z_\perp^N \\ y_\perp^N \end{pmatrix}. \tag{2.55}$$

This decomposition of the amplitudes z^N and y^N implies, via (2.50), a decomposition of Z and Y and thus, via (2.22), a decomposition of the electric and of the magnetic component of our approximate-plane-wave family.

In terms of the decomposition (2.55), the inductive scheme for the amplitudes is given by the following proposition.

Proposition 2.4.3. *Let S be a solution of the eikonal equation and fix an integer $N_0 \geq 0$. Then the one-parameter family (2.50) is an N_0^{th} order asymptotic solution of Maxwell's equations if and only if the amplitudes z^N and y^N satisfy*

$$\begin{pmatrix} z_\parallel^0 \\ y_\parallel^0 \end{pmatrix} = \begin{pmatrix} 0 \\ 0 \end{pmatrix} \tag{2.56}$$

and

$$(1 - P_S)(L^a \partial_a + M) \begin{pmatrix} z^N \\ y^N \end{pmatrix} = -i\,\partial_a S\, L^a \begin{pmatrix} z_\parallel^{N+1} \\ y_\parallel^{N+1} \end{pmatrix} \tag{2.57}$$

$$P_S\, L^a\, \partial_a \begin{pmatrix} z_\perp^N \\ y_\perp^N \end{pmatrix} + P_S\, M \begin{pmatrix} z_\perp^N \\ y_\perp^N \end{pmatrix} = -P_S\, (L^a \partial_a + M) \begin{pmatrix} z_\parallel^N \\ y_\parallel^N \end{pmatrix} \tag{2.58}$$

for $0 \leq N \leq N_0$. (2.56) is called the 0th order polarization condition, (2.57) is called the $(N+1)^{\text{th}}$ order polarization condition and (2.58) is called the Nth order transport equation.

Proof. (2.56) is obviously equivalent to (2.51). To prove that (2.57) and (2.58) together are equivalent to (2.52), we decompose (2.52) into two equations by applying P_S and $1 - P_S$ respectively. The first equation gives (2.57), the second equation gives (2.58). This is readily verified with the help of the equations $\partial_a S\, L^a\, P_S = 0$ and $\partial_a S\, P_S\, L^a = 0$. (The first equation is trivial and the second follows from the fact that $\partial_a S\, L^a$ is symmetric with respect to the scalar product (2.29).) $\qquad\square$

Since (2.57) can be solved for $z_\parallel^{N+1}$ and $y_\parallel^{N+1}$, by this equation the $\parallel$-components of z^{N+1} and y^{N+1} are algebraically determined through the lower order amplitudes z^N and y^N. This gives a restriction on the allowed directions of the electric and magnetic field vectors which justifies the name "polarization condition".

If $z_\parallel^N$ and $y_\parallel^N$ are known, (2.58) gives a system of first order differential equations for $z_\perp^N$ and $y_\perp^N$. Later we shall associate solutions of the eikonal equation with congruences of rays. The name "transport equation" refers to the fact that (2.58) gives us ordinary differential equations (i.e., "transport laws") for the components of $z_\perp^N$ and $y_\perp^N$ along each ray, as will be shown in Sect. 2.4 below.

In spacetime regions where $\boldsymbol{P}_S$ has constant rank, (2.58) admits a well-posed initial value problem in the following sense. If

$$1 \leq \operatorname{rank} \boldsymbol{P}_S = k = \text{const.} \tag{2.59}$$

we can choose k basis vector fields $a_1, ..., a_k$ (complex six-tuples depending on x), orthonormal with respect to the scalar product (2.29), such that

$$\boldsymbol{P}_S = \sum_{A=1}^{k} a_A \otimes a_A , \tag{2.60}$$

where $\otimes$ denotes the standard tensor product on $\mathbb{C}^6$. Hence, $z_\perp^N$ and $y_\perp^N$ are of the form

$$\begin{pmatrix} z_\perp^N \\ y_\perp^N \end{pmatrix} = \sum_{A=1}^{k} \xi_A^N a_A \tag{2.61}$$

with some $\mathbb{C}$-valued functions ξ_A^N. Then the Nth order transport equation (2.58) gives a system of k differential equations for the k coefficients ξ_A^N which is symmetric hyperbolic. (This follows from the facts that each matrix $\boldsymbol{L}^a$ is symmetric with respect to the scalar product (2.29) and that $\boldsymbol{L}^4$ is close to **1**.) Hence, local existence and uniqueness of solutions $\xi_1^N, ..., \xi_k^N$ is guaranteed for arbitrary initial values given on a hypersurface $x^4 = $ const. By solving the transport equations in this way at each level N, we determine that part of the polarization direction which is not fixed already by the polarization condition, and we determine the intensity of our approximate plane wave.

Now it is clear how, for a solution S of the eikonal function that satisfies the rank condition (2.59), the amplitudes z^N and y^N can be determined inductively to construct an N_0^{th} order asymptotic solution of Maxwell's equations.

1. The induction starts with setting $z_\parallel^0 = y_\parallel^0 = 0$.
2. The Nth step of the induction, $0 \leq N \leq N_0$, is given by the following prescription. With $z_\parallel^N$ and $y_\parallel^N$ known, determine $z_\perp^N$ and $y_\perp^N$ by solving (2.58) with arbitrary initial values. (The only restriction on the initial values is that $z_\perp^0$ and $y_\perp^0$ must not vanish simultaneously.) Then, determine $z_\parallel^{N+1}$ and $y_\parallel^{N+1}$ with the help of (2.57).

The other amplitudes (i.e, $z_\perp^N$, $y_\perp^N$, $z_\parallel^{N+1}$, $y_\parallel^{N+1}$ for $N \geq N_0 + 1$) and the $O(\alpha^{N_0+2})$ term can be chosen arbitrarily. (E.g., they could be set equal to zero.) Then (2.50) gives an approximate-plane-wave family that satisfies Maxwell's equations asymptotically to order N_0.

This construction can be carried through for arbitrarily large N_0, i.e., it can be used to construct (non-convergent) infinite asymptotic series solutions of Maxwell's equations. In the very special case that the induction yields $z_\parallel^N = y_\parallel^N = 0$ for some $N \geq 1$ we can set z^M and y^M equal to zero for $M \geq N$ to get an approximate-plane-wave family that satisfies Maxwell's equations exactly for all $\alpha \in \mathbb{R}^+$.

The results of this section show how to construct, locally around any spacetime point, an approximate-plane-wave family that satisfies Maxwell's equations in a linear dielectric and permeable medium asymptotically to some order $N \geq 0$. The physical relevance of those one-parameter families is in the fact that they can be interpreted as *approximate* solutions of Maxwell's equations as well. This will be proven in Sect. 2.7 below. Already now we emphasize that this is not true for asymptotic solutions of lowest non-trivial order $N = -1$. In other words, if it is our goal to set up a viable approximation scheme for exact Maxwell fields we have to consider approximate-plane-wave families that satisfy Maxwell's equations asymptotically to order $N = 0$ at least. In this order we get polarization conditions that fix $z_\parallel^0$, $y_\parallel^0$, $z_\parallel^1$, $y_\parallel^1$, and we get transport equations for $z_\perp^0$ and $y_\perp^0$. This $N = 0$ theory is often called the *geometric optics approximation* of Maxwell fields.

2.5 Discussion of the eikonal equation

In the preceding section we have derived the eikonal equation of our medium, locally around an arbitrarily chosen point, in a special coordinate system. It is now our goal to analyze the structure of this equation and, in particular, to rewrite the eikonal equation in covariant form.

In a coordinate system adapted to the rest system of the medium, the eikonal equation was given by (2.54) supplemented with the condition that $\partial_4 S$ has no zeros. Clearly, the characteristic matrix $\boldsymbol{L}(x,p) = p_a \boldsymbol{L}^a(x)$ is a real 6×6 matrix, symmetric with respect to the scalar product (2.29). Hence, it has six real eigenvalues and the characteristic determinant $\det(p_a \, \boldsymbol{L}^a(x))$ is the product of these eigenvalues. If we want to bring the eikonal equation in a more explicit form we have to determine these six eigenvalues.

First we reduce this six-dimensional eigenvalue problem to a three-dimensional eigenvalue problem. To that end we introduce, for all x in the spacetime neighborhood considered and for all $p = (p_1, p_2, p_3, p_4) \in \mathbb{R}^4$, the real 3×3 matrix

$$\boldsymbol{W}(x,p) = \frac{1}{\sqrt{-g_{44}(x)}} \left(p_4 \, \boldsymbol{Q}(x) + p_\rho \, \boldsymbol{A}^\rho(x) \right) \qquad (2.62)$$

which, by (2.25), enters into the characteristic matrix according to

$$p_a \mathbf{L}^a(x) = p_4 \begin{pmatrix} 1 & 0 \\ 0 & 1 \end{pmatrix} + \sqrt{-g_{44}(x)} \begin{pmatrix} 0 & \mathbf{W}(x,p)^T \\ \mathbf{W}(x,p) & 0 \end{pmatrix}. \tag{2.63}$$

The (strictly positive) factor $\sqrt{-g_{44}(x)}$ was introduced in (2.62) for later convenience. Then the 3×3 matrix $\mathbf{W}(x,p)^T \mathbf{W}(x,p)$ is obviously positive semidefinite and symmetric with respect to the scalar product (2.29). Hence, it has three real eigenvectors $u_1(x,p), u_2(x,p), u_3(x,p)$ which are orthonormal with respect to the scalar product (2.29), and the pertaining eigenvalues are real and non-negative. We denote these eigenvalues by $h_1(x,p)^2$, $h_2(x,p)^2$, $h_3(x,p)^2$ with $h_A(x,p) \geq 0$ for $A = 1,2,3$. Similarly, the 3×3 matrix $\mathbf{W}(x,p)\,\mathbf{W}(x,p)^T$ has three real eigenvectors $v_1(x,p), v_2(x,p), v_3(x,p)$ which are orthonormal with respect to the scalar product (2.29), and the pertaining eigenvalues are the same as for $\mathbf{W}(x,p)^T \mathbf{W}(x,p)$, i.e.,

$$\begin{aligned} \mathbf{W}(x,p)^T \, \mathbf{W}(x,p)\, u_A(x,p) &= h_A(x,p)^2 \, u_A(x,p)\,, \\ \mathbf{W}(x,p)\, \mathbf{W}(x,p)^T \, v_A(x,p) &= h_A(x,p)^2 \, v_A(x,p)\,, \end{aligned} \tag{2.64}$$

for $A = 1,2,3$. The bases of eigenvectors can be chosen in such a way that

$$\begin{aligned} \mathbf{W}(x,p)\, u_A(x,p) &= h_A(x,p)\, v_A(x,p)\,, \\ \mathbf{W}(x,p)^T \, v_A(x,p) &= h_A(x,p)\, u_A(x,p)\,, \end{aligned} \tag{2.65}$$

for $A = 1,2,3$. (In the non-degenerate case, i.e., if the eigenvalues $h_1(x,p)^2$, $h_2(x,p)^2$, $h_3(x,p)^2$ are mutually different, the eigenvectors $u_A(x,p)$ and $v_A(x,p)$ are unique up to sign and the equations (2.65) are automatically true up to sign.) These equations imply that the characteristic matrix (2.63) satisfies

$$p_a \mathbf{L}^a(x) \begin{pmatrix} u_A(x,p) \\ \pm v_A(x,p) \end{pmatrix} = \left(p_4 \pm \sqrt{-g_{44}(x)}\, h_A(x,p) \right) \begin{pmatrix} u_A(x,p) \\ \pm v_A(x,p) \end{pmatrix} \tag{2.66}$$

for $A = 1,2,3$. This equation gives us six (real) eigenvalues of the 6×6 matrix $p_a \mathbf{L}^a$ and pertaining eigenvectors in terms of the eigenvalues and eigenvectors of the 3×3 matrices $\mathbf{W}(x,p)^T \, \mathbf{W}(x,p)$ and $\mathbf{W}(x,p)\,\mathbf{W}(x,p)^T$. As the characteristic determinant is the product of these six eigenvalues, the eikonal equation (2.54) takes the form

$$\prod_{A=1}^{3} \left((\partial_4 S)^2 + g_{44}\, h_A(\,\cdot\,,\partial S)^2 \right) = 0 \tag{2.67}$$

supplemented with the condition that $\partial_4 S$ has no zeros. To get a more explicit form of the eikonal equation, we have to calculate the eigenvalues $h_A(x,p)^2$ of the matrix $W(x,p)^T W(x,p)$. If we insert the general expressions (2.26) and (2.27) for the components of the matrices $Q(x)$ and $A^\rho(x)$ into the definition (2.62) of $W(x,p)$, we find that the components of the matrix $R(x,p) = W(x,p)^T W(x,p)$ are

$$R_\sigma{}^\tau(x,p) = R^{ab}{}_\sigma{}^\tau(x)\, p_a\, p_b \qquad (2.68)$$

with

$$R^{ab}{}_\sigma{}^\tau(x) = \frac{1}{g_{44}}\, v_\sigma{}^\rho(x)\, \eta_{\rho 4}{}^{a\gamma}(x)\, w_\gamma{}^\lambda(x)\, w_\lambda{}^\nu(x)\, \eta_{\nu 4}{}^{b\kappa}(x)\, v_\kappa{}^\tau(x)\,. \qquad (2.69)$$

The three eigenvalues $h_1(x,p)^2$, $h_2(x,p)^2$ and $h_3(x,p)^2$ of the matrix $R(x,p)$ are then given by

$$h_{1/2}(x,p)^2 = \tfrac{1}{2}\, R^{ab}{}_\sigma{}^\sigma(x)\, p_a\, p_b \pm$$

$$\sqrt{\left(\tfrac{1}{2}\, R^{ab}{}_\rho{}^\mu(x)\, R^{cd}{}_\mu{}^\rho(x) - \tfrac{1}{4}\, R^{ab}{}_\sigma{}^\sigma(x)\, R^{cd}{}_\tau{}^\tau(x)\right) p_a\, p_b\, p_c\, p_d}\,, \qquad (2.70)$$

$$h_3(x,p)^2 = 0\,.$$

The appearance of the square root in (2.5) has the unpleasant consequence that h_1 and h_2 might fail to be differentiable at some points even if all input functions are C^∞ as tacitly assumed throughout Part I. In the following we assume that h_1 and h_2 are C^∞ functions at all points with $(p_1, p_2, p_3, p_4) \neq (0,0,0,0)$.

The whole calculation was done around an arbitrarily chosen spacetime point x_0, in a coordinate system adapted to the rest system U^a of the medium. From Sect. 2.1 we know that such a coordinate system is unique, locally near x_0, to within coordinate transformations of the special form (2.18). If we perform such a coordinate change, viewing $p = (p_1, p_2, p_3, p_4)$ as canonical momentum coordinates conjugate to $x = (x^1, x^2, x^3, x^4)$ which transform as

$$p_a \longmapsto p'_a = \frac{\partial x^b}{\partial x'^a}\, p_b\,, \qquad (2.71)$$

the components of the matrix $R(x,p) = W(x,p)^T W(x,p)$ transform according to

$$R'_\lambda{}^\tau(x',p') = \frac{\partial x^\sigma}{\partial x'^\lambda}\, \frac{\partial x'^\tau}{\partial x^\rho}\, R_\sigma{}^\rho(x,p)\,, \qquad (2.72)$$

as can be read from (2.68) and (2.69). (That is the reason why we introduced the factor $\sqrt{-g_{44}}$ in (2.62).) The eigenvalues of the matrix $R(x,p)$ are, thus, invariant under coordinate transformations of the form (2.18), i.e., $h'_A(x',p')^2 = h_A(x,p)^2$. In other words, h_1 and h_2 are uniquely determined (global and invariant) functions on the cotangent bundle over spacetime.

Hence, for $A = 1$ and $A = 2$, the function

$$H_A(x,p) = \tfrac{1}{2}\Big(h_A(x,p)^2 - U^a(x)\,U^b(x)\,p_a\,p_b\Big) \tag{2.73}$$

is a uniquely determined (global and invariant) function on the cotangent bundle over spacetime. We refer to H_1 and H_2 as to the *partial Hamiltonians* of our linear dielectric and permeable medium. The eikonal equation can then be formulated in the following way.

Proposition 2.5.1. *A real-valued function S, defined on some open space-time region $\mathcal{U}$, is a solution of the eikonal equation if and only if*

$$H_1\big(x, \partial S(x)\big)\, H_2\big(x, \partial S(x)\big) = 0 \tag{2.74}$$

and $\partial S(x) \neq (0,0,0,0)$ for all $x \in \mathcal{U}$. Here H_1 and H_2 denote the partial Hamiltonians introduced in (2.73).

Proof. S is a solution of the eikonal equation near any spacetime point if and only if, in adapted coordinates near this point, (2.67) holds and $\partial_4 S$ has no zeros. Since, by (2.5), h_3 vanishes, this is true if and only if

$$\Big(h_1(\cdot,\partial S)^2 + \tfrac{1}{g_{44}}(\partial_4 S)^2\Big)\Big(h_2(\cdot,\partial S)^2 + \tfrac{1}{g_{44}}(\partial_4 S)^2\Big) = 0 \tag{2.75}$$

holds and $\partial_4 S$ has no zeros. From (2.5) we read that $h_1(x,p)$ and $h_2(x,p)$ are non-zero at points (x,p) with $p_4 = 0$ but $(p_1,p_2,p_3) \neq (0,0,0)$. (This follows from the fact that $(v_\sigma{}^\rho(x))$ and $(w_\sigma{}^\rho(x))$ are invertible 3×3 matrices and that the kernel of the matrix $(\eta_{\sigma 4}{}^{\mu\rho}(x)p_\mu)$ is exactly one-dimensional if $(p_1,p_2,p_3) \neq (0,0,0)$.) Hence, for a solution of (2.75) the condition $\partial_4 S \neq 0$ is equivalent to $\partial S \neq (0,0,0,0)$. With the help of (2.73) we can rewrite (2.75) in the coordinate invariant form (2.74). $\qquad\square$

We shall refer to the equations

$$H_A\big(x, \partial S(x)\big) = 0 \tag{2.76}$$

for $A = 1$ and $A = 2$ as to the *partial eikonal equations*. A solution of the eikonal equation has to satisfy at each point at least one of the two partial eikonal equations. In the terminology of classical mechanics, (2.76) is called the *Hamilton-Jacobi equation* of the Hamiltonian H_A.

The set of all (x,p) with $p \neq (0,0,0,0)$ and

$$H_A(x,p) = 0 \tag{2.77}$$

is called the *A-branch of the characteristic variety* and the equation (2.77) is called the *A-dispersion relation* of our medium. The following proposition gives some information on the geometry of the A-branch of the characteristic variety.

Proposition 2.5.2. *For $A = 1$ and $A = 2$, the partial Hamiltonian H_A introduced in (2.73) has the following properties.*

(a) H_A *is homogeneous of degree two with respect to the momentum coordinates,*

$$H_A(x, tp) = t^2 H_A(x, p) \tag{2.78}$$

for all real numbers t.

(b) H_A *satisfies the differential equation*

$$U_a(x) \frac{\partial H_A(x, p)}{\partial p_a} = \frac{1}{2} U^b(x) p_b. \tag{2.79}$$

(c) *At all points (x, p) with $p \neq (0, 0, 0, 0)$ but $U^b(x) p_b = 0$ the partial Hamiltonian is strictly positive,*

$$H_A(x, p) > 0. \tag{2.80}$$

In (b) and (c), U^a denotes the rest system of the medium.

Proof. (a) is obviously true in the special coordinates where $h_A(x, p)^2$ is given by (2.5). As a consequence, it is true in any coordinates since the conjugate momenta transform homogeneously according to (2.71). To prove (b), we read from (2.5) that, in the special coordinates used there, the momentum coordinates enter into $h_A(x, p)^2$ only in terms of the combination $g_{44}(x) p_\sigma - g_{4\sigma}(x) p_4$. Thus, the coordinate invariant differential equation $U_a \frac{\partial}{\partial p_a} \left(h_A(x, p)^2 \right) = 0$ holds true. To prove (c) it suffices to verify from (2.5) that $h_A(x, p)^2$ is non-zero if, in the coordinates used there, $p_4 = 0$ but $(p_1, p_2, p_3) \neq (0, 0, 0)$. This follows from the fact that $(v_\sigma{}^\rho(x))$ and $(w_\sigma{}^\rho(x))$ are invertible 3×3 matrices and that the kernel of the matrix $(\eta_{\sigma4}{}^{\mu\rho}(x)p_\mu)$ is exactly one-dimensional if $(p_1, p_2, p_3) \neq (0, 0, 0)$. $\square$

By differentiating (2.78) with respect to t and setting $t = 1$ afterwards, part (a) of this proposition implies that H_A satisfies the equations

$$p_a \frac{\partial H_A(x, p)}{\partial p_a} = 2 H_A(x, p), \tag{2.81}$$

$$\frac{1}{2} p_a \, p_b \frac{\partial^2 H_A(x, p)}{\partial p_a \partial p_b} = H_A(x, p). \tag{2.82}$$

Thus, the Hamiltonian H_A is similar to the quadratic form of metric tensor, $H(x, p) = \frac{1}{2} g^{ab}(x) p_a p_b$, but with a metric tensor that depends not only on x but also homogeneously on p. Such generalized metrics are usually called *Finsler metrics*; we may thus say that each partial Hamiltonian H_A defines a Finsler metric on the cotangent bundle over spacetime. Note, however, that some authors include the assumption of positive-definiteness into the definition of the term "Finsler metric", whereas our metric $(\partial^2 H_A(x, p)/\partial p_a \partial_b)$ cannot be positive definite. This follows from differentiating (2.79) with respect to p_b which demonstrates that $U_a U_b \partial^2 H_A(x, p)/\partial p_a \partial_b < 0$. For literature on Finsler structures we refer to Rund [124] and to Asanov [10].

From part (b) and (c) of Proposition 2.5.2 we read that

$$\left(\frac{\partial H_A}{\partial p_1}, \ldots, \frac{\partial H_A}{\partial p_4}\right) \neq (0,0,0,0) \tag{2.83}$$

on the A-branch of the characteristic variety. Hence, this branch is a codimension-one submanifold of the cotangent bundle which is transverse to the fibers. By part (a) of Proposition 2.5.2, the intersection of this manifold with each fiber has a "conic" structure.

In general, the union of the 1-branch and of the 2-branch of the characteristic variety need not be a manifold. The two branches might intersect or coalesce. It is, of course, also possible that the two branches coincide completely. (This is necessarily true if the medium is isotropic, as we shall verify soon.) Whenever the two branches do not coincide, the medium is called *birefringent* or *double-refractive*.

The fact that the two branches can intersect or coalesce is related to the following unpleasant feature. Whereas (2.83) guarantees that either partial eikonal equation (2.76) can be solved, locally around any one point, for one of the partial derivatives $\partial_1 S$, $\partial_2 S$, $\partial_3 S$, $\partial_4 S$, this is not necessarily true for the full eikonal equation (2.74). Hence it is not guaranteed that we can find a hypersurface through each point such that initial data for S on that hypersurface determine a solution of (2.74) *uniquely*.

The term "birefringence" refers to the fact that a light wave that enters into such a medium from vacuum will split up, in general, into two waves. In the appproximate-plane-wave setting considered here, one of the two waves has an eikonal function that solves (2.76) with $A = 1$, the other one with $A = 2$. In general, a solution of the full eikonal equation (2.74) can satisfy (2.76) with $A = 1$ at some points and with $A = 2$ at other points. Moreover, there might be solutions of the full eikonal equation which solve (2.74) with $A = 1$ and with $A = 2$ simultaneously. If the two branches of the characteristic variety coincide, this is true for all solutions of the full eikonal equation.

It is worthwhile to note that the partial Hamiltonians can be changed according to

$$H_A(x,p) \longmapsto \tilde{H}_A(x,p) = F_A(x,p)\, H_A(x,p) \tag{2.84}$$

for $A = 1, 2$, where F_A is any real-valued function that has no zeros on the A-branch of the characteristic variety. Clearly, such a transformation does not affect the solutions of the partial eikonal equations. In this sense, the dynamics of wave surfaces in our medium determines two *equivalence classes* of Hamiltonians. A transformation of the form (2.84) does, of course, not preserve the degree-two homogeneity of H_A with respect to the momentum coordinates. Thus, it will lead to a representation in which the Finsler structure is "hidden".

Finally, we illustrate the results of this section by considering two more special kinds of media.

Example 2.5.1. The Isotropic Case

If our linear dielectric and permeable medium is isotropic in the sense of Definition 2.1.2, the two branches of the characteristic variety coincide and are given by the null cone bundle of a Lorentzian metric. In particular, there is no birefringence in an isotropic medium. To verify these well-known facts with the help of our general results, we first observe that, in the isotropic case, (2.21) simplifies to

$$v_\sigma{}^\tau = \frac{1}{\sqrt{\mu}}\,\delta_\sigma^\tau \quad \text{and} \quad w_\sigma{}^\tau = \frac{1}{\sqrt{\varepsilon}}\,\delta_\sigma^\tau\,. \tag{2.85}$$

Upon inserting this into (2.69) and using the identity (2.7) of the Levi-Civita tensor field, the non-zero eigenvalues in (2.5) take the form

$$h_1(x,p)^2 = h_2(x,p)^2 = \frac{g^{ab}(x) + U^a(x)\,U^b(x)}{\varepsilon(x)\mu(x)}\,p_a\,p_b\,. \tag{2.86}$$

Thus, the partial Hamiltonians (2.73) coincide and are given by

$$H(x,p) = H_1(x,p) = H_2(x,p) = \tfrac{1}{2}\,g_o^{ab}(x)\,p_a\,p_b\,, \tag{2.87}$$

where

$$g_o^{ab} = \frac{1}{\varepsilon\mu}\left(g^{ab} + U^a\,U^b\right) - U^a\,U^b \tag{2.88}$$

are the contravariant components of a Lorentzian metric which is called the *optical metric* of the isotropic medium. Please note that $g_o^{ab}\,U_a\,U_b = -1$ and that $g_o^{ab}\,U_a\,X_b = 0$ implies $g_o^{ab}\,X_a\,X_b = \frac{1}{\varepsilon\mu}\,g^{ab}\,X_a\,X_b$. Both observations together imply that the optical metric is, indeed, of Lorentzian signature $(+,+,+,-)$.

The strictly positive function

$$n = \sqrt{\varepsilon\mu} \tag{2.89}$$

is called the *index of refraction* of the isotropic medium. If $n = 1$ (and, in particular, in the vacuum case $\varepsilon = \mu = 1$) the optical metric and the spacetime metric coincide, $g_o^{ab} = g^{ab}$. Hence, the eikonal equation in an isotropic medium has exactly the same structure as in vacuum; we just have to replace the spacetime metric with the optical metric. This is a well-known result. It was derived, with increasing mathematical rigor, by Gordon [50], Pham Mau Quan [117] and Ehlers [38].

Example 2.5.2. The Uniaxial Case

Now we specialize from a general linear dielectric and permeable medium to the case that the permeability tensor field has the same form as for vacuum, $\mu^{ab} = g^{ab} + U^a\,U^b$. Moreover, we assume that the dielectricity tensor field, which can be written in the form $\varepsilon^{ab} = \varepsilon_1\,X_1^a\,X_1^b + \varepsilon_2\,X_2^a\,X_2^b + \varepsilon_3\,X_3^a\,X_3^b$ with

$g_{ab} X^a_\rho X^b_\sigma = \delta_{\rho\sigma}$, $g_{ab} X^a_\sigma U^b = 0$ and $\varepsilon_\sigma > 0$ for $\sigma = 1, 2, 3$, has a double eigenvalue, say $\varepsilon_1 = \varepsilon_2$. In this case the functions $h_1(x,p)^2$ and $h_2(x,p)^2$ of (2.5) are bilinear with respect to the momentum coordinates. An example for such a medium is a *uniaxial crystal*. For the partial Hamiltonians (2.73) we find in this special case after a quick calculation

$$H_A(x,p) = \tfrac{1}{2} g^{ab}_{oA}(x)\, p_a\, p_b \qquad (2.90)$$

for $A = 1$ and $A = 2$, where

$$g^{ab}_{o1} = \frac{1}{\varepsilon_1}\left(g^{ab} + U^a\, U^b\right) - U^a\, U^b\,,$$

$$g^{ab}_{o2} = \frac{1}{\varepsilon_3}\left(X^a_1\, X^b_1 + X^a_2\, X^b_2\right) + \frac{1}{\varepsilon_1}\, X^a_3\, X^b_3 - U^a\, U^b\,. \qquad (2.91)$$

Hence, either branch of the characteristic variety is the null cone bundle of a Lorentzian metric. Generalizing the terminology from the isotropic case, the two metrics (2.91) can be called the *optical metrics* of the medium. The first optical metric does not distinguish a spatial direction, i.e., it is of the same kind as the optical metric in an isotropic medium. The second optical metric, however, reflects the fact that the X_3-direction is distinguished in the medium considered. In a situation like this the 1-branch of the characteristic variety is usually called the *ordinary branch* whereas the 2-branch is called the *extraordinary branch*. In this terminology solutions of the partial eikonal equation (2.76) with $A = 1$ are associated with *ordinary waves* and solutions with $A = 2$ are associated with *extraordinary waves*.

If the eigenvalues ε_1, ε_2, ε_3 of the dielectricity tensor field are mutually different, the two characteristic varieties are no longer the null cone bundles of Lorentzian metrics. An example for such a medium is a *biaxial crystal*. If we want to speak of "optical metrics" in such a medium, we have to understand the term "metric" in the Finslerian sense. Both these optical Finsler metrics display the anisotropy of the medium in a symmetrical way, i.e., it is not justified to distinguish one of them by the attribute "ordinary". For this reason, we prefer to speak of 1-waves and 2-waves (rather than of ordinary waves and extraordinary waves) in general anisotropic media.

2.6 Discussion of transport equations and the introduction of rays

In this section we associate solutions of the eikonal equation in a linear dielectric and permeable medium with congruences of rays. The guiding idea is to introduce the notion of rays in such a way that the transport equations (2.58) can be reinterpreted as ordinary differential equations along rays.

According to Proposition 2.5.1 the left-hand side of the eikonal equation has a product structure. This suggests to introduce, for solutions S of the eikonal equation, the following terminology. S is called a *solution of multiplicity two* iff it satisfies both partial eikonal equations (2.76), and it is called a *solution of multiplicity one* iff it satisfies exactly one of the two partial eikonal equations. The multiplicity can, of course, change from point to point.

In the following we consider solutions of the eikonal equation on neighborhoods where the multiplicity is constant. Note that, for any given solution, there exists such a neighborhood near almost all spacetime points. We begin our discussion with solutions of multiplicity one, later we consider the somewhat more complicated case of solutions of multiplicity two. We introduce the following definition.

Definition 2.6.1. *Let S be a solution of the eikonal equation according to Proposition 2.5.1. Assume that, on some open spacetime region $\mathcal{U}$, S is of multiplicity one, i.e., that the partial eikonal equation (2.76) is satisfied for $A = 1$, say, but not for $A = 2$ at all points of $\mathcal{U}$. Then the vector field*

$$K^a(x) = \frac{\partial H_1}{\partial p_a}\left(x, \partial S(x)\right) \tag{2.92}$$

on $\mathcal{U}$ is called the transport vector field *and its integral curves are called the* rays *associated with S.*

In the theory of partial differential equations the rays are also known as (*bi-*)*characteristic curves.*

Owing to (2.83) the transport vector field has no zeros, i.e., the rays are immersed curves. Changing the partial Hamiltonian according to (2.84) corresponds to reparametrizing the rays. Please note that

$$\partial_a S(x)\, K^a(x) = 0 \,. \tag{2.93}$$

This follows from the fact that H_1 satisfies equation (2.81) which was a consequence of the homogeneity property established in Proposition 2.5.2 (a). Thus, the transport vector field is tangent to the hypersurfaces $S = \text{const}$.

We want to show now that the transport equations can be viewed as ordinary differential equations along rays. We do that locally around any point x_0 of the neighborhood $\mathcal{U}$ mentioned in Definition 2.6.1. To that end we introduce a coordinate system adapted to the rest system U^a of the medium near x_0. (Please recall Definition 2.1.3.) In such a coordinate system, the partial Hamiltonians (2.73) take the form

$$H_A(x,p) = \frac{1}{2}\left(h_A(x,p) + \frac{p_4}{\sqrt{-g_{44}(x)}}\right)\left(h_A(x,p) - \frac{p_4}{\sqrt{-g_{44}(x)}}\right), \tag{2.94}$$

i.e., the partial eikonal equations (2.76) are equivalent to

$$h_A\big(x, \partial S(x)\big) \pm \frac{\partial_4 S}{\sqrt{-g_{44}(x)}} = 0 \qquad (2.95)$$

for $A = 1, 2$. Here and in the following, the upper sign corresponds to negative frequency solutions, $\partial_4 S < 0$, and the lower sign corresponds to positive frequency solutions, $\partial_4 S > 0$. With (2.95) the partial derivative of H_A takes the form

$$\frac{\partial H_A}{\partial p_a}\big(x, \partial S(x)\big) = \frac{\partial_4 S}{\sqrt{-g_{44}(x)}}\left(\mp \frac{\partial h_A}{\partial p_a}\big(x, \partial S(x)\big) - \frac{\delta_4^a}{\sqrt{-g_{44}(x)}}\right). \qquad (2.96)$$

With these informations at hand, we take a closer look at the Nth order transport equation, i.e., at equation (2.58) viewed as a differential equation for $z_\perp^N$ and $y_\perp^N$ with $z_\parallel^N$ and $y_\parallel^N$ assumed known. In the situation of Definition 2.6.1, the projection operator $\boldsymbol{P}_S(x)$ onto the kernel of $\partial_a S(x)\,\boldsymbol{L}^a(x)$ is given, in terms of the eigenvectors u_A and v_A of (2.66), by

$$\boldsymbol{P}_S(x) = \begin{pmatrix} u_1\big(x, \partial S(x)\big) \\ \pm v_1\big(x, \partial S(x)\big) \end{pmatrix} \otimes \begin{pmatrix} u_1\big(x, \partial S(x)\big) \\ \pm v_1\big(x, \partial S(x)\big) \end{pmatrix}, \qquad (2.97)$$

and $z_\perp^N$ and $y_\perp^N$ are necessarily of the form

$$\begin{pmatrix} z_\perp^N(x) \\ y_\perp^N(x) \end{pmatrix} = \xi^N(x) \begin{pmatrix} u_1\big(x, \partial S(x)\big) \\ \pm v_1\big(x, \partial S(x)\big) \end{pmatrix} \qquad (2.98)$$

with a $\mathbb{C}$-valued function ξ^N. After multiplication with the non-vanishing factor $\partial_4 S(x)/g_{44}(x)$, the Nth order transport equation (2.58) reduces to a differential equation for the function ξ^N of the form

$$K^a(x)\,\partial_a \xi^N(x) + f(x)\,\xi^N(x) + k^N(x) = 0. \qquad (2.99)$$

Here K^a is an abbreviation for

$$K^a(x) = \frac{\partial_4 S(x)}{g_{44}(x)}\begin{pmatrix} u_1\big(x, \partial S(x)\big) \\ \pm v_1\big(x, \partial S(x)\big) \end{pmatrix} \cdot \boldsymbol{L}^a(x)\begin{pmatrix} u_1\big(x, \partial S(x)\big) \\ \pm v_1\big(x, \partial S(x)\big) \end{pmatrix} \qquad (2.100)$$

and $f(x)$ and $k^N(x)$ are known $\mathbb{C}$-valued functions. Clearly, (2.99) gives an ordinary differential equation for ξ^N along each integral curve of the vector field K^a. We now show that K^a is, indeed, the transport vector field defined through (2.92). We first observe that (2.66) implies

$$\begin{pmatrix} u_A(x, p) \\ \pm v_A(x, p) \end{pmatrix} \cdot p_a \boldsymbol{L}^a(x)\begin{pmatrix} u_B(x, p) \\ \pm v_B(x, p) \end{pmatrix} = $$

$$\Big(p_4 \pm \sqrt{-g_{44}(x)}\, h_A(x, p)\Big)\delta_{AB} \qquad (2.101)$$

for $A, B = 1, 2$. Upon differentiation with respect to p_b, (2.101) yields

$$\begin{pmatrix} u_A(x,p) \\ \pm v_A(x,p) \end{pmatrix} \cdot L^b(x) \begin{pmatrix} u_B(x,p) \\ \pm v_B(x,p) \end{pmatrix} \pm$$

$$\sqrt{-g_{44}(x)} \left(h_A(x,p) - h_B(x,p) \right) \begin{pmatrix} u_A(x,p) \\ \pm v_A(x,p) \end{pmatrix} \cdot \frac{\partial}{\partial p_b} \begin{pmatrix} u_B(x,p) \\ \pm v_B(x,p) \end{pmatrix} =$$

$$\left(\delta_4^b \pm \sqrt{-g_{44}(x)} \frac{\partial h_A}{\partial p_b}(x,p) \right) \delta_{AB} . \tag{2.102}$$

If we evaluate this equation with $A = B = 1$ along $p = \partial S(x)$, we see that the right-hand side of (2.100) coincides with the right-hand side of (2.96) for $A = 1$. This proves that the vector field K^a in (2.99) is, indeed, the transport vector field associated with S.

Now let us investigate to what extent these results carry over to the case of solutions of multiplicity two. In analogy to Definition 2.6.1, we introduce the following notions.

Definition 2.6.2. *Let S be a solution of the eikonal equation according to Proposition 2.5.1. Assume that, on some open spacetime region $\mathcal{U}$, S is of multiplicity two, i.e., that the partial eikonal equation (2.76) is satisfied for $A = 1$ and $A = 2$ at all points of $\mathcal{U}$. Then for $A = 1$ and $A = 2$ the vector field*

$$K_A^a(x) = \frac{\partial H_A}{\partial p_a}\left(x, \partial S(x)\right) \tag{2.103}$$

on $\mathcal{U}$ is called the A-transport vector field and its integral curves are called the A-rays associated with S. We shall also refer to $K_1^a(x)$ and $K_2^a(x)$ as to the partial transport vector fields associated with S.

For a solution of multiplicity two, K_1^a and K_2^a may or may not coincide. (If they are collinear, they can be made equal by a transformation of the form (2.84).) If the two branches of the characteristic variety coincide, all solutions are of multiplicity two with $K_1^a = K_2^a$. This is the case for an isotropic medium where, by (2.87),

$$K_1^a(x) = K_2^a(x) = g_o^{ab}(x) \partial_b S(x) . \tag{2.104}$$

Hence, there is only one congruence of rays associated with each solution of the eikonal equation in an isotropic medium.

In the anisotropic case we have to live with the situation that solutions of the eikonal equation might be associated with two different congruences of rays. Clearly, this makes it more complicated to interpret the transport equations as ordinary differential equations along rays. We are now going to work out the details.

In the situation of Definition 2.6.2, (2.97) is to be replaced with

$$P_S(x) = \sum_{A=1}^{2} \begin{pmatrix} u_A(x,p) \\ \pm v_A(x,p) \end{pmatrix} \otimes \begin{pmatrix} u_A(x,p) \\ \pm v_A(x,p) \end{pmatrix} \tag{2.105}$$

and $z_\perp^N$ and $y_\perp^N$ are of the form

$$\begin{pmatrix} z_\perp^N(x) \\ y_\perp^N(x) \end{pmatrix} = \sum_{A=1}^{2} \xi_A^N(x) \begin{pmatrix} u_A(x,\partial S(x)) \\ \pm v_A(x,\partial S(x)) \end{pmatrix} \tag{2.106}$$

with two $\mathbb{C}$-valued functions ξ_1^N and ξ_2^N to be determined. Upon multiplication with the non-vanishing factor $\partial_4 S(x)/g_{44}(x)$, the transport equation (2.58) gives a system of two coupled differential equations for ξ_1^N and ξ_2^N of the form

$$K_A^a(x)\,\partial_a \xi_A^N(x) + \sum_{B=1}^{2} f_{AB}(x)\,\xi_B^N(x) + k_A^N(x) = 0 , \quad A = 1,2 . \tag{2.107}$$

Here K_A^a is an abbreviation for

$$K_A^a(x) = \frac{\partial_4 S(x)}{g_{44}(x)} \begin{pmatrix} u_A(x,\partial S(x)) \\ \pm v_A(x,\partial S(x)) \end{pmatrix} \cdot L^a(x) \begin{pmatrix} u_A(x,\partial S(x)) \\ \pm v_A(x,\partial S(x)) \end{pmatrix} \tag{2.108}$$

for $A = 1,2$ and f_{AB}, k_A^N are known $\mathbb{C}$-valued functions. To put the transport equation into the form (2.107), we made use of the fact that, by (2.6), our multiplicity-two solution satisfies

$$\begin{pmatrix} u_1(x,\partial S(x)) \\ \pm v_1(x,\partial S(x)) \end{pmatrix} \cdot L^a(x) \begin{pmatrix} u_2(x,\partial S(x)) \\ \pm v_2(x,\partial S(x)) \end{pmatrix} =$$
$$\begin{pmatrix} u_2(x,\partial S(x)) \\ \pm v_2(x,\partial S(x)) \end{pmatrix} \cdot L^a(x) \begin{pmatrix} u_1(x,\partial S(x)) \\ \pm v_1(x,\partial S(x)) \end{pmatrix} = 0 . \tag{2.109}$$

To verify that the K_A^a given by (2.108) are, indeed, the partial transport vector fields defined in (2.103), we consider (2.6) with $A = B$. This shows that the right-hand side of (2.108) coincides with the right-hand side of (2.96).

If the two partial transport vector fields coincide, (2.107) gives an ordinary differential equation for the tuple (ξ_1^N, ξ_2^N) along the integral curves of $K_1^a = K_2^a$. In the general case, the situation is more complicated. (2.107) with $A = 1$ gives an ordinary differential equation for ξ_1^N along the integral curves of K_1^a that involves ξ_2^N, and (2.107) with $A = 2$ gives an ordinary differential equation for ξ_2^N along the integral curves of K_2^a that involves ξ_1^N.

We summarize our discussion in the following way. For a solution of the eikonal equation in a linear dielectric and permeable medium, Definition 2.6.1 gives a transport vector field and, thus, a congruence of rays on open subsets on which the multiplicity is one, and Definition 2.6.2 gives two partial transport vector fields and, thus, two congruences of rays on open subsets on which the multiplicity is two. What is left out is the set of all points where the multiplicity changes. By continuous extension into such point we might get pathologies such as bifurcating rays.

Any ray is an integral curve of a vector field K_A^a given by (2.103) with $A = 1$ and/or $A = 2$. For any such integral curve $s \longmapsto x(s)$ we can define a map $s \longmapsto p(s)$ by $p(s) = \partial S(x(s))$, thereby getting a solution of *Hamilton's equations*

$$H_A\Big(x(s), p(s)\Big) = 0 \,,$$

$$\dot{x}^a(s) = \frac{\partial H_A}{\partial p_a}\Big(x(s), p(s)\Big) \,, \tag{2.110}$$

$$\dot{p}_a(s) = -\frac{\partial H_A}{\partial x^a}\Big(x(s), p(s)\Big) \,.$$

We call any immersed curve $s \longmapsto x(s)$ for which (2.6) is satisfied, with some $s \longmapsto p(s)$, an *A-ray* for short, $A = 1, 2$.

If the partial Hamiltonian H_A is changed into $\tilde{H}_A$ by a transformation of the form (2.84), the A-rays undergo a reparametrization but they are unchanged otherwise. In other words, the A-rays are determined, up to their parametrization, by the A-branch of the characteristic variety.

In the uniaxial case discussed in Example 2.5.2, the 1-rays are called *ordinary rays* and the 2-rays are called *extraordinary rays* . If we solve Hamilton's equations (2.6) with the partial Hamiltonians given by (2.90), we find that the ordinary and extraordinary rays are the light-like geodesics of the first and the second optical metric (2.91), respectively.

In the isotropic case there is only one optical metric and the notions of 1-rays and 2-rays coincide. By solving Hamilton's equations (2.6) with $H_1 = H_2$ given by (2.87), we find that the rays are exactly the light-like geodesics of the optical metric. This implies, of course, in particular the familiar textbook result that in vacuum the rays are exactly the light-like geodesics of the spacetime metric.

2.7 Ray optics as an approximation scheme

From the preceding sections we know that rays are associated with *asymptotic solutions* of Maxwell's equations. We are now going to show that they are associated, moreover, with *approximate solutions* of Maxwell's equations. For the physical interpretation of ray optics this is a crucial point.

Let us start with a solution S of the eikonal equation which, in a medium of the kind under consideration, is given by (2.74) with partial Hamiltonians (2.73). As always, we assume that S is given on some open neighborhood of spacetime and that its gradient has no zeros. Moreover, we have to assume in the following that S is associated with a unique congruence of rays. In other words, we have to assume that either S is a solution of multiplicity one or that S is a solution of multiplicity two for which the two partial transport vector fields coincide.

It is our goal to associate such an eikonal function S with an approximate solution of Maxwell's equations. To that end, we fix a spacetime point and, on a neighborhood of that point, a coordinate system adapted to the rest system of the medium in the sense of Definition 2.1.3. We use the inductive method of Sect. 2.5 to construct an Nth order asymptotic solution

$$\begin{pmatrix} Z(\alpha, x) \\ Y(\alpha, x) \end{pmatrix} = \text{Re}\left\{ e^{iS(x)/\alpha} \sum_{M=0}^{N+1} \alpha^M \begin{pmatrix} z^M(x) \\ y^M(x) \end{pmatrix} + O(\alpha^{N+2}) \right\} \qquad (2.111)$$

of the evolution equations (2.23), where N can be chosen as large as we want. This leaves the $O(\alpha^{N+2})$ term arbitrary. For any choice of this term, (2.111) is automatically an Nth order asymptotic solution of the constraints as well. It is not difficult to check that the $O(\alpha^{N+2})$ term can be chosen in such a way that the constraints are *exactly* satisfied on the initial hypersurface $x^4 = \text{const}$. These initial values determine a unique exact solution of the evolution equations (2.23) for each α, thereby giving us a one-parameter family of exact solutions that will be denoted by $Z^*(\alpha, \cdot)$, $Y^*(\alpha, \cdot)$. Now the difference

$$\Delta Z(\alpha, \cdot) = Z(\alpha, \cdot) - Z^*(\alpha, \cdot)\,,$$
$$\Delta Y(\alpha, \cdot) = Y(\alpha, \cdot) - Y^*(\alpha, \cdot)\,, \qquad (2.112)$$

satisfies

$$\left(L^a \partial_a + M \right) \begin{pmatrix} \Delta Z(\alpha, \cdot) \\ \Delta Y(\alpha, \cdot) \end{pmatrix} = O(\alpha^{N+1}) \qquad (2.113)$$

and vanishes on the initial hypersurface. We have already stressed in Sect. 2.1 that the differential operator on the left-hand side of (2.113) is symmetric hyperbolic with respect to the scalar product (2.29). Hence, we have the so-called energy inequalities at our disposal (see, e.g., Theorem 4.3 in Chazarain and Piriou [26] or Theorem 2.63 in Egorov and Shubin [36]). As a consequence, (2.113) implies the existence of a constant C such that the inequality

$$\begin{pmatrix} \Delta Z(\alpha, \cdot) \\ \Delta Y(\alpha, \cdot) \end{pmatrix} \cdot \begin{pmatrix} \Delta Z(\alpha, \cdot) \\ \Delta Y(\alpha, \cdot) \end{pmatrix} \leq C^2 \alpha^{2(N+1)} \qquad (2.114)$$

holds on an appropriately chosen (relatively compact) neighborhood. The constant C can be written as an integral over this spacetime neighborhood where the integrand involves the (known) tensor fields g_{ab}, U^c, $\varepsilon_d{}^e$, and $\mu_f{}^k$. Actually, the energy inequalities allow to estimate ΔZ and ΔY not only in the pointwise sense as in (2.114) but even in terms of *Sobolev norms* involving arbitrarily high derivatives. For our purpose, however, (2.114) will do.

(2.114) can be rewritten in terms of the field strengths

$$B = \begin{pmatrix} B_1 \\ B_2 \\ B_3 \end{pmatrix} \quad \text{and} \quad E = \begin{pmatrix} E_1 \\ E_2 \\ E_3 \end{pmatrix} \tag{2.115}$$

rather than in terms of our dynamical variables Z and Y. Since the constitutive equations are linear, and since the dielectricity and permeability tensor fields can be uniformly estimated on compact subsets of spacetime, we get an inequality of the form

$$\begin{pmatrix} \Delta B(\alpha, \cdot) \\ \Delta E(\alpha, \cdot) \end{pmatrix} \cdot \begin{pmatrix} \Delta B(\alpha, \cdot) \\ \Delta E(\alpha, \cdot) \end{pmatrix} \le \tilde{C}^2 \alpha^{2(N+1)} , \tag{2.116}$$

where $\tilde{C}$ is another constant. This shows that for α sufficiently small $B(\alpha, \cdot)$ and $E(\alpha, \cdot)$ are arbitrarily close to the exact solution $B^*(\alpha, \cdot)$ and $E^*(\alpha, \cdot)$, i.e., that our N^{th} order asymptotic solution is indeed an approximate solution, recall Figure 2.1. The higher N, the faster $\Delta B(\alpha, \cdot)$ and $\Delta E(\alpha, \cdot)$ converge to zero for $\alpha \to 0$.

In physical terms, the possibility to measure electric and magnetic field strengths is limited by some measuring accuracy δ. If α is so small that the right-hand side of (2.116) is smaller than δ^2, (2.116) implies that an observer moving along an x^4-line cannot distinguish, by way of measurement, the approximate solution from the exact solution. It is important to realize that this is true only for observers moving along an x^4-line (or at a small velocity with respect to the x^4-lines). If we exclude the case that approximate solution and exact solution coincide, we can always find observers, moving at a high velocity with respect to the x^4-lines, who measure an arbitrarily large difference between them. This follows immediately from the transformation behavior of electric and magnetic field strength under a Lorentz transformation, given in any textbook on special relativity. In other words, the question of whether or not our N^{th} order asymptotic solution, for some finite value of α, can be viewed as a valid approximation for some specific exact solution depends on the observer field with respect to which electric and magnetic field strengths are to be measured.

A similar observation, based on a different argument, was brought forward by Mashhoon [92]) who only considered light propagation in vacuum. He came to the conclusion that the equations of general relativistic ray optics have a meaning only in the limit of infinite frequency but not in the sense of a physically reasonable approximation for any finite frequency value. We do not share this radical point of view. Our results show that the ray method does give a viable approximation scheme for light propagation in a medium of the kind under consideration in the following sense. Any solution S of the eikonal equation which is associated with a unique congruence of rays can be viewed locally as the eikonal function of an approximate-plane-wave family that satisfies Maxwell's equations asymptotically to order N. This was shown in Sect. 2.4 for arbitrary $N \geq 0$. Moreover, we can find a one-parameter family of exact solutions of Maxwell's equations such that the difference between asymptotic solution and exact solution goes to zero for $\alpha \to 0$ like α^{N+1}. This follows from (2.114) or from the equivalent result (2.116). We just have to keep in mind that the constant C in (2.114) and the constant $\tilde{C}$ in (2.116) depend on the observer field U^a; it is impossible to find error bounds that are valid with respect to all observer fields simultaneously.

Having thus associated a congruence of rays with a one-parameter family of exact solutions of Maxwell's equations $F_{ab}^*(\alpha, \cdot)$, it is natural to ask if the rays are related, at least in the sense of an approximation, to the *energy flux* of $F_{ab}^*(\alpha, \cdot)$. After all, the intuitive idea behind ray optics is to view light propagation as a sort of energy transfer along rays. We need some more calculations to prove that this idea is, indeed, correct for media of the kind under consideration.

We start again with a solution S of the eikonal equation and assume that it is associated with a unique congruence of rays. We construct, locally around any point as outlined above, an approximate-plane-wave family $F_{ab}(\alpha, \cdot)$ with eikonal function S and a one-parameter family of exact solutions $F_{ab}^*(\alpha, \cdot)$ such that (2.114) holds for $N = 0$ at least. Then the energy flux of $F_{ab}^*(\alpha, \cdot)$ in the rest system U^a of the medium is given by

$$S^{*a}(\alpha, \cdot) = U^b(x)\, T^*{}_b{}^a(\alpha, x) \tag{2.117}$$

where $T^*{}_b{}^a(\alpha, \cdot)$ is the *Minkowski energy-momentum* tensor of $F_{cd}^*(\alpha, \cdot)$,

$$T^*{}_b{}^a(\alpha, \cdot) = F_{bc}^*(\alpha, \cdot)\, G^{*ac}(\alpha, \cdot) - \tfrac{1}{4}\delta_b^a\, F_{cd}^*(\alpha, \cdot)\, G^{*cd}(\alpha, \cdot)\,. \tag{2.118}$$

The component of the energy flux four-vector (2.117) orthogonal to U^a gives the familiar *Poynting vector*, whereas the component parallel to U^a gives the *energy density* of the electromagnetic field.

In a coordinate system adapted to U^a, in the sense of Definition 2.1.3, F_{ab}^* and G_{cd}^* can be expressed in terms of our dynamical variables Z_ρ^* and Y_ρ^* as in (2.22). Then (2.117) takes the form

$$\sqrt{-g_{44}(x)}\, S^{*a}(\alpha, \cdot) = g^{a\nu}(x)\, \eta^\sigma{}_{4\nu}{}^\rho(x)\, v_\sigma{}^\lambda(x)\, w_\rho{}^\mu(x)\, Z_\lambda^*(\alpha, \cdot)\, Y_\mu^*(\alpha, \cdot) -$$
$$\tfrac{1}{2}\delta_4^a\, g^{\sigma\tau}(x)\Big(Z_\sigma^*(\alpha, \cdot)\, Z_\tau^*(\alpha, \cdot) + Y_\sigma^*(\alpha, \cdot)\, Y_\tau^*(\alpha, \cdot)\Big)\,. \tag{2.119}$$

Since (2.114) holds with $N = 0$, (2.119) can be rewritten in terms of our approximate-plane-wave family $Z_\rho(\alpha, \cdot)$, $Y_\rho(\alpha, \cdot)$ in the form

$$\sqrt{-g_{44}(x)}\, S^{*a}(\alpha, \cdot) = g^{a\nu}(x)\, \eta^\sigma{}_{4\nu}{}^\rho(x)\, v_\sigma{}^\lambda(x)\, w_\rho{}^\mu(x)\, Z_\lambda(\alpha, \cdot)\, Y_\mu(\alpha, \cdot) -$$
$$\tfrac{1}{2}\delta_4^a\, g^{\sigma\tau}(x)\Big(Z_\sigma(\alpha, \cdot)\, Z_\tau(\alpha, \cdot) + Y_\sigma(\alpha, \cdot)\, Y_\tau(\alpha, \cdot)\Big) + O(\alpha)\,. \tag{2.120}$$

Since $Z_\rho(\alpha, \cdot)$ and $Y_\rho(\alpha, \cdot)$ are given by (2.50) with $N_0 = 0$, we have

$$Z_\lambda(\alpha, \cdot) Y_\mu(\alpha, \cdot) = \tfrac{1}{2} \operatorname{Re}\left\{ e^{2iS(x)/\alpha}\, z_\lambda^0(x)\, y_\mu^0(x) + \overline{z_\lambda^0}(x)\, y_\mu^0(x) \right\} + O(\alpha) \ . \tag{2.121}$$

Let us denote by $< f > (x)$ the average of a spacetime function f taken over a neighborhood of x on which the gradient of S and the amplitudes z_λ^0, y_μ^0 can be viewed as approximately constant. (Please recall our discussion of approximate-plane-wave families in Sect. 2.2.) For α sufficiently small, the first term on the right-hand side of (2.121) gives an average arbitrarily close to zero. We may thus write

$$< Z_\lambda(\alpha, \cdot) Y_\mu(\alpha, \cdot) > \approx \tfrac{1}{2} \operatorname{Re}\left\{ z_\lambda^0 \overline{y_\mu^0} \right\} , \tag{2.122}$$

where $x \approx y$ means that the difference between x and y can be made arbitrarily small by choosing α sufficiently small. Similar expressions hold for the averaged products $< Z_\lambda(\alpha, \cdot) Z_\mu(\alpha, \cdot) >$ and $< Y_\lambda(\alpha, \cdot) Y_\mu(\alpha, \cdot) >$. With these equations at hand, we can calculate the averaged energy flux from (2.119). If we assume that the background fields (i.e., the spacetime metric and the tensor fields that characterize the medium) do not vary significantly over the neighborhood used for the averaging procedure, we find

$$-2\sqrt{-g_{44}} < S^{*a}(\alpha, \cdot) > \approx \delta_\rho^a \operatorname{Re}\{z^0 \cdot A^\rho y^0\} + \\ \delta_4^a \operatorname{Re}\{z^0 \cdot Q\, y^0\} + \delta_4^a \left(z^0 \cdot z^0 + y^0 \cdot y^0 \right) \tag{2.123}$$

where the 3×3 matrices Q and A^ρ from (2.26) and (2.27) are used.

We shall now show that the right-hand side of (2.123) is, indeed, proportional to the transport vector field of our eikonal function. To that end, we recall that $Z(\alpha, \cdot), Y(\alpha, \cdot)$ is an Nth order asymptotic solution of Maxwell's equations for $N = 0$ at least. Thus, z^0 and y^0 have to satisfy the 0th order polarization condition (2.56). This implies

$$\begin{pmatrix} z^0(x) \\ y^0(x) \end{pmatrix} = \xi^0(x) \begin{pmatrix} u_1(x, \partial S(x)) \\ -v_1(x, \partial S(x)) \end{pmatrix} \tag{2.124}$$

with a $\mathbb{C}$-valued function ξ^0 if S is of multiplicity one, and

$$\begin{pmatrix} z^0(x) \\ y^0(x) \end{pmatrix} = \sum_{A=1}^{2} \xi_A^0(x) \begin{pmatrix} u_A(x, \partial S(x)) \\ -v_A(x, \partial S(x)) \end{pmatrix} \tag{2.125}$$

with $\mathbb{C}$-valued functions ξ_1^0 and ξ_2^0 if S is of multiplicity two. In the first case the transport vector field is given by (2.100), and (2.124) implies

$$K^a = u \begin{pmatrix} z^0 \\ y^0 \end{pmatrix} \cdot L^a \begin{pmatrix} z^0 \\ y^0 \end{pmatrix} \tag{2.126}$$

with some $\mathbb{R}$-valued function u. In the second case the partial transport vector fields are given by (2.108). Since we assume that our eikonal equation is associated with a unique congruence of rays, the two partial transport vector fields coincide, $K_1^a = K_2^a = K^a$, and (2.126) holds in this case as well. With L^a given by (2.25), (2.126) takes the form

$$2K^a = u\left(\delta_\rho^a \operatorname{Re}\{z^0 \cdot A^\rho y^0\} + \delta_4^a \operatorname{Re}\{z^0 \cdot Q\, y^0\} + \delta_4^a \left(z^0 \cdot z^0 + y^0 \cdot y^0\right)\right).$$

$$(2.127)$$

Comparison of (2.123) and (2.127) shows that

$$< S^{*a}(\alpha, \, \cdot \,) > \approx v\, K^a \qquad (2.128)$$

with some $\mathbb{R}$-valued function v. In other words, the averaged energy flux of the exact Maxwell field follows the rays up to terms that can be made arbitrarily small by choosing α sufficiently small. Please note that we have considered the energy flux only in the rest system of the medium. This is important unless in the vacuum case where there is no distinguished rest system of the medium and (2.128) holds for the energy flux with respect to any observer field.

With these findings we have completed our discussion of light propagation in a linear dielectric and permeable medium. In particular, we have now established the missing link between asymptotic solutions and approximate solutions. Let us emphasize the main point again. For the mathematical derivation of eikonal equation and transport equations through a mathematically well-defined limit procedure it is necessary to consider approximate-plane-wave families that satisfy Maxwell's equations asymptotically for $\alpha \to 0$. From a physical point of view, however, this limit $\alpha \to 0$ is a purely formal device. The physical meaning of the method is in the fact that the resulting approximate-plane-wave families give approximate solutions of Maxwell's equations for (sufficiently small but) finite values of α.

3. Light propagation in other kinds of media

In Chap. 2 we considered the homogeneous Maxwell equations (2.1) supplemented by linear constitutive equations (2.12). This ansatz does, of course, not cover all sorts of media with relevance to physics. Modifications of the following kind are possible. First, we could replace ansatz (2.12) with more complicated relations between field strengths and excitations. Second, we could introduce a current, i.e., a source term in the second Maxwell equation (and, similarly, in the first Maxwell equation if the hypothetical existence of magnetic monopoles is to be taken into account). In the latter case, the current must be specified by additional equations. E.g., we could assume, in analogy to (2.12), a linear relation between 3-current and electric field strength in the rest system of the medium, thereby generalizing *Ohm's law*.

For any such specification of the medium we can investigate if the resulting system of equations determines reasonable dynamics for the electromagnetic field. Here, a dynamical law is to be viewed as "reasonable" if it is governed by a set of evolution equations characterized by a local existence and uniqueness theorem. This set of evolution equations might be supplemented by a set of constraints that are preserved by the evolution equations. If the medium under consideration gives rise to a dynamical law of this kind, it is reasonable to proceed along the lines of Chap. 2, i.e., to consider approximate-plane-wave families (2.33) that satisfy evolution equations and constraints asymptotically for $\alpha \to 0$ to some order N. The passage to ray optics has been achieved if it is possible to derive, on this assumption, an eikonal equation of the form $H(x, \partial S(x)) = 0$, where H can be chosen as a product, $H = H_1 \cdot \ldots \cdot H_k$ with each partial Hamiltonian H_A, $A = 1, \ldots, k$, satisfying condition (2.83) on its characteristic variety. This guarantees that each solution S of a partial eikonal equation $H_A(x, \partial S(x)) = 0$ is associated with a nowhere vanishing transport vector field (2.103) whose integral curves give a congruence of A-rays, i.e., of solutions to Hamilton's equations (2.6) projected to spacetime.

Even if all this works out nicely, it is of course not guaranteed that ray optics in the medium under consideration can be viewed as a valid approximation scheme for exact Maxwell fields. This has to be checked, along the lines of Sect. 2.7, in each case individually.

We are now going to discuss the question of if and how such a treatment of media more general than considered in Chap. 2 is able to cover the phenomenon of *dispersion*.

3.1 Methodological remarks on dispersive media

The most important motivation to go beyond the kind of media considered in Chap. 2 is the following. We have found that the media considered in Chap. 2 are characterized by an eikonal equation of the form $H\big(x, \partial S(x)\big) = 0$ where the Hamiltonian H can be chosen homogeneous with respect to the momenta. In other words, if a 4-momentum $p = (p_1, \ldots, p_4)$ satisfies the dispersion relation at some spacetime point x, then any multiple $t\, p = (t\, p_1, \ldots, t\, p_4)$ also satisfies the dispersion relation at this spacetime point. Whenever this homogeneity property is satisfied the medium is called *dispersion-free* or *non-dispersive*; otherwise, it is called *dispersive*. In a non-dispersive medium, a ray is fixed by giving an initial event and an initial direction for the spatial wave covector (with respect to any normalized time-like vector field); in a dispersive medium, one has to give the length of the spatial covector in addition. This definition can be rephrased in terms of phase velocities and group velocities to yield the familiar physics textbook definition of dispersive and non-dispersive media, see Sect. 6.2 in Part II below.

Hence, we have to ask ourselves what sort of modified ansatz for the medium could be able to cover the phenomenon of dispersion. This is an important question not only from a theoretical point of view but also in view of applications to astropyhsics. Dispersion plays a role for light propagation in planetary and stellar atmospheres and in interstellar plasma clouds.

A closer look at the treatment of Chap. 2 shows that the following features are causative for the homogeneity of the eikonal equation.

(a) Evolution equations and constraints give a system of linear differential equations for the electromagnetic field strength.

(b) The limit $\alpha \to 0$ is taken on a fixed background, i.e., neither evolution equations nor constraints involve the parameter α.

As a matter of fact, it is easy to check that, whenever (a) and (b) are satisfied, the eikonal equation arises in the form $H(x, \partial S(x)) = 0$ where the Hamiltonian H is a homogeneous polynomial with respect to the momenta. (Afterwards, we are free to change the Hamiltonian according to transformations of the form given in (2.84) for $H = H_A$. This leaves, of course, the homogeneity of the eikonal equation unchanged.) If we want to treat dispersive media we have, thus, to modify the method of Chap. 2 by violating at least one of the two properties (a) and (b).

The most obvious idea to violate property (a) is to modify the linear constitutive equations (2.12) by adding terms quadratic in the field strengths.

This is common practice in ordinary optics where it gives rise to interesting effects with relevance to strong electromagnetic wave fields. However, it is quite evident that such non-linear terms are not typically the origin of dispersion in crystals, gases, or plasmas. As a matter of fact, dispersion is frequently observed in situations where the field strengths are far too weak to make non-linear modifications of the constitutive equations necessary. Moreover, there are several technical problems associated with the ray method if evolution equations and/or constraints are non-linear. Contrary to the linear case, our assumption that the approximate-plane-wave family (2.33) satisfies the differential equations asymptotically cannot be evaluated inductively in general. The reason is that even in lowest non-trivial order amplitudes f_{ab}^N with arbitrarily large N may show up, i.e., we do not get an eikonal equation for S alone. In other words, in a non-linear medium the propagation of wave surfaces in the high frequency limit and, thus, the corresponding propagation of rays depends on the amplitudes of the wave fields. In this sense, there is no self-contained theory of ray optics for such media. To be sure, there are some non-linear equations that do give an eikonal equation for S alone. This is true, in particular, of *semi-linear* equations, i.e., of equations which are linear in the highest order derivatives of the dynamical variables (field strengths) with coefficients independent of these variables. However, for the inductive method of determining the amplitudes f_{ab}^N to carry over we need nothing less than linearity. For this reason, only in the linear case is it possible to check, along the lines of Sect. 2.7, whether or not ray optics gives a viable approximation scheme.

It is worthwile to mention another problem with non-linear equations. Suppose we know that some approximate-plane-wave family (2.33) stays close to exact solutions, for $0 < \alpha < a_0$, within some given error bounds. Then it is still possible that a generalized Fourier integral (2.40), formed with this family over a real interval $[\alpha_1, \alpha_2] \subseteq [0, \alpha_0]$, deviates from all exact solutions by an arbitrarily large amount. In this sense, studying approximate-plane-wave families of the form (2.33) is of limited usefulness in a non-linear medium since it gives no information on non-monochromatic waves.

These arguments show that it is somewhat problematic to apply the ray method to non-linear differential equations. As a matter of fact, the existing literature on this topic is much more "heuristic" than in the linear case. A typical reference is the book by Jeffrey and Kawahara [66] where many applications to physics are mentioned. Those applications refer mainly to fluid mechanics where nonlinear effects are more important than in optics. In our context, the following strategy is advisable. When dealing with a medium for electromagnetic fields that gives non-linear evolution equations and/or non-linear constraints, it is reasonable to linearize these equations around a ("background") solution and to apply the ray method to the linearized equations. The resulting theory is valid for all wave fields which are sufficiently weak such that their self-interactions, caused by the non-linearities of the full equations, can be ignored. Interactions with the background field are, of course, taken into account.

Following this line of thought, the only possibility to treat dispersive media is by violating the above-mentioned property (b). At first sight, the idea to smuggle the parameter α into the differential equations seems alien to optics. (This is a major difference to the JWKB method in wave mechanics. In the latter case, the role of α is played by Planck's constant $\hbar$ which, evidently, appears in Schrödinger's equation.) Nonetheless, there is a sound method of achieving this goal. Strictly speaking, this method comes in various different variants. The common feature is that one considers asymptotic behavior of approximate-plane-wave families on a one-parameter family of background geometries, rather than on a fixed background geometry. Circumstances permitted, this gives an eikonal equation (and transport equations) in close analogy to the treatment of Chap. 2. The crucial point is that even in the case of linear differential equations the eikonal equation need not be homogeneous with respect to ∂S, i.e., dispersion is not excluded. The physical meaning of an eikonal equation derived that way depends, of course, on the way in which the background geometries depend on the parameter. In the following section we demonstrate the method by way of a special example.

3.2 Light propagation in a non-magnetized plasma

In this section we consider a simple plasma model as a medium for electromagnetic waves and we perform the passage to ray optics in such a way that dispersion is taken into account. Apart from some modifications, our treatment follows Breuer and Ehlers [18] [19]. For earlier references on the same subject we refer to Madore [90], to Bičák and Hadrava [14] and to Anile and Pantano [5] [6].

We restrict ourselves to the most simple plasma model, viz., to a two-fluid model with vanishing pressure. Then the dynamical system to be considered is governed by the equations

$$\partial_{[a} F_{bc]} = 0 \,, \tag{3.1}$$

$$\nabla_b F^{ab} = J^a + e\,n\,U^a \,, \tag{3.2}$$

$$m\,U^b\,\nabla_b U^a = e\,F^a{}_b\,U^b \,, \tag{3.3}$$

$$\nabla_a\!\left(n\,U^a\right) = 0 \,, \tag{3.4}$$

$$g_{ab}\,U^a\,U^b = -1 \,. \tag{3.5}$$

(3.1) and (3.2) are the Maxwell equations for the electromagnetic field strength tensor F_{ab}, where square brackets around indices mean antisymmetrization. In (3.2), the ionic current is denoted by J^a, whereas the electronic current is written as the product of electron charge e, electron particle density n, and electron 4-velocity U^a. In mathematical terms, e is a negative constant, n is a nonnegative scalar function, and U^a is a vector field normalized by (3.5).

(3.3) is the equation of motion for the electron fluid (Euler equation plus Lorentz force), where m is a positive constant with the meaning of the electron mass. Here we assume, as already mentioned, that the pressure of the electron fluid vanishes. This is a legitimate approximation as long as the plasma is sufficiently cold.

(3.4) is the equation of charge conservation of the electron component. Please note that (3.2) already implies conservation of the total charge, $\nabla_a(J^a + e n U^a) = 0$, but not of the electron component alone.

We want to view (3.1)–(3.5) as a system of non-linear first order differential equations for F_{ab}, n and U^a with the metric g_{ab} and the ionic current J^a assumed known. Viewed in this sense, (3.1)–(3.5) give us $4+4+4+1+1 = 14$ component equations for $6+1+4 = 11$ unknown functions. In a local coordinate system with time-like x^4-lines and space-like hypersurfaces $x^4 = \text{const.}$ our 14 equations split up into 11 evolution equations and 3 constraints. It is easy to verify that the evolution equations preserve the constraints. Moreover, Breuer and Ehlers [18] were able to show that the system of evolution equations admits a locally well-posed initial value problem, and that the equations (3.1)–(3.5) are linearization stable. The latter property guarantees that solutions of the linearized equations are close to solutions of the full equations, i.e., that linearization gives a meaningful approximation.

This is of particular relevance for us since, following the strategy outlined in Sect. 3.1, we are now going to linearize (3.1)–(3.5) around some ("background") solution. For simplicity we restrict ourselves to the case of a background solution with vanishing electromagnetic field. In other words, our background solution is given by a nonnegative scalar function $\overset{\circ}{n}$ and a vector field $\overset{\circ}{U}{}^a$ that satisfy the following set of equations.

$$0 = J^a + e\overset{\circ}{n}\overset{\circ}{U}{}^a , \tag{3.6}$$

$$\overset{\circ}{U}{}^b \nabla_b \overset{\circ}{U}{}^a = 0 , \tag{3.7}$$

$$\nabla_a(\overset{\circ}{n}\overset{\circ}{U}{}^a) = 0 , \tag{3.8}$$

$$g_{ab} \overset{\circ}{U}{}^a \overset{\circ}{U}{}^b = -1 . \tag{3.9}$$

Now we linearize the equations (3.1)–(3.5) around this background solution, i.e., we consider these equations for perturbed fields

$$F_{ab} = 0 + \hat{F}_{ab} , \tag{3.10}$$

$$n = \overset{\circ}{n} + \hat{n} , \tag{3.11}$$

$$U^a = \overset{\circ}{U}{}^a + \hat{U}{}^a , \tag{3.12}$$

and we drop all terms of second and higher order with respect to the perturbations $\hat{F}_{ab}$, $\hat{n}$, $\hat{U}^a$. The resulting equations govern the dynamics of sufficiently weak electromagnetic waves $\hat{F}_{ab}$ in our plasma which, according to $\overset{\circ}{F}_{ab} = 0$, is assumed non-magnetized. We shall presuppose that the metric g_{ab} and the ionic current J^a are unperturbed. The first assumption is in agreement with our general stipulation to work on a fixed metric background, i.e., to disregard the back-reaction, governed by Einstein's field equations, of matter and electromagnetic fields on the metric. The second assumption means that the effect of the electromagnetic wave on the ions is ignored. This is a reasonable approximation since the inertia of the ions is much bigger than that of the electrons. On these assumptions, the linearized system of equations for the perturbations takes the following form.

$$\partial_{[a}\hat{F}_{bc]} = 0 \,, \tag{3.13}$$

$$\nabla_b \hat{F}^{ab} = e\overset{\circ}{n}\hat{U}^a + e\hat{n}\overset{\circ}{U}^a \,, \tag{3.14}$$

$$m\overset{\circ}{U}^b\nabla_b\hat{U}^a + m\hat{U}^b\nabla_b\overset{\circ}{U}^a = e\hat{F}^a{}_b\overset{\circ}{U}^b \,, \tag{3.15}$$

$$\nabla_a\big(\overset{\circ}{n}\hat{U}^a + \hat{n}\overset{\circ}{U}^a\big) = 0 \,, \tag{3.16}$$

$$g_{ab}\overset{\circ}{U}^a\hat{U}^b = 0 \,. \tag{3.17}$$

With g_{ab}, $\overset{\circ}{n}$ and $\overset{\circ}{U}^a$ known, (3.13)–(3.17) is a system of first order linear differential equations for $\hat{F}_{ab}$, $\hat{n}$ and $\hat{U}^a$. It is our goal to find dynamical equations for $\hat{F}_{ab}$ alone, i.e., to eliminate $\hat{n}$ and $\hat{U}^a$. This is indeed possible provided that the background density $\overset{\circ}{n}$ has no zeros,

$$\overset{\circ}{n} > 0 \,, \tag{3.18}$$

in the spacetime region considered. If this condition is satisfied, we can proceed in the following way.

From (3.14) we find, with the help of (3.9) and (3.17),

$$e\hat{n} = -\overset{\circ}{U}_a\nabla_b\hat{F}^{ab} \,, \tag{3.19}$$

$$e\overset{\circ}{n}\hat{U}^a = \nabla_b\hat{F}^{cb}\big(\delta^a_c + \overset{\circ}{U}^a\overset{\circ}{U}_c\big) \,. \tag{3.20}$$

Since we can divide by $\overset{\circ}{n}$, (3.20) can be used to eliminate $\hat{U}^a$ from (3.15). This results in the following linear second order differential equation for $\hat{F}_{ab}$:

$$\overset{\circ}{U}^b\big(\delta^a_c + \overset{\circ}{U}^a\overset{\circ}{U}_c\big)\nabla_b\nabla_d\hat{F}^{cd} + \Big(\nabla_b\overset{\circ}{U}^b\big(\delta^a_c + \overset{\circ}{U}^a\overset{\circ}{U}_c\big) + \nabla_c\overset{\circ}{U}^a\Big)\nabla_d\hat{F}^{cd} -$$

$$\frac{e^2}{m}\overset{\circ}{n}\overset{\circ}{U}^b\hat{F}^a{}_b = 0 \,. \tag{3.21}$$

If we have a solution $\hat{F}_{ab}$ of (3.13) and (3.21), we can define $\hat{n}$ and $\hat{U}^a$ by (3.19) and (3.20), respectively. It is easy to check that then the full system of equations (3.13)–(3.17) is satisfied. In other words, we have reduced this system (3.13)–(3.17) to dynamical equations for $\hat{F}_{ab}$ alone, given by (3.13) and (3.21).

To rewrite (3.13) and (3.21) in a more convenient form, we express $\hat{F}_{ab}$ in terms of a *potential* $\hat{A}_a$,

$$\hat{F}_{ab} = \partial_{[a}\hat{A}_{b]} = \nabla_{[a}\hat{A}_{b]} \,, \tag{3.22}$$

and we assume that $\hat{A}_a$ satisfies the *Landau gauge condition* in the rest system of the background electron fluid,

$$\hat{A}_a \overset{\circ}{U}{}^a = 0 \,. \tag{3.23}$$

It is a standard exercise in Maxwell theory to verify that any antisymmetric tensor field $\hat{F}_{ab}$ that satisfies (3.13) can be locally represented in this way, and that $\hat{A}_a$ is (locally) uniquely determined by $\hat{F}_{ab}$ up to *gauge transformations*

$$\hat{A}_a \longmapsto \hat{A}_a + \partial_a h \tag{3.24}$$

where h is any spacetime function that is constant along the flow lines of $\overset{\circ}{U}{}^a$. In other words, h can be freely prescribed on a hypersurface transverse to those flow lines.

With (3.22), (3.13) is automatically satisfied and (3.21) takes the form

$$\mathcal{D}^{af}\hat{A}_f = 0 \tag{3.25}$$

where the differential operator $\mathcal{D}^{af}$ is defined by

$$\mathcal{D}^{af}\hat{A}_f = \overset{\circ}{U}{}^b(\delta^a_c + \overset{\circ}{U}{}^a\overset{\circ}{U}_c)\nabla_b(\nabla^f\nabla^c - g^{fc}\nabla^d\nabla_d)\hat{A}_f +$$
$$\left(\nabla_b\overset{\circ}{U}{}^b(\delta^a_c + \overset{\circ}{U}{}^a\overset{\circ}{U}_c) + \nabla_c\overset{\circ}{U}{}^a\right)\left(\nabla^f\nabla^c - g^{fc}\nabla^d\nabla_d\right)\hat{A}_f -$$
$$\frac{e^2}{m}\overset{\circ}{n}\left(\overset{\circ}{U}{}^f\nabla^a - g^{af}\overset{\circ}{U}{}^b\nabla_b\right)\hat{A}_f \,. \tag{3.26}$$

(3.25) determines the dynamics of electromagnetic waves in our plasma. (3.25) consists of four component equations, but only three of them are independent since the equation

$$\overset{\circ}{U}_a\mathcal{D}^{af}\hat{A}_f = 0 \tag{3.27}$$

is identically satisfied for any $\hat{A}_f$. By the Landau gauge condition (3.23), $\hat{A}_f$ has three independent components. Hence, we have as many equations as unknown functions. In this sense, (3.25) gives a *determined* system of linear third order differential equations for the electromagnetic potential. To make this explicit, one can choose, on an appropriate open subset of spacetime, an orthonormal tetrad field E_1, E_2, E_3, E_4 with $E_4^a = \overset{\circ}{U}{}^a$. By (3.23), $\hat{A}_f$ is of the form $\hat{A}_f = g_{fk}\,\hat{A}^\mu\,E_\mu^k$ with some scalar functions $\hat{A}^1$, $\hat{A}^2$, $\hat{A}^3$ on that domain. Multiplication of (3.25) with $g_{ac}\,E_\mu^c$ gives us three equations (numbered by $\mu = 1,2,3$) for the three functions $\hat{A}^1$, $\hat{A}^2$, $\hat{A}^3$. It is shown in Breuer and Ehlers [18] [19] that this system of linear differential equations admits a local existence and uniqueness theorem for any data $\hat{A}^\mu$, $\overset{\circ}{U}{}^a\,\partial_a\hat{A}^\mu$, $\overset{\circ}{U}{}^a\,\overset{\circ}{U}{}^b\,\partial_a\partial_b\hat{A}^\mu$ prescribed on a space-like hypersurface.

Viewed in this sense, (3.25) is the system of evolution equations for electromagnetic waves in our plasma. Those evolution equations are of second order in the field strengths, and they are not supplemented by constraints. They are, thus, quite different from the evolution equations (2.23) in a linear dielectric and permeable medium. Unfortunately, (3.25) is not of the kind for which standard theorems guarantee the validity of energy inequalities.

With the dynamical law (3.25) at hand, we can now perform the passage to ray optics. Since it is our goal to take dispersion into account, we proceed in a way different from Chap. 2. As outlined in Sect. 3.1, it will be crucial to consider one-parameter families of background fields rather than fixed background fields. The background fields that enter into the differential operator $\mathcal{D}^{af}$ are the metric g_{ab}, the electron number density $\overset{\circ}{n}$ and the electron 4-velocity $\overset{\circ}{U}{}^a$. Let us fix such a set of background fields which have to satisfy (3.6)–(3.9) and (3.18). Further, let us fix a spacetime point and a coordinate system around this point. We assume that the chosen point is represented by the coordinates $x_0 = (x_0^1, x_0^2, x_0^3, x_0^4)$ and that the considered coordinate domain is star-shaped with respect to x_0 in $\mathbb{R}^4$. The latter condition means that for any point x in this domain the straight line between x and x_0 is completely contained in this domain. Refering to this fixed coordinate system, we define new background fields, depending on a real parameter β, by

$$g_{ab}(\beta, x) = g_{ab}(x_0 + \beta(x - x_0)) \,, \tag{3.28}$$

$$\overset{\circ}{n}(\beta, x) = \overset{\circ}{n}(x_0 + \beta(x - x_0)) \,, \tag{3.29}$$

$$\overset{\circ}{U}{}^a(\beta, x) = \overset{\circ}{U}{}^a(x_0 + \beta(x - x_0)) \,. \tag{3.30}$$

For $0 \le \beta \le 1$, the new background fields $g_{ab}(\beta, \cdot)$, $\overset{\circ}{n}(\beta, \cdot)$, and $\overset{\circ}{U}{}^a(\beta, \cdot)$ are well defined on the star-shaped domain considered, and they satisfy again equations (3.6)–(3.9) and condition (3.18). (This observation does not carry over if an electromagnetic background field $\overset{\circ}{F}_{ab} \ne 0$ is to be taken into account. For a magnetized plasma, one cannot assume the same β-dependence for all background fields g_{ab}, $\overset{\circ}{n}$, $\overset{\circ}{U}{}^a$, and $\overset{\circ}{F}_{ab}$.)

For $\beta \to 0$, the components of the background fields become constant in the coordinate system under consideration. In this sense, $g_{ab}(0, \cdot)$, $\overset{\circ}{n}(0, \cdot)$

and $\overset{\circ}{U}{}^a(0, \cdot)$ are homogeneous fields. In particular, $g_{ab}(0, \cdot)$ is a flat metric and $\overset{\circ}{U}{}^a(0, \cdot)$ is covariantly constant, i.e., an inertial system, with respect to this metric. For this reason, we shall refer to the limit $\beta \to 0$ as to the *homogeneous background limit*.

If we replace in (3.2) the original background fields g_{ab}, $\overset{\circ}{n}$ and $\overset{\circ}{U}{}^a$ by $g_{ab}(\beta, \cdot)$, $\overset{\circ}{n}(\beta, \cdot)$ and $\overset{\circ}{U}{}^a(\beta, \cdot)$, respectively, we get a one-parameter family of differential operators $\mathcal{D}^{af}(\beta, \cdot)$. It is our plan to enter into the differential equation $\mathcal{D}^{af}(\beta, \cdot)\hat{A}_f(\beta, \cdot) = 0$ with an approximate-plane-wave ansatz for the potential $\hat{A}_f(\beta, \cdot)$. Hence, we consider two-parameter families of the form

$$\hat{A}_f(\alpha, \beta, x) = \tag{3.31}$$

$$\tfrac{\alpha}{\beta} \operatorname{Re}\left\{ e^{i\, S(x_0 + \beta(x - x_0))/\alpha}\, \hat{a}_f(\alpha, x_0 + \beta(x - x_0)) \right\}$$

which satisfy the Landau gauge condition

$$\overset{\circ}{U}{}^f(\beta, x)\, \hat{A}_f(\alpha, \beta, x) = 0 \,. \tag{3.32}$$

We assume that the complex amplitudes are of the form

$$\hat{a}_f(\alpha, \cdot) = \sum_{N=0}^{N_0+1} \hat{a}_f^N(\cdot)\, \alpha^N + O(\alpha^{N_0+2}) \tag{3.33}$$

for all integers $N_0 \geq -1$ and that

$$\hat{F}_{ab}(\alpha, \beta, x) = \partial_{[a}\hat{A}_{b]}(\alpha, \beta, x) = \tag{3.34}$$

$$\operatorname{Re}\left\{ e^{i\, S(x_0 + \beta(x - x_0))/\alpha}\, i\, \left(\partial_{[a}S\, \hat{a}_{b]}^0 \right)(x_0 + \beta(x - x_0)) + O(\alpha) \right\}$$

is an approximate-plane-wave family, in the sense of Sect. 2.2, for any fixed β with $0 < \beta \leq 1$. For an approximate plane wave in this family, the frequency function with respect to the background electron rest system (3.30) is then given by

$$\omega(\alpha, \beta, x) = \tag{3.35}$$

$$\tfrac{\beta}{\alpha} \overset{\circ}{U}{}^a\left(x_0 + \beta(x - x_0)\right) \partial_a S\left(x_0 + \beta(x - x_0)\right) \,.$$

To perform the passage to ray optics, we have to assume that our approximate-plane-wave family satisfies the dynamical equations asymptotically. Since we have two parameters α and β at our disposal, we can consider asymptotic behavior with respect to different kinds of limits.

The first possibility is to keep β fixed and to consider the condition

$$\lim_{\alpha \to 0} \left(\tfrac{1}{\alpha^N} \mathcal{D}^{af}(\beta, \cdot)\hat{A}_f(\alpha, \beta, \cdot) \right) = 0 \tag{3.36}$$

for $N \in \mathbb{Z}$. This is essentially the same kind of limit as considered in Chap. 2. It can be characterized as the *high frequency limit on a fixed background*. In the case at hand, the lowest non-trivial order is $N = -3$. We leave it to the reader to compute from (3.36) with $N = -3$ that the resulting eikonal equation equals the vacuum eikonal equation in the background metric $g_{ab}(\beta, \cdot)$, i.e., that the corresponding rays are exactly the light-like geodesics of this background metric. In other words, if the high-frequency limit is taken on a fixed background, the plasma has no influence on the rays. In particular, there is no dispersion. (If this kind of limit is to be considered, one can, of course, stick to the case $\beta = 1$ throughout, i.e., there is no need to introduce the parameter β at all.)

Now we want to consider a different kind of limit, namely to let β and α go to zero simultaneously with the quotient $\frac{\alpha}{\beta}$ kept fixed. We can then simply put $\alpha = \beta$ and consider the condition

$$\lim_{\alpha \to 0} \left(\tfrac{1}{\alpha^N} \mathcal{D}^{af}(\alpha, \cdot) \hat{A}_f(\alpha, \alpha, \cdot) \right) = 0 \tag{3.37}$$

for $N \in \mathbb{Z}$. Keeping $\frac{\alpha}{\beta}$ fixed implies that the frequency function (3.35) is kept fixed at the point x_0. Therefore, this kind of limit can be characterized as the *homogeneous background limit with fixed frequency* at x_0. We shall now prove that this limit gives, indeed, a different eikonal equation. To that end, we have to assume that (3.37) holds in lowest non-trivial order which is now given by $N = 0$. This is true if and only if the equation

$$Q_a{}^f \, \hat{a}_f^0 = 0 \tag{3.38}$$

holds at x_0, where $Q_a{}^f$ is an abbreviation for

$$Q_a{}^f = \tag{3.39}$$

$$\overset{\circ}{U}{}^b \, \partial_b S \left(- \partial_a S \, \partial^f S - \overset{\circ}{U}_a \overset{\circ}{U}{}^c \, \partial_c S \, \partial^f S + \delta_a^f \, \partial^d S \, \partial_d S + \delta_a^f \, \tfrac{e^2}{m} \overset{\circ}{n} \right) .$$

Here we have used the equation

$$\overset{\circ}{U}{}^f(x_0) \, \hat{a}_f^0(x_0) = 0 \tag{3.40}$$

which follows from the Landau gauge condition (3.32). Since (3.34) is supposed to be an approximate-plane-wave family, $\hat{a}_f^0$ must be non-zero and linearly independent of $\partial_f S$. The condition that (3.38) admits a solution $\hat{a}_f^0$ of this kind at x_o gives the desired eikonal equation at x_0 for S. We have, thus, to solve the eigenvalue problem of $Q_a{}^f$ restricted to the orthocomplement of $\overset{\circ}{U}{}^f$. We find that there are three real eigenvalues, viz.

$$\lambda_1 = \overset{\circ}{U}{}^b \, \partial_b S \left(- \left(\overset{\circ}{U}{}^c \, \partial_c S \right)^2 + \tfrac{e^2}{m} \overset{\circ}{n} \right) , \tag{3.41}$$

$$\lambda_2 = \lambda_3 = \overset{\circ}{U}{}^b \, \partial_b S \left(\partial^d S \, \partial_d S + \tfrac{e^2}{m} \overset{\circ}{n} \right) . \tag{3.42}$$

If either $\overset{\circ}{U}{}^b \partial_b S = 0$ or $\partial_a S = \pm \sqrt{\frac{e^2}{m} \overset{\circ}{n}} \, \overset{\circ}{U}_a$, all three eigenvalues coincide and (3.38) is satisfied by any $\hat{a}_f^0$. Otherwise, we find $\lambda_1 \neq \lambda_2 = \lambda_3$. In the latter case, the eigenspace pertaining to λ_1 is one-dimensional and spanned by $\partial_f S + \overset{\circ}{U}_f \overset{\circ}{U}{}^b \partial_b S$ whereas the eigenspace pertaining to $\lambda_2 = \lambda_3$ is two-dimensional and consists of all X_f with $\overset{\circ}{U}{}^f X_f = \partial^f S \, X_f = 0$.

Equation (3.38) admits a non-trivial solution $\hat{a}_f^0$ which is perpendicular to $\overset{\circ}{U}{}^f$ if and only if one of the eigenvalues λ_1, λ_2, λ_3 is zero. From the form of the eigenspaces we see that in any such case $\hat{a}_f^0$ can be chosen linearly independent of $\partial_f S$. Hence, the eikonal equation takes the form $\lambda_1 \lambda_2 \lambda_3 = 0$ which is equivalent to

$$\overset{\circ}{U}{}^b \partial_b S \left(- \left(\overset{\circ}{U}{}^c \partial_c S \right)^2 + \frac{e^2}{m} \overset{\circ}{n} \right) \left(\partial^d S \, \partial_d S + \frac{e^2}{m} \overset{\circ}{n} \right) = 0 \, . \qquad (3.43)$$

Let us be precise about this result. Our assumption that the asymptotic condition (3.37) holds in lowest non-trivial order requires that S satisfies (3.43) at the point x_0 around which the construction was done.

Although we have used a fixed coordinate system around the chosen space-time point to perform the homogeneous background limit, the eikonal equation is a covariant equation (i.e., independent of this coordinate system). If S satisfies this covariant equation (3.43) on an open spacetime domain $\mathcal{U}$, it is associated with an asymptotic solution of lowest non-trivial order, in the homogeneous-background sense, around any point of $\mathcal{U}$. That is to say, to any such S we can find a non-trivial amplitude $\hat{a}_f(\alpha, \cdot)$ on $\mathcal{U}$ such that the following holds. If we choose any coordinate system around any point of $\mathcal{U}$, thereby defining the one-parameter family of operators $\mathcal{D}^{af}(\beta, \cdot)$ and the two-parameter family (3.31) of electromagnetic fields, the asymptotic condition (3.37) is satisfied for $N = 0$. As a matter of fact, a similar statement is true for any N. However, this more general result does not follow from our reasoning so far.

Owing to the terms proportional to $\overset{\circ}{n}$, the eikonal equation (3.43) is not homogeneous with respect to ∂S. This indicates dispersion.

The product structure of the eikonal equation (3.43) suggests to introduce three partial Hamiltonians

$$H_1(x, p) = \overset{\circ}{U}{}^b(x) \, p_b \, , \qquad (3.44)$$

$$H_2(x, p) = \tfrac{1}{2} \left(- \overset{\circ}{U}{}^a(x) \, \overset{\circ}{U}{}^b(x) \, p_a \, p_b + \frac{e^2}{m} \overset{\circ}{n}(x) \right) , \qquad (3.45)$$

$$H_3(x, p) = \tfrac{1}{2} \left(g^{ab}(x) \, p_a \, p_b + \frac{e^2}{m} \overset{\circ}{n}(x) \right) . \qquad (3.46)$$

Our assumptions guarantee that each partial Hamiltonian satisfies condition (2.83) on its characteristic variety. We are, of course, free to change each partial Hamiltonian by a transformation of the form (2.84).

The three partial Hamiltonians determine three branches of the dispersion relation. The branches defined by H_2 and H_3 have an intersection given by the equation $p_a = \pm \sqrt{\frac{e^2}{m} \overset{\circ}{n}(x)} \overset{\circ}{U}_a(x)$. At all points of phase space where this equation does not hold, at most one of the three partial dispersion relations can be satisfied. (This is true as long as our assumption (3.18) is valid.)

In analogy to Chap. 2 we assign to each solution S of the partial eikonal equation

$$H_A(x, \partial S(x)) = 0 , \quad A = 1, 2, 3, \tag{3.47}$$

a (partial) transport vector field K^a defined by

$$K^a(x) = \frac{\partial H_A}{\partial p_a} (x, \partial S(x)) . \tag{3.48}$$

The integral curves of K^a are, again, called the $(A\text{-})rays$ associated with S. The totality of all A-rays, associated with any solution of (3.47), is found by solving Hamilton's equations (2.6) for $A = 1, 2, 3$, respectively.

It is worthwhile to mention that this definition associates a unique congruence of rays to each solution S of the full eikonal equation (3.43). This can be verified in the following way. In almost all cases, a solution of the full eikonal equation satisfies exactly one of the three partial eikonal equations (3.47). The only exception occurs if, at some point x, the equation $\partial_a S(x) = \pm \sqrt{\frac{e^2}{m} \overset{\circ}{n}(x)} \overset{\circ}{U}_a(x)$ holds such that (3.47) is satisfied for $A = 2$ and for $A = 3$ simultaneously. At such points we have two partial transport vectors, given by (3.48) with $A = 2$ and with $A = 3$, respectively. Luckily enough, we find from (3.45) and (3.46) that these two partial transport vectors coincide.

Let us consider the three partial Hamiltonians one by one. Solutions of the partial eikonal equation (3.47) with $A = 1$ are pathological insofar as they have vanishing frequency in the background rest system of the electron fluid, $\overset{\circ}{U}{}^a(x)\partial_a S(x) = 0$. Hence, $\overset{\circ}{U}{}^a$ is not an "admissible reference system" for the approximate-plane-wave interpretation. The transport vector field (3.48) associated with such a solution is given by

$$K^a(x) = \overset{\circ}{U}{}^a(x) , \tag{3.49}$$

i.e, the rays are the integral curves of $\overset{\circ}{U}{}^a$. Note that $H_1(\cdot, \partial S(\cdot)) = 0$ implies that the eigenvalues (3.41) and (3.42) coincide, $\lambda_1 = \lambda_2 = \lambda_3 = 0$, and that equation (3.38) is identically satisfied for all $\hat{a}^0_f$. In other words, the amplitude $\hat{F}^0_{ab} = i\,\hat{a}^0_{[a}\partial_{b]}S$ is not restricted by any polarization condition.

For a solution of the second partial eikonal equation $H_2(x, \partial S(x)) = 0$ the frequency function with respect to the background rest system of the electron fluid is determined by the equation

$$\overset{\circ}{U}{}^a(x)\,\partial_a S(x) = \pm\,\omega_p(x)\,, \tag{3.50}$$

where ω_p denotes the *plasma frequency* defined by

$$\omega_p(x)^2 = \frac{e^2}{m}\,\overset{\circ}{n}(x)\,. \tag{3.51}$$

For the transport vector field (3.48) associated with such a solution S we find

$$K^a(x) = \pm\,\omega_p(x)\,\overset{\circ}{U}{}^a(x) \tag{3.52}$$

such that the rays coincide, again, with the integral curves of $\overset{\circ}{U}{}^a$. (Please recall that the parametrization of rays is arbitrary.) The case $\partial_a S = \pm\,\omega_p\,\overset{\circ}{U}{}^a$ plays a special role since in this case S satisfies the partial eikonal equation (3.47) not only for $A = 2$ but also for $A = 3$. For this special solution we have again $\lambda_1 = \lambda_2 = \lambda_3$ and, thus, no polarization condition of 0th order. For all other solutions of $H_2\big(x,\partial S(x)\big) = 0$, (3.38) requires that $\hat{a}^0_f$ is in the eigenspace pertaining to the eigenvalue λ_1 given by (3.41), i.e., that $\hat{a}^0_f$ is a multiple of $\partial_f S + \overset{\circ}{U}_f\,\overset{\circ}{U}{}^c\,\partial_c S$. This condition implies that the electric component of $\hat{f}^0_{ab} = i\,\hat{a}^0_{[a}\partial_{b]}S$ with respect to $\overset{\circ}{U}{}^b$ is a linear combination of $\overset{\circ}{U}_a$ and $\partial_a S$ and that the corresponding magnetic component vanishes. This is tantamount to a longitudinal polarization condition in the sense that the electric field strength is parallel to the spatial wave covector, i.e., $\hat{f}^0_{ab}\,\overset{\circ}{U}{}^b = u\big(\partial_a S + \overset{\circ}{U}{}^b\,\partial_b S\,\overset{\circ}{U}_a\big)$ with some real-valued function u. Those longitudinal modes described by the partial Hamiltonian H_2 are known as *plasma oscillations*.

Now let us turn to the third partial Hamiltonian H_3. For $A = 3$, formula (3.48) yields the same expression for the transport vector field as in vacuum, viz.

$$K^a(x) = g^{ab}(x)\,\partial_b S(x)\,. \tag{3.53}$$

Using our assumption that $\overset{\circ}{n}$ has no zeros, we find that the 3-rays (i.e., the rays determined by the partial Hamiltonian H_3) are exactly the time-like geodesics of the metric $\omega_p^2\,g_{ab}$ which is conformally equivalent to g_{ab}. The easiest way to verify this result is by changing H_3 according to

$$H_3(x,p) = \tfrac{1}{2}\Big(g^{ab}(x)\,p_a\,p_b + \omega_p(x)^2\Big) \;\longmapsto \tag{3.54}$$

$$\tilde{H}_3(x,p) = \tfrac{1}{\omega_p(x)^2}\,H_3(x,p) = \tfrac{1}{2}\Big(\tfrac{1}{\omega_p(x)^2}\,g^{ab}(x)\,p_a\,p_b + 1\Big)\,.$$

Since this transformation is of the form (2.84), it leaves the rays unchanged up to reparametrization, i.e., we can use $\tilde{H}_3$ instead of H_3 for the determination of the 3-rays. Solving Hamilton's equations (2.6) with this transformed Hamiltonian gives, of course, the time-like geodesics of the conformally rescaled metric $\tilde{g}_{ab} = \omega_p^2 \, g_{ab}$ parametrized by $\tilde{g}_{ab}$-proper time.

To further analyze the third partial Hamiltonian we consider a solution S of $H_3(\,\cdot\,,\partial S(\,\cdot\,)) = 0$ but exclude the case $\partial_a S = \pm \omega_p \overset{\circ}{U}{}^a$ which was already considered above. Then (3.38) requires that $\hat{a}_f^0$ is in the eigenspace pertaining to the eigenvalue $\lambda_2 = \lambda_3$ given by (3.42), i.e., that $\hat{a}_f^0$ satisfies the condition

$$\hat{a}_f^0 \, \partial^f S = 0 \tag{3.55}$$

in addition to the Landau gauge condition $\hat{a}_f^0 \, U^f = 0$. This is tantamount to a transverse polarization condition for the 0th order amplitude $\hat{f}_{ab}^0 = i \, \hat{a}_{[a}^0 \, \partial_{b]} S$ in the following sense. (3.55) and the Landau gauge condition imply that the electric and magnetic components of $\hat{f}_{ab}^0$ with respect to $\overset{\circ}{U}{}^a$ are perpendicular to the gradient of S, i.e., that $\hat{f}_{ab}^0 \, \overset{\circ}{U}{}^b \, \partial^a S = 0$ and $\eta^{abcd} \, \overset{\circ}{U}_b \, \hat{f}_{cd}^0 \, \partial_a S = 0$. By (3.53), this implies that the electric and magnetic components of the 0th order amplitude are perpendicular to the rays.

From this analysis of the three partial Hamiltonians we see that, for transverse modes with non-zero frequency, the eikonal equation reduces to the form

$$\partial^a S(x) \, \partial_a S(x) + \frac{e^2}{m} \, \overset{\circ}{n}(x) = 0 \,. \tag{3.56}$$

i.e., to the partial eikonal equation determined by H_3. On a flat spacetime, the eikonal equation (3.56) is discussed in any textbook on plasma physics, see, e.g., Stix [134]. On a curved spacetime, it was derived, with increasing rigor, by Madore [90], Bičák and Hadrava [14], Anile and Pantano [5] [6] and Breuer and Ehlers [18] [19].

If we consider the limit $\overset{\circ}{n} \to 0$ we reobtain the familiar eikonal equation for light propagation in vacuum from (3.56). (Note that this is not the case for the partial eikonal equation (3.47) with $A = 1$ or $A = 2$.) It is, thus, admissible to consider (3.56) for any spacetime function $\overset{\circ}{n}$ which is non-negative (but not necessarily strictly positive). Spacetime regions on which $\overset{\circ}{n} > 0$ are to be interpreted as occupied by a plasma whereas spacetime regions on which $\overset{\circ}{n} = 0$ are to be interpreted as vacuum. To stick with our general stipulations, we assume that $\overset{\circ}{n}$ is a C^∞ function everywhere. We can then find (C^∞) solutions S of (3.56) which give us wave surfaces traveling partly through vacuum and partly through plasma clouds. An analogous treatment based on H_1 or H_2 rather than on H_3 is impossible. This indicates that H_1 and H_2 have nothing to do with electromagnetic waves *passing through* our plasma. (A full discussion of this topic requires replacing our C^∞ condition with a *piecewise* C^∞ condition and deriving *junction conditions* for $\partial_a S$ and for the amplitudes $\hat{a}_f^N$ from the asymptotic condition (3.37). We shall not embark upon such an investigation here.)

It is, therefore, justified to concentrate the discussion of light propagation in our plasma on the partial Hamiltonian H_3. Using the eikonal equation (3.56) and the 0th order polarization condition (3.55) as the starting point, we could now go on to evaluate (3.37) inductively for $N = 1, 2, 3$ etc. This would result in transport equations and polarization conditions for the amplitudes $\hat{a}_f^N$ of arbitrarily high order. We leave it to the reader to verify that, proceeding along the lines of Sect. 2.4, this hierarchy of equations can be solved inductively to construct solutions of the asymptotic condition (3.37) for arbitrarily large N. Unfortunately, it is a difficult problem, apparently unsolved so far, to prove that those asymptotic solutions are *approximate solutions* as well. The method of Sect. 2.7 does not carry over since the differential equation (3.25) is not of the kind for which energy inequalities are known to hold true. Therefore, it is hard to see how the difference between our asymptotic solutions and appropriate exact solutions could be estimated. If such error estimates do exist, they are, of course, quite different depending on which sort of limit is considered. For the homogeneous background limit with fixed frequency considered here, the error bounds must go to zero if the background fields $\overset{\circ}{g}_{ab}$, $\overset{\circ}{n}$ and $\overset{\circ}{U}{}^a$ become homogeneous. In other words, our asymptotic solutions yield good approximations if the background fields are sufficiently homogeneous. Clearly, this is not necessarily the case if the high-frequency limit on a fixed background is considered. Either limit yields a reasonable eikonal equation, reasonable transport equations and reasonable polarization conditions. The difference is in the range of validity as an approximation scheme for exact electromagnetic wave fields (providing this validity can be established, in terms of error estimates, at all).

From the results of this section we can draw the following lesson. The eikonal equation for light propagation in a plasma does not only depend on the plasma model (two-fluid model, infinite inertia of the ion component, vanishing pressure of the electron component, linearization around background with vanishing electromagnetic field, etc.); it also depends on the kind of asymptotic limit considered. For the high frequency limit on a fixed background, the eikonal equation is exactly the same as for light propagation in vacuum, i.e., there is no effect of the plasma on the rays. For the homogeneous background limit with fixed frequency, on the other hand, the eikonal equation is given by (3.43), i.e., there is an effect of the plasma on the rays which causes, in particular, dispersion. Although the eikonal equation (3.43) has a product structure associated with three partial Hamiltonians, one should not speak of "multiple refraction" in this case. The reason is that only the transverse modes described by H_3 can be linked to solutions of the vacuum eikonal equation in the way indicated above. In other words, rays that enter into our plasma from an adjacent vacuum region have to proceed as 3-rays, i.e., they are not multiply refracted. In a *magnetized* plasma, however, the background electromagnetic field causes the branch of the dispersion relation associated with H_3 to split into two branches. Then the medium becomes double-refractive. On a special-relativistic background this is a standard result of plasma physics. For a general-relativistic treatment of this case we refer to Breuer and Ehlers [18] [19].

In addition to the high frequency limit on a fixed background and the homogeneous background limit with fixed frequency there are many other possibilities. To mention just one further example, we could modify ansatz (3.28)–(3.30) by assuming a different scaling behavior for the one-parameter family of background fields. In this way it is possible to derive, e.g., an eikonal equation such that the rays are directly affected by the rotation of the background rest system $\overset{\circ}{U}{}^{a}$. An eikonal equation of this kind was brought forward by Heintzmann and Schrüfer [61], based on earlier work by Heintzmann, Kundt and Lasota [60] in the context of special relativity. For each eikonal equation derived that way, the range of validity as an approximation scheme (if any) must be checked individually.

Part II

A mathematical framework for ray optics

4. Introduction to Part II

In Part II we treat ray optics as a theory in its own right. In Chap. 5 we presuppose an arbitrary finite-dimensional manifold $\mathcal{M}$ and set up a Hamiltonian formalism for ray optics in the cotangent bundle over $\mathcal{M}$. In Chap. 6 we assume that, in addition, a Lorentzian metric is given on $\mathcal{M}$. Specialized to the case $\dim(\mathcal{M}) = 4$, $(\mathcal{M}, g)$ can then be interpreted as a spacetime in the sense of general relativity and our formalism covers ray optics in arbitrary media on such a spacetime. This procedure has the advantage that the results of Chap. 5 apply equally well to spacetime theories other than general relativity and to the case that $\mathcal{M}$ is to be interpreted as space, rather than as spacetime, in any kind of theory where such a notion makes sense. Chapter 7 will then be devoted to variational principles for rays and Chap. 8 presents applications of the general formalism to astrophysics and astronomy.

The results of Part I will often be used for the sake of motivation, and they will provide us with illustrative examples. However, the mathematical formalism developed in Part II is completely self-contained.

4.1 A brief guide to the literature

In the following we make extensive use of Hamiltonian formalism. For the most part we use coordinate free notation since it is our goal to also treat some global questions. We assume that the reader is familiar with differential calculus and with symplectic geometry as it is used in the modern treatment of classical mechanics. Our standard reference for background material is the textbook by Abraham and Marsden [1]. In addition, we also refer to Arnold [8] and to Woodhouse [150]. More particularly in view of optics, it might be helpful to consult the textbook by Guillemin and Sternberg [55] where applications of the Hamiltonian formalism and of symplectic geometry to optics are given in modern mathematical terminology. In traditional notation, applications of the Hamiltonian formalism to optics can be found, e.g., in the classical work of Carathéodory [25] and of Luneburg [88] [89]. Readers interested in the historical roots of "Hamiltonian optics" should go back to the original work of Sir William Rowan Hamilton who established this formalism in the 1820s, see vol. 1 of the collected papers of Hamilton edited by Conway and Synge [29]. Next to the work of Hamilton, the most fundamental contribution to the mathematical theory of ray optics is due to Bruns [23] who introduced the socalled *eikonal function*. The relation of Bruns's eikonal function to Hamilton's *characteristic function* is controversially discussed in articles by Herzberger [63] and Synge [139].

Textbooks on general relativity do not usually treat ray optics in detail. Most of them are restricted to light propagation in vacuo where the light rays are just the light-like geodesics of the spacetime metric. An important exception to this rule is the book by Synge [142] where a Hamiltonian formalism for ray optics in isotropic media is discussed in some detail. It is worthwile to compare this to earlier work on ray optics by the same author, see Synge [138] [140] [141]. More recent work by Miron and Kawaguchi [97], also see Kawaguchi and Miron [68] [69], is strongly influenced by ideas of Synge but uses a more modern mathematical terminology. It is the main purpose of their work to develop a differential geometric formalism for light propagation in isotropic dispersive media. Miron and Kawaguchi repeatedly claim that standard symplectic geometry does not provide an appropriate framework for the treatment of such media. We do not share this point of view .

Having set up a Hamiltonian formalism for general-relativistic ray optics, the way is paved for characterizing rays by a variational principle. Some of these variational principles can be interpreted as general-relativistic versions of *Fermat's principle*. The oldest versions, which hold for vacuum rays in static or stationary spacetimes, date back to Weyl [149] and Levi-Civita [81]. Related material can be found in Levi-Civita [80] [82] [83] and Synge [137]. (The reader is cautioned that the latter paper does not meet the standard of Synge's later work.) These versions of Fermat's principle are also discussed in several modern textbooks and review articles, see, e.g., Frankel [43] or Straumann [136] for the static case and Landau and Lifshitz [76] or Brill [21] for the stationary case. For a discussion from a mathematical point of view we refer to Masiello [93]. Generalizations from vacuum to an isotropic medium, but still assuming stationarity, were first considered by Pham Mau Quan [116] [117] [118] [119]. On the other hand, Uhlenbeck [144] found the first variational principle for (vacuum) light rays in general-relativistic spacetimes without symmetries. Whereas for the work of Uhlenbeck it was crucial that the spacetime be globally hyperbolic, Kovner [74] was able to formulate a Fermat principle for vacuum light rays in an arbitrary Lorentzian manifold. A rigorous proof that the solution curves of Kovner's variational principle are, indeed, the light-like geodesics was given in Perlick [108]. Kovner's variational principle was further discussed, both from a physical and from a mathematical point of view, e.g., by Faraoni [42], Nityananda and Samuel [101], Schneider, Ehlers and Falco [128], Bel and Martín [12], Perlick [110] and in several articles by Giannoni, Masiello and Piccione, see, e.g., Giannoni and Masiello [46] or Giannoni, Masiello and Piccione [47] [48].

4.2 Assumptions and notations

Throughout Part II we presuppose a finite-dimensional real C^∞ manifold $\mathcal{M}$ whose topology satisfies Hausdorff's axiom and the second axiom of countability. This implies that $\mathcal{M}$ is paracompact. The terms "manifold" and "submanifold" always mean manifold and submanifold without boundary. The physical interpretation we have in mind refers to $\mathcal{M}$ as to a spacetime model in the sense of general relativity. However, for the basic concepts of ray optics, to be introduced and discussed in Chap. 5, we need no additional structure on $\mathcal{M}$ and $n = \dim(\mathcal{M})$ need not be specified. In Chap. 6 we shall assume that there is a C^∞ Lorentzian metric g given on $\mathcal{M}$, whereas $n = \dim(\mathcal{M})$ will still be an unspecified positive integer except for the restriction that, in Chap. 6, we assume $n > 2$ to exclude some pathologies. In all applications to relativity we use units making the vacuum velocity c of light equal to one.

At a point $q \in \mathcal{M}$, we denote the tangent space by $T_q\mathcal{M}$ and its dual, the cotangent space, by $T_q^*\mathcal{M}$. The tangent bundle will be denoted by $\tau_\mathcal{M} \colon T\mathcal{M} \longrightarrow \mathcal{M}$ and the cotangent bundle by $\tau_\mathcal{M}^* \colon T^*\mathcal{M} \longrightarrow \mathcal{M}$. It will often be necessary to remove the zero section from $T\mathcal{M}$ and from $T^*\mathcal{M}$; what is left will be denoted by $\overset{o}{T}\mathcal{M}$ and $\overset{o}{T}{}^*\mathcal{M}$, respectively.

By a "Lorentzian metric" we always mean a covariant symmetric second rank tensor field with signature $(+, \ldots, +, -)$. With respect to a Lorentzian metric g, a linear subspace of the tangent space $T_q\mathcal{M}$ is called *space-like* if on this subspace the metric g is positive definite, *light-like* if it is positive semidefinite but not positive definite, and *time-like* otherwise. A vector $X \in T_q\mathcal{M}$ is called space-like, light-like, or time-like whenever the linear subspace spanned by this vector has the respective property. As a consequence, X is space-like if $X = 0$ or $g(X, X) > 0$, light-like if $X \neq 0$ and $g(X, X) = 0$, and time-like if $g(X, X) < 0$. If X is space-like, light-like, or time-like, the same property is assigned to the covector $g(X, \cdot)$. Finally, a submanifold of $\mathcal{M}$ is called space-like, light-like, or time-like whenever its tangent space has the respective property at all points.

With a Lorentzian metric (or, more generally, a pseudo-Riemannian metric of any signature) on $\mathcal{M}$ there is associated its Levi-Civita connection ∇. This defines the notions of parallel transport and of geodesics. By a *geodesic* we always mean a map $\lambda \colon I \longrightarrow \mathcal{M}$ from a real interval into $\mathcal{M}$ such that $\nabla_{\dot\lambda}\dot\lambda$ is parallel to $\dot\lambda$. Such a curve is called an *affinely parametrized geodesic* if, more specifically, the equation $\nabla_{\dot\lambda}\dot\lambda = 0$ is satisfied.

We assume that the reader is familiar with exterior calculus. As to the definition of antisymmetric tensor product, exterior derivative, etc., our sign and factor conventions follow Abraham and Marsden [1]. Whenever refering to a local chart $(x^1, \ldots, x^n)$ on $\mathcal{M}$, we use Einstein's summation convention with latin indices running from 1 to n and greek indices running from 1 to $n-1$. With respect to such a local chart, elements of $T\mathcal{M}$ can be represented in the form $v^a \partial/\partial x^a$, and elements of $T^*\mathcal{M}$ can be represented in the form

$p_a dx^a$. In this way we get a local chart $(x^1, \ldots, x^n, v^1, \ldots, v^n)$ on TM, and a local chart $(x^1, \ldots, x^n, p_1, \ldots, p_n)$ on T^*M. Following Abraham and Marsden [1] we refer to charts constructed in this way as *natural charts* induced by $(x^1, \ldots, x^n)$. It is also usual to refer to the v^a as to *velocity coordinates* and to the p_a as to *momentum coordinates* conjugate to the x^a. Occasionally we also refer to elements of TM as to *velocity vectors* and to elements of T^*M as to *momentum covectors*.

It is well known and easily verified that there is a (unique and globally well-defined) one-form θ on T^*M such that $\theta = p_a dx^a$ in any natural chart. θ is known as the *canonical one-form* on T^*M. It can be characterized in a coordinate-free manner as the unique one-form on T^*M such that $\beta^*\theta = \beta$ for all C^∞ sections $\beta: M \longrightarrow T^*M$, see Abraham and Marsden [1], p. 179. (Here, $\beta^*\theta$ denotes the pull-back of θ with β.) The two-form $\Omega = -d\theta$, which is known as the *canonical two-form* on T^*M, takes the form $\Omega = dx^a \wedge dp_a$ in any natural chart. More generally, any chart on T^*M in which Ω takes this special form is called a *canonical chart*. It is obvious that Ω is closed (i.e., $d\Omega = 0$) and non-degenerate (i.e., the equation $\Omega(X, \cdot) = 0$ implies $X = 0$). Hence, Ω makes T^*M into a *symplectic manifold*. The restrictions of θ and Ω to $\overset{\circ}{T}^*M$ will again be denoted by θ and Ω for the sake of simplicity.

Ω can be used to assign to each C^∞ function $H: T^*M \longrightarrow \mathbb{R}$ a *Hamiltonian vector field* X_H on T^*M by the formula

$$\Omega(X_H, \cdot) = dH. \tag{4.1}$$

The non-degeneracy of Ω guarantees that the assignment $H \longmapsto X_H$ is, indeed, well defined. In a natural chart, the Hamiltonian vector field X_H takes the form

$$X_H = \frac{\partial H}{\partial p_a} \frac{\partial}{\partial x^a} - \frac{\partial H}{\partial x^a} \frac{\partial}{\partial p_a}. \tag{4.2}$$

A C^∞ curve $\xi: I \longrightarrow T^*M$, defined on a real interval I, is called a solution of *Hamilton's equations* iff it is an integral curve of the Hamiltonian vector field X_H, i.e., iff

$$\Omega_{\xi(s)}\big(\dot\xi(s), \cdot\big) = (dH)_{\xi(s)} \tag{4.3}$$

for all $s \in I$. If ξ is represented in a natural chart as a map $s \longmapsto \big(x(s), p(s)\big)$, (4.3) takes the familiar canonical form of Hamilton's equations

$$\dot{x}^a(s) = \frac{\partial H}{\partial p_a}\big(x(s), p(s)\big), \tag{4.4}$$

$$\dot{p}_a(s) = -\frac{\partial H}{\partial x^a}\big(x(s), p(s)\big). \tag{4.5}$$

Equation (4.4), which gives the velocity coordinates as functions of the position and momentum coordinates, is properly called the *vertical part* of

Hamilton's equations since it corresponds to equation (4.3) applied to vectors tangent to the fibers of $T^*\mathcal{M}$. If (4.4) can be solved for the momentum coordinates p_a, the result can be inserted into (4.5). This leaves us with a set of second order equations for the position coordinates x^a. Locally around some point $u \in T^*\mathcal{M}$, (4.4) can be solved for the momentum coordinates p_a if and only if the condition

$$\det(H^{ab}) \neq 0 \qquad (4.6)$$

holds at u. Here and in the following, we use the abbreviations

$$H^a = \frac{\partial H}{\partial p_a}, \quad H^{ab} = \frac{\partial^2 H}{\partial p_a \partial p_b}, \quad \text{etc.} \qquad (4.7)$$

A Hamiltonian H that satisfies (4.6) at u is called *regular* or *non-degenerate* at u. (It is easy to check that (4.6) holds in any natural chart if it holds in just one natural chart.) Hence, locally Hamilton's equations can be viewed as a system of second order differential equations on $\mathcal{M}$ if and only if H is everywhere regular.

To express regularity in invariant notation, without refering to a natural chart, one introduces the *fiber derivative* of a function $H : T^*\mathcal{M} \longrightarrow \mathbb{R}$ in the following way. For $q \in \mathcal{M}$, we denote the restriction of H to $T_q^*\mathcal{M}$ by H_q; then, for each $u \in T_q^*\mathcal{M}$, the differential $(dH_q)_u : T_q^*\mathcal{M} \longrightarrow \mathbb{R}$ being a linear map can be viewed as an element of $(T_q^*\mathcal{M})^* \cong T_q\mathcal{M}$. The fiber derivative $\mathbb{F}H : T^*\mathcal{M} \longrightarrow T\mathcal{M}$ of H is defined by the equation $(\mathbb{F}H)(u) = (dH_q)_u$ where $q = \tau_\mathcal{M}^*(u)$. Using natural charts on $T^*\mathcal{M}$ and $T\mathcal{M}$, induced by one and the same chart $x = (x^1, \ldots, x^n)$ on $\mathcal{M}$, the fiber derivative takes the form $(x, p) \longmapsto (x, v = \partial H/\partial p)$. With the help of the map $\mathbb{F}H$ the vertical part of Hamilton's equations (4.3) can be written in the form $(\tau_\mathcal{M}^* \circ \xi)^\cdot = \mathbb{F}H \circ \xi$, where the ring denotes composition of maps and the dot stands for derivative with respect to the curve parameter. Hence the desired invariant characterization of regularity can be given in the following way. A Hamiltonian is regular at a point $u \in T^*\mathcal{M}$ if and only if its fiber derivative $\mathbb{F}H$ maps a neighborhood of u diffeomorphically onto an open subset of $T\mathcal{M}$. If $\mathbb{F}H : T^*\mathcal{M} \longrightarrow T\mathcal{M}$ is even a global diffeomorphism, H is called *hyperregular*.

5. Ray-optical structures on arbitrary manifolds

In Part I we have seen that the fundamental object on which all of ray optics can be based is a dispersion relation, i.e., a characteristic variety in a cotangent bundle. We make this into the general definition of ray optical structures on our arbitrary n-dimensional manifold.

5.1 Definition and basic properties of ray-optical structures

When we performed the passage from Maxwell's equations to ray optics in Part I, the characteristic variety came about as the zero-level surface of a Hamiltonian function H, where H was a smooth function on the punctured cotangent bundle whose derivative with respect to the momentum coordinates had no zeros. It is therefore natural to define a ray-optical structure on our arbitrary n-dimensional manifold $\mathcal{M}$ as a closed codimension-one submanifold $\mathcal{N}$ of $\overset{\circ}{T}{}^*\mathcal{M}$ which is everywhere transverse to the fibers. If we add the condition that $\mathcal{N}$ covers all points of $\mathcal{M}$, we are led to the following definition which is fundamental for all of Part II.

Definition 5.1.1. *A ray-optical structure on $\mathcal{M}$ is a $(2n-1)$-dimensional closed embedded C^∞ submanifold $\mathcal{N}$ of $\overset{\circ}{T}{}^*\mathcal{M}$ such that $\tau^*_\mathcal{M}|_\mathcal{N}: \mathcal{N} \longrightarrow \mathcal{M}$ is a surjective submersion.*

Here $\tau^*_\mathcal{M}|_\mathcal{N}$ denotes the restriction of the bundle projection $\tau^*_\mathcal{M}: T^*\mathcal{M} \longrightarrow \mathcal{M}$ to $\mathcal{N}$. The condition of $\tau^*_\mathcal{M}|_\mathcal{N}$ being a submersion guarantees that $\mathcal{N}$ is transverse to the fibers of $\overset{\circ}{T}{}^*\mathcal{M}$, whereas the condition of $\tau^*_\mathcal{M}|_\mathcal{N}$ being surjective guarantees that $\mathcal{N}$ covers all of $\mathcal{M}$. Both conditions together imply that the set $\mathcal{N}_q = \mathcal{N} \cap \overset{\circ}{T}{}^*_q\mathcal{M}$ is a codimension-one submanifold of the punctured cotangent space $\overset{\circ}{T}{}^*_q\mathcal{M}$ for all $q \in \mathcal{M}$. As $\mathcal{N}$ is closed in $\overset{\circ}{T}{}^*\mathcal{M}$, so is $\mathcal{N}_q$ in $\overset{\circ}{T}{}^*_q\mathcal{M}$. Note that, for two different points q and q', the manifolds $\mathcal{N}_q$ and $\mathcal{N}_{q'}$ are not necessarily diffeomorphic. In particular, $\mathcal{N}$ need not be a fiber bundle over $\mathcal{M}$. This will be exemplified below.

According to Definition 5.1.1, $\mathcal{N}$ need not be closed in $T^*\mathcal{M}$ and its closure in $T^*\mathcal{M}$ might fail to be a smooth manifold at the zero section. In view of the examples we have in mind, it is indeed necessary to keep Definition 5.1.1 as general as that. In particular, we do not want to exclude the case that $\mathcal{N}$ is the null cone bundle of a Lorentzian metric.

If $\mathcal{N}_1$ and $\mathcal{N}_2$ are two ray-optical structures on $\mathcal{M}$ with $\mathcal{N}_1 \cap \mathcal{N}_2 = \emptyset$, then $\mathcal{N} = \mathcal{N}_1 \cup \mathcal{N}_2$ is again a ray-optical structure on $\mathcal{M}$. Conversely, each connected component of a ray-optical structure is a ray-optical structure in its own right.

Ray-optical structures that come about as level surfaces of Hamiltonian functions are characterized by the following proposition.

Proposition 5.1.1. *Fix a C^∞ function $H \colon \overset{\circ}{T}{}^*\mathcal{M} \longrightarrow \mathbb{R}$ and let $\mathcal{N}$ be the zero-level surface of H, i.e., $\mathcal{N} = \{\, u \in \overset{\circ}{T}{}^*\mathcal{M} \,|\, H(u) = 0 \,\}$. Then $\mathcal{N}$ is a ray-optical structure on $\mathcal{M}$, provided that H satisfies the following two properties.*

(a) *For all $q \in \mathcal{M}$, the set $\mathcal{N}_q = \{\, u \in \overset{\circ}{T}{}_q^*\mathcal{M} \,|\, H(u) = 0 \,\}$ is non-empty.*
(b) *For all $u \in \mathcal{N}$, the fiber derivative of H satisfies $(\mathbb{F}H)(u) \neq 0$.*

Proof. Condition (a) guarantees that the map $\tau_\mathcal{M}^*|_\mathcal{N}$ is surjective. Condition (b), which is the coordinate-free way of saying that the derivative of H with respect to the momentum coordinates has no zeros, guarantees that $\mathcal{N}$ is a closed embedded codimension-one submanifold of $\overset{\circ}{T}{}^*\mathcal{M}$ and that $\tau_\mathcal{M}^*|_\mathcal{N}$ is a submersion.   □

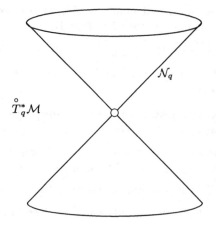

Fig. 5.1. For the ray-optical structure of Example 5.1.1, $\mathcal{N}_q$ is the null cone of the metric g_o at each point $q \in \mathcal{M}$.

A ray-optical structure need not be generated globally by a function H in this way. In general, such a function H exists only locally. (Later in this section we shall investigate this question in more detail.) Insofar, Definition 5.1.1 applies to situations more general than those encountered in Part I. Such a generalization seems to be reasonable since the treatment of Part I was local throughout.

Next we mention some examples of ray-optical structures. They will serve as our standard examples for the discussion of all properties of ray-optical structures. Therefore the reader is requested to commit them to his or her memory.

Example 5.1.1. Let g_o be a C^∞ Lorentzian metric on $\mathcal{M}$ and denote the induced fiber metric on $T^*\mathcal{M}$ by $g_o^\#$. (In other words, if the components of g_o are denoted by $(g_o)_{ab}$, the components of $g_o^\#$ are given by g_o^{ab} with $(g_o)_{ac}\, g_o^{cd} = \delta_a^d$.) Define $H: \overset{\circ}{T}{}^*\mathcal{M} \longrightarrow \mathbb{R}$ by $H(u) = \tfrac{1}{2} g_o^\#(u,u)$, i.e.,

$$H(x,p) = \tfrac{1}{2} g_o^{ab}(x)\, p_a\, p_b \tag{5.1}$$

in terms of natural coordinates. Then $\mathcal{N} = \{ u \in \overset{\circ}{T}{}^*\mathcal{M} \mid H(u) = 0 \}$ is a ray-optical structure on $\mathcal{M}$.

For the case $\dim(\mathcal{M}) = 4$ this example admits several different interpretations on the basis of general relativity. The first possibility is to interpret g_o as the spacetime metric such that $\mathcal{N}$ gives light propagation in vacuum. The second possibility is to interpret g_o as the optical metric (2.88) in a linear dielectric and permeable medium which is isotropic. The third possibility is to interpret g_o as one of the two optical metrics (2.91) in a linear dielectric and permeable medium which is anisotropic with the special features typical of a uniaxial crystal. In the latter case it is important to realize that, in general, the two optical metrics must be treated separately; the union of the two "light cone bundles" is not a ray-optical structure in the sense of Definition 5.1.1 unless they are disjoint.

We should keep in mind that Example 5.1.1 is not general enough to cover light propagation in arbitrary linear dielectric and permeable media. According to Sect. 2.5 this would require a generalization to Finslerian metrics.

Here is another example of a ray-optical structure.

Example 5.1.2. Let g_o and $g_o^\#$ be as in Example 5.1.1 and define a function $H: \overset{\circ}{T}{}^*\mathcal{M} \longrightarrow \mathbb{R}$ by $H(u) = \tfrac{1}{2}\big(g_o^\#(u,u) + 1\big)$, i.e.,

$$H(x,p) = \tfrac{1}{2}\big(g_0^{ab}(x)p_a\, p_b + 1\big) \tag{5.2}$$

in terms of natural coordinates. Then $\mathcal{N} = \{ u \in \overset{\circ}{T}{}^*\mathcal{M} \mid H(u) = 0 \}$ is a ray-optical structure on $\mathcal{M}$.

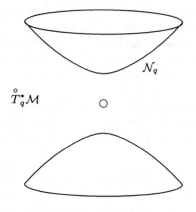

Fig. 5.2. For the ray-optical structure of Example 5.1.2, $\mathcal{N}_q$ is a two-shell hyperboloid at each point $q \in \mathcal{M}$.

According to Chap. 3, light propagation in a non-magnetized plasma is given by a ray-optical structure of this sort, generated by the (partial) Hamiltonian (3.46). Here we have to put $g_o = \omega_p^2 \, g$, where g is the spacetime metric and ω_p is the plasma frequency (3.51), and we have to restrict ourselves to regions where ω_p has no zeros.

Example 5.1.1 and Example 5.1.2 are special cases of the following more general construction. Let g be a C^∞ Lorentzian metric on $\mathcal{M}$ and denote the induced fiber metric on $T^*\mathcal{M}$ by $g^\#$. Fix a C^∞ function $h \colon \mathcal{M} \longrightarrow \mathbb{R}$ and define $H \colon \overset{\circ}{T}{}^*\mathcal{M} \longrightarrow \mathbb{R}$ by $H(u) = \frac{1}{2} \big(g^\#(u, u) + h(\tau^*_\mathcal{M}(u)) \big)$. Then $\mathcal{N} = \{ u \in \overset{\circ}{T}{}^*\mathcal{M} \mid H(u) = 0 \}$ is a ray-optical structure on $\mathcal{M}$. In regions where h vanishes this leads us back to Example 5.1.1 with $g_o = g$. In regions where h is strictly positive this leads us back to Example 5.1.2 with $g_o = h \, g$. This generalized example has relevance for light propagation in a non-magnetized plasma which occupies only part of the spacetime region considered. Mathematically it exemplifies our earlier remark that, for an arbitrary ray-optical structure $\mathcal{N}$ on $\mathcal{M}$, the manifolds $\mathcal{N}_q$ and $\mathcal{N}_{q'}$ need not be diffeomorphic for two points q and q' in $\mathcal{M}$. If $h(q) = 0$ and $h(q') > 0$, $\mathcal{N}_q$ is a double cone whereas $\mathcal{N}_{q'}$ is a two-shell hyperboloid, i.e., $\mathcal{N}_q$ and $\mathcal{N}_{q'}$ are not even homeomorphic, let alone diffeomorphic .

We now turn to a different kind of examples for ray-optical structures.

Example 5.1.3. Fix a C^∞ vector field U on $\mathcal{M}$ that has no zeros and define a function $H \colon \overset{\circ}{T}{}^*\mathcal{M} \longrightarrow \mathbb{R}$ by $H(u) = u(U_q)$ for all $q \in \mathcal{M}$ and $u \in \overset{\circ}{T}{}^*_q\mathcal{M}$, i.e.,

$$H(x, p) = U^a(x) \, p_a \tag{5.3}$$

in terms of natural coordinates. Then the set $\mathcal{N} = \{\, u \in \overset{\circ}{T}{}^*\mathcal{M} \mid H(u) = 0 \,\}$ is a ray-optical structure on $\mathcal{M}$.

In our discussion of light propagation in a non-magnetized plasma we encountered the (partial) Hamiltonian (3.44) which generates a ray-optical structure of this sort. This example will be useful in the following to demonstrate some possible pathologies.

Example 5.1.3 admits the following generalization which is mathematically interesting although somewhat contrived in view of physical applications. Instead of a C^∞ vector field U without zeros, it suffices to have a C^∞ line field L, i.e., a map that assigns to each point $q \in \mathcal{M}$ a one-dimensional subspace L_q of the tangent space $T_q\mathcal{M}$. Then we define $\mathcal{N}_q$ as the set of all covectors in $\overset{\circ}{T}{}^*_q\mathcal{M}$ that annihilate all vectors in L_q. It is easy to check that $\mathcal{N} = \{\, u \in \mathcal{N}_q \mid q \in \mathcal{M} \,\}$ is indeed a ray-optical structure on $\mathcal{M}$. If L is not globally spanned by a C^∞ vector field without zeros, this ray-optical structure $\mathcal{N}$ is not globally generated by a Hamiltonian function, i.e., it is not of the kind considered in Proposition 5.1.1.

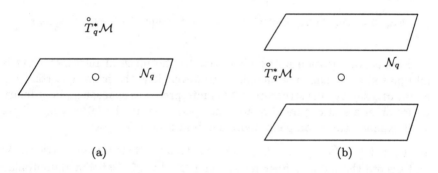

Fig. 5.3. For the ray-optical structure of Example 5.1.3, $\mathcal{N}_q$ is a punctured hyperplane (a), whereas for Example 5.1.4 it is a pair of hyperplanes (b).

The following example is related to Example 5.1.3 in a similar way as Example 5.1.2 is related to Example 5.1.1.

Example 5.1.4. Let U be a C^∞ vector field on $\mathcal{M}$ that has no zeros and define a function $H \colon \overset{\circ}{T}{}^*\mathcal{M} \longrightarrow \mathbb{R}$ by $H(u) = \frac{1}{2}\big(u(U_q)^2 - 1\big)$ for all points $q \in \mathcal{M}$ and for all covectors $u \in \overset{\circ}{T}{}^*_q\mathcal{M}$, i.e.,

$$H(x,p) = \tfrac{1}{2}\big(U^a(x)U^b(x)p_a\,p_b - 1\big) \tag{5.4}$$

in terms of natural coordinates. Then the set $\mathcal{N} = \{\, u \in \overset{\circ}{T}{}^*\mathcal{M} \mid H(u) = 0 \,\}$ is a ray-optical structure on $\mathcal{M}$.

The partial Hamiltonian (3.45), which determines the plasma oscillations in a non-magnetized plasma, generates a ray-optical structure of this kind. Here we have to divide the vector field $\overset{\circ}{U}$ of (3.45) by the plasma frequency ω_p, given by (3.51), to get the vector field U, and we have to restrict to regions where the plasma frequency has no zeros.

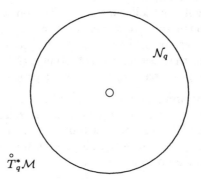

Fig. 5.4. For the ray-optical structure of Example 5.1.5, $\mathcal{N}_q$ is a sphere.

Finally, we mention an example that has no physical relevance if $\mathcal{M}$ is interpreted as a spacetime manifold. However, it is the most important example of a ray-optical structure if $\mathcal{M}$ is interpreted as space, e.g., in ordinary optics or in a static general-relativistic spacetime. In the latter context, we shall examine this example in more detail in Sect. 6.5 below.

Example 5.1.5. Let g_+ be a C^∞ (positive definite) Riemannian metric on $\mathcal{M}$ and denote the induced fiber metric on $T^*\mathcal{M}$ by $g_+^\#$. Define a Hamiltonian $H: \overset{\circ}{T}^*\mathcal{M} \longrightarrow \mathbb{R}$ by $H(u) = \frac{1}{2}\big(g_+^\#(u,u) - 1\big)$, i.e.,

$$H(x,p) = \tfrac{1}{2}\big(g_+^{ab} p_a p_b - 1\big) \tag{5.5}$$

in terms of natural coordinates. Then $\mathcal{N} = \{\, u \in \overset{\circ}{T}^*\mathcal{M} \,|\, H(u) = 0 \,\}$ is a ray-optical structure on $\mathcal{M}$.

This ends our list of examples to which we shall come back frequently.

It is now our goal to justify the term "ray-optical structure" by showing that, indeed, any such structure gives rise to the notions of rays. If $\mathcal{N}$ is a ray-optical structure on $\mathcal{M}$, we can use the inclusion map $i: \mathcal{N} \longrightarrow \overset{\circ}{T}^*\mathcal{M}$ to pull back the canonical one-form θ and the canonical two-form Ω to $\mathcal{N}$. The resulting forms will be denoted by

$$\theta_\mathcal{N} = i^*\theta \quad \text{and} \quad \Omega_\mathcal{N} = i^*\Omega \,. \tag{5.6}$$

Since the exterior derivative d commutes with the pull-back operation i^*, these two forms are related by the equation $\Omega_{\mathcal{N}} = d\theta_{\mathcal{N}}$. In particular, this implies that $\Omega_{\mathcal{N}}$ is a closed two-form, $d\Omega_{\mathcal{N}} = 0$. Moreover, at each point $u \in \mathcal{N}$ the kernel of $\Omega_{\mathcal{N}}$ is one-dimensional since Ω is non-degenerate and $\mathcal{N}$ has codimension one. This shows that $(\mathcal{N}, \Omega_{\mathcal{N}})$ is a *contact manifold*, see Abraham and Marsden [1], Definition 5.1.4 .

A vector field X on $\mathcal{N}$ is called a *characteristic vector field* iff it satisfies the equation $\Omega_{\mathcal{N}}(X, \cdot) = 0$. As the kernel of $\Omega_{\mathcal{N}}$ is one-dimensional, any two characteristic vector fields are linearly dependent. Integral curves of characteristic vector fields (or their projections to $\mathcal{M}$) are often called *characteristic curves*. They give us the rays of $\mathcal{N}$, according to the following definition.

Definition 5.1.2. *Let $\mathcal{N}$ be a ray-optical structure on $\mathcal{M}$ and let $\Omega_{\mathcal{N}}$ be the contact two-form on $\mathcal{N}$ defined by (5.6). A C^∞ immersion $\xi \colon I \longrightarrow \mathcal{N}$ from a real interval I into $\mathcal{N}$ is called a* lifted ray *iff*

$$(\Omega_{\mathcal{N}})_{\xi(s)}\big(\dot{\xi}(s), \cdot\big) = 0 \tag{5.7}$$

for all $s \in I$. Then the projected curve $\tau_{\mathcal{M}}^* \circ \xi \colon I \longrightarrow \mathcal{M}$ is called a ray.

Clearly, lifted rays satisfy the following existence and (non-)uniqueness properties.

(a) A lifted ray remains a lifted ray under an arbitrary reparametrization (which need not be orientation-preserving).
(b) Through each point $u \in \mathcal{N}$ there is a lifted ray, and it is unique up to reparametrization and extension.

Moreover, it is important to realize that a lifted ray is nowhere tangent to a fiber of $T^*\mathcal{M}$. (This follows from the fact that $\mathcal{N}$ is everywhere transverse to the fibers.) Hence, every ray is an immersed curve in $\mathcal{M}$. In other words, "rays do not stand still in $\mathcal{M}$".

In the case of Example 5.1.1 the rays are the light-like geodesics of the metric g_o whereas in the case of Example 5.1.2 they are the time-like geodesics of the metric g_o. In the case of Example 5.1.3 and 5.1.4 an immersed curve in $\mathcal{M}$ is a ray iff it is an integral curve of the vector field U. In the case of Example 5.1.5 the rays are the immersed geodesics of the metric g_+.

It follows immediately from the definitions that two ray-optical structures on $\mathcal{M}$ are equal if and only if they determine the same set of lifted rays. However, it is very well possible for two different ray-optical structures to have the same rays. As an example we may consider a ray-optical structure constructed from a (positive definite) Riemannian metric g_+ as in Example 5.1.5 such that the rays are the geodesics of the metric g_+. If we change g_+ by multiplication with a positive constant $c \neq 1$ we get a different ray-optical structure but the set of rays remains unchanged. This is obvious since g_+ and $c\,g_+$ have the same Levi-Civita connection.

So far our notion of rays made no use of Hamiltonian functions. To cast the ray equation (5.7) into Hamiltonian form we need the following proposition which can be viewed as the converse of Proposition 5.1.1.

Proposition 5.1.2. *Let $\mathcal{N}$ be a ray-optical structure on $\mathcal{M}$ and fix a point $u \in \mathcal{N}$. Then there is an open neighborhood W of u in $\overset{\circ}{T}{}^*\mathcal{M}$ and a C^∞ function $H \colon W \longrightarrow \mathbb{R}$ with the following properties.*

(a) $\mathcal{N} \cap W = \{ w \in W \mid H(w) = 0 \}$;
(b) dH has no zeros on $\mathcal{N} \cap W$.

Any such function H is called a local Hamiltonian *for $\mathcal{N}$. H is called a* global Hamiltonian *for $\mathcal{N}$ if all of $\mathcal{N}$ is covered by W, i.e., if $\mathcal{N} \subset W$.*

Proof. This is an immediate consequence of the fact that, by Definition 5.1.1, $\mathcal{N}$ is an embedded codimension-one C^∞ submanifold of $T^*\mathcal{M}$.   □

Later we shall give criteria for the existence of global Hamiltonians, see Proposition 5.1.4 below.

We shall now rewrite the ray equation (5.7) in Hamiltonian form. To begin with, let us consider the special case of a ray-optical structure $\mathcal{N}$ on $\mathcal{M}$ that admits a global Hamiltonian $H \colon W \longrightarrow \mathbb{R}$ and let us denote the Hamiltonian vector field of H by X_H. Then, at all points $u \in \mathcal{N}$, the vector $(X_H)_u$ is non-zero and tangent to $\mathcal{N}$. Hence, $X_H|_{\mathcal{N}}$ gives a nowhere vanishing vector field on $\mathcal{N}$. If the defining equation (4.1) of X_H is pulled back to $\mathcal{N}$, we find $\Omega_{\mathcal{N}}(X_H|_{\mathcal{N}}, \cdot) = 0$, i.e., $X_H|_{\mathcal{N}}$ is a characteristic vector field. Thus, any other characteristic vector field on $\mathcal{N}$ must be a multiple of $X_H|_{\mathcal{N}}$. This implies that an immersion $\xi \colon I \longrightarrow \mathcal{N}$ is a lifted ray if and only if its tangent field is everywhere a multiple of X_H. Thus, lifted rays ξ are characterized by the equations

$$H\big(\xi(s)\big) = 0 , \tag{5.8}$$

$$\Omega_{\xi(s)}\big(\dot{\xi}(s), \cdot \big) = k(s)\,(dH)_{\xi(s)} , \tag{5.9}$$

where k is a nowhere vanishing but otherwise arbitrary function. The freedom to choose this function at will corresponds to the fact that lifted rays can be arbitrarily reparametrized.

If H is a local rather than a global Hamiltonian, this result remains true on that part of $\mathcal{N}$ which is covered by W. This implies that, with respect to a natural chart and a local Hamiltonian, lifted rays are characterized by the equations

$$H\Big(x(s), p(s)\Big) = 0\,, \tag{5.10}$$

$$\dot{x}^a(s) = k(s)\,\frac{\partial H}{\partial p_a}\Big(x(s), p(s)\Big)\,, \tag{5.11}$$

$$\dot{p}_a(s) = -k(s)\,\frac{\partial H}{\partial x^a}\Big(x(s), p(s)\Big)\,. \tag{5.12}$$

These considerations can be summarized in the following way. If we are lucky enough to have a global Hamiltonian H for $\mathcal{N}$, the lifted rays of $\mathcal{N}$ can be found by (i) solving Hamilton's equations with this function H; (ii) singling out those solutions that lie in $\mathcal{N}$; (iii) allowing for arbitrary reparametrizations. If there is no global Hamiltonian, the same procedure can be carried through on the domain of each local Hamiltonian. On the mutual overlaps of those domains the lifted rays have to be patched up.

From a geometrical point of view it is quite satisfactory to work with the contact manifold $(\mathcal{N}, \Omega_{\mathcal{N}})$ without refering to Hamiltonians. On the other hand, the use of (local) Hamiltonians leads to a formalism that looks more familiar to physicists. For that reason we will often refer to Hamiltonians.

We end this section with two propositions that are helpful when working with local Hamiltonians. The first proposition clarifies the relation between two local Hamiltonians for one and the same ray-optical structure, the second proposition gives criteria for the existence of a global Hamiltonian.

Proposition 5.1.3. *Let $\mathcal{N}$ be a ray-optical structure on $\mathcal{M}$ and assume that the C^∞ function $H: \mathcal{W} \longrightarrow \mathbb{R}$ is a local Hamiltonian for $\mathcal{N}$, defined on some open subset $\mathcal{W}$ of $\overset{\circ}{T}{}^*\mathcal{M}$ with $\mathcal{N} \cap \mathcal{W} \neq \emptyset$. Then another C^∞ function $\tilde{H}: \mathcal{W} \longrightarrow \mathbb{R}$, defined on the same open subset $\mathcal{W}$ of $\overset{\circ}{T}{}^*\mathcal{M}$, is again a local Hamiltonian for $\mathcal{N}$ if and only if there is a C^∞ function $F: \mathcal{W} \longrightarrow \mathbb{R} \setminus \{0\}$ such that the equation $\tilde{H} = FH$ holds on $\mathcal{W}$. (By continuity, the function F must be either everywhere positive or everywhere negative.)*

Proof. Since the "if" part is trivial, we just have to prove the "only if" part. So let us assume that both H and $\tilde{H}$ are local Hamiltonians for $\mathcal{N}$. Then $F_0(u) = \tilde{H}(u)/H(u)$ defines a C^∞ function $F_0: \mathcal{W} \setminus \mathcal{N} \longrightarrow \mathbb{R} \setminus \{0\}$ since both H and $\tilde{H}$ are non-zero on $\mathcal{W} \setminus \mathcal{N}$. As dH and $d\tilde{H}$ have no zeros on $\mathcal{N} \cap \mathcal{W}$, the Bernoulli-l'Hôpital rule guarantees that F_0 has a continuous extension $F: \mathcal{W} \longrightarrow \mathbb{R} \setminus \{0\}$. At a point $u \in \mathcal{N} \cap \mathcal{W}$, the value of F is given by $F(u) = (d\tilde{H})_u(X)/(dH)_u(X)$, where X is any vector in $T_u(T^*\mathcal{M})$ which is non-tangential to $\mathcal{N}$. What remains to be shown is that at all points of $\mathcal{N} \cap \mathcal{W}$ the function F is, indeed, of class C^r for all $r \in \mathbb{N}$. This can be verified by induction over r where again the Bernoulli-l'Hôpital rule has to be applied. $\qquad\square$

Proposition 5.1.4. *Let $\mathcal{N}$ be a ray-optical structure on $\mathcal{M}$. Then the following properties are mutually equivalent.*

(a) $\mathcal{N}$ admits a global Hamiltonian H.
(b) There is a nowhere vanishing characteristic C^∞ vector field X on $\mathcal{N}$.
(c) There is a nowhere vanishing C^∞ one-form β on $\mathcal{N}$ such that the characteristic direction on $\mathcal{N}$ is transverse to the kernel of β at all points of $\mathcal{N}$.
(d) $\mathcal{N}$ is orientable, i.e., there is a nowhere vanishing C^∞ $(2n - 1)$-form ε on $\mathcal{N}$.

Proof. The implication "(a)$\Rightarrow$(b)" is trivial since we can choose $X = X_H|_{\mathcal{N}}$. To prove the implication "(b)$\Rightarrow$(c)" we put a C^∞ (positive definite) Riemannian metric g_+ on $\mathcal{N}$. Such a metric exists since $\mathcal{N}$ is a finite-dimensional paracompact manifold; for a proof of this well-known fact see Proposition 2.5.13 in Abraham and Marsden [1]. Then $\beta = g_+(X, \cdot)$ will do the job. To prove the implication "(c)$\Rightarrow$(d)" we can put $\varepsilon = \beta \wedge (\Omega_{\mathcal{N}})^{(n-1)}$, where $(\Omega_{\mathcal{N}})^{(n-1)} = \Omega_{\mathcal{N}} \wedge \cdots \wedge \Omega_{\mathcal{N}}$ with $(n - 1)$ factors on the right-hand side. Finally, we prove the implication "(d)$\Rightarrow$(a)". By assumption, $\mathcal{N}$ is an orientable codimension-one submanifold of $\overset{\circ}{T}^*\mathcal{M}$. We choose an orientation for $\mathcal{N}$ and put a (positive definite) C^∞ Riemannian metric G_+ on $\overset{\circ}{T}^*\mathcal{M}$. (The existence of such a metric is guaranteed by the same argument as above.) This defines an outward unit normal vector at each point $u \in \mathcal{N}$. Let $t \longmapsto \phi(u,t)$ denote the affinely parametrized G_+-geodesic tangent to this unit vector at $t = 0$. Then a global Hamiltonian $H: \mathcal{W} \longrightarrow \mathbb{R}$ for $\mathcal{N}$ is well-defined on some neighborhood $\mathcal{W}$ of $\mathcal{N}$ by setting $H(w) = t$ if and only if there is a $u \in \mathcal{N}$ such that $w = \phi(u,t)$. $\qquad\qquad\qquad\qquad\qquad\qquad\qquad\qquad\qquad\qquad\quad\square$

From this proposition we should keep in mind, in particular, that orientability of $\mathcal{N}$ is equivalent to the existence of a global Hamiltonian. We are thus in agreement with usual terminology if we define the choice of an orientation for $\mathcal{N}$ in the following way.

Definition 5.1.3. An orientation *for a ray-optical structure $\mathcal{N}$ is an equivalence class $[H]$ of global Hamiltonians for $\mathcal{N}$. Here two global Hamiltonians for $\mathcal{N}$ are called* equivalent *if they are related, according to Proposition 5.1.3, by a positive function F. After an orientation $[H]$ for $\mathcal{N}$ has been chosen, the parametrization of a lifted ray is called* positively oriented *if (5.9) holds with a positive function k for any $H \in [H]$.*

5.2 Regularity notions for ray-optical structures

In Proposition 5.1.3 we have seen that two local Hamiltonians H and $\tilde{H}$ for one and the same ray-optical structure are related, on their common domain of definition, by an equation of the form $\tilde{H} = F H$ where F is a nowhere vanishing function. As a consequence, their derivatives with respect to the momentum coordinates in any natural chart have to satisfy the equations

$$\tilde{H}^a = F\,H^a \quad \text{and} \quad \tilde{H}^{ab} = F\,H^{ab} + F^a\,H^b + F^b\,H^a \tag{5.13}$$

on $\mathcal{N}$, where the abbreviations (4.7) have been used for the functions H, $\tilde{H}$, and F. This implies, in particular, that the usual regularity condition (4.6) cannot be satisfied at a point $u \in \mathcal{N}$ by all Hamiltonians for $\mathcal{N}$. If this regularity condition is satisfied by one Hamiltonian for $\mathcal{N}$ it can always be spoiled by switching to another Hamiltonian with the help of an appropriate function F, according to (5.13). Therefore we define regularity of ray-optical structures in the following way.

Definition 5.2.1. *A ray-optical structure $\mathcal{N}$ on $\mathcal{M}$ is called* regular *at a point $u \in \mathcal{N}$ if there is a local Hamiltonian H for $\mathcal{N}$, defined on some neighborhood W of u, such that the fiber derivative $\mathbb{F}H$ of H maps $\mathcal{N} \cap W$ diffeomorphically onto its image in $\overset{\circ}{T}\mathcal{M}$. This is true if and only if H satisfies condition (4.6) at u in any natural chart. A ray-optical structure $\mathcal{N}$ on $\mathcal{M}$ is called* hyperregular *if there is a global Hamiltonian H for $\mathcal{N}$ such that the fiber derivative $\mathbb{F}H$ of H maps $\mathcal{N}$ diffeomorphically onto its image in $\overset{\circ}{T}\mathcal{M}$.*

If we take a look at our five examples of ray-optical structures mentioned above, we find that Example 5.1.1, Example 5.1.2 and Example 5.1.5 give hyperregular ray-optical structures. In each of these cases the fiber derivative of the given Hamiltonian is a global diffeomorphism $\mathbb{F}H \colon \overset{\circ}{T}{}^*\mathcal{M} \longrightarrow \overset{\circ}{T}\mathcal{M}$. The ray-optical structures of Example 5.1.3 and Example 5.1.4, on the other hand, are nowhere regular, provided that $\dim(\mathcal{M}) > 2$. It is easy to check that on a two-dimensional manifold all ray-optical structures are everywhere regular.

The best strategy to verify results of this kind is the following. To find out whether a ray-optical structure $\mathcal{N}$ is regular at some point $u \in \mathcal{N}$, we choose a local Hamiltonian H and a natural chart around u. As before, we use the abbreviations (4.7). If $\det\!\big(H^{ab}(u)\big) \neq 0$, it is obvious that $\mathcal{N}$ is regular at u. If $\det\!\big(H^{ab}(u)\big) = 0$, we consider the second-order polynomial

$$P_u(X^1, \ldots, X^n) = \det\!\big(H^{ab}(u) + H^a(u)\,X^b + X^a\,H^b(u)\big). \tag{5.14}$$

If this is the zero polynomial, i.e., if the equation $P_u(X^1, \ldots, X^n) = 0$ is satisfied by all $(X^1, \ldots, X^n) \in \mathbb{R}^n$, we know that $\mathcal{N}$ cannot be regular at u. This follows immediately from the fact that any other local Hamiltonian $\tilde{H}$ for $\mathcal{N}$ must be related to H by the transformation formulae (5.13). If, on the other hand, there is an $(X^1, \ldots, X^n) \in \mathbb{R}^n$ with $P_u(X^1, \ldots, X^n) \neq 0$, $\mathcal{N}$ must be regular at u. In order to prove this we switch to a new Hamiltonian $\tilde{H} = FH$, choosing the function F in such a way that $F(u) = 1$ and $F^a(u) = X^a$. Then we can read from (5.14) that $\det\!\big(\tilde{H}^{ab}(u)\big)$ is equal to $P_u(X^1, \ldots, X^n)$ which, by assumption, is different from zero.

If $\mathcal{N}$ is regular at u, an appropriate choice of a Hamiltonian near u gives us a local one-to-one correspondence between momentum covectors and velocity vectors. It is clear that a ray-optical structure alone, without choosing a particular Hamiltonian, cannot give such a correspondence since lifted rays can be arbitrarily reparametrized. Under such a reparametrization the momentum covectors remain unchanged whereas the velocity vectors are multiplied by a scalar factor. This observation suggests a different regularity notion for ray-optical structures. We want to call a ray-optical structure "strongly regular" if it gives a local one-to-one correspondence between momentum covectors and *directions* of velocity vectors. To make this precise we consider the vertical part of the ray equation together with the dispersion relation in a natural chart, i.e., the equations (5.10) and (5.11). The desired local one-to-one correspondence holds if and only if this system of equations can be solved for the momentum coordinates $p_1(s), \ldots, p_n(s)$ and for the stretching factor $k(s)$. This solvability condition can be written in the form

$$
\det \begin{pmatrix} (H^{ab}) & (H^a) \\ (H^b) & 0 \end{pmatrix} \neq 0 , \tag{5.15}
$$

where a is an index numbering rows and b is an index numbering columns such that we get an $(n+1) \times (n+1)$ matrix on the left-hand side of (5.15). With the help of (5.13) it is easy to check that (5.15) is, indeed, independent of the Hamiltonian chosen. Switching back to coordinate-free notation, this leads to the following definition.

Definition 5.2.2. *A ray-optical structure $\mathcal{N}$ on $\mathcal{M}$ is called* strongly regular *at a point $u \in \mathcal{N}$ if for one and hence for any Hamiltonian $H \colon W \longrightarrow \mathbb{R}$, defined on a sufficiently small neighborhood W of u in $\overset{\circ}{T}{}^*\mathcal{M}$, the map $\sigma_H \colon \mathcal{N} \cap W \times \mathbb{R}^+ \longrightarrow T\mathcal{M}$ defined by*

$$
\sigma_H(w, c) = c\,\mathbb{F}H(w) \tag{5.16}
$$

is a diffeomorphism onto its image. This is the case if and only if in any natural chart condition (5.15) holds at u. A ray-optical structure $\mathcal{N}$ on $\mathcal{M}$ is called strongly hyperregular *if there is a global Hamiltonian H for $\mathcal{N}$ such that the map $\sigma_H \colon \mathcal{N} \times \mathbb{R}^+ \longrightarrow T\mathcal{M}$ defined by (5.16) is a diffeomorphism onto its image. Here $\mathbb{F}H$ denotes the fiber derivative of H and $\mathbb{R}^+$ denotes the set of strictly positive real numbers.*

Strong regularity is easier to check than regularity; we just have to calculate the left-hand side of (5.15) with any Hamiltonian for $\mathcal{N}$.

If we look at our standard examples, we find that Example 5.1.2 and Example 5.1.5 give strongly hyperregular ray-optical structures. On the other hand, Example 5.1.1 gives ray-optical structures which are nowhere strongly regular. The same is true of Example 5.1.3 and Example 5.1.4 if we exclude the trivial case $\dim(\mathcal{M}) = 1$. This shows that strong regularity is violated for several physically interesting ray-optical structures on spacetimes. (In the next section we shall see that a ray-optical structure that describes light propagation in a non-dispersive medium on a spacetime cannot be strongly regular.) Nonetheless strong regularity is a physically useful notion, in particular in the case that $\mathcal{M}$ is to be interpreted as space rather than as spacetime.

We have not yet justified our terminology by showing that, indeed, strong regularity implies regularity. This follows from the next proposition.

Proposition 5.2.1. *A ray-optical structure $\mathcal{N}$ on $\mathcal{M}$ is strongly regular at a point $u \in \mathcal{N}$ if and only if there exists a local Hamiltonian H for $\mathcal{N}$ on some neighborhood of u such that in any natural chart*

(a) $\det\left(H^{ab}\right) \neq 0$ *and*

(b) $H_{ab}\, H^a\, H^b \neq 0$

at u. Here we use again the abbreviations (4.7) and H_{ab} is defined through $H_{ab}H^{bc} = \delta_a^c$.

Proof. The "if" part is a trivial exercise in linear algebra. To prove the "only if" part, we assume that $\mathcal{N}$ is strongly regular at u and fix a local Hamiltonian H for $\mathcal{N}$ on a neighborhood $\mathcal{W}$ of u. If H satisfies (a) we are done since in this case, by (5.15), (b) is also satisfied. So let us assume that H does not satisfy (a). It is our goal to find another Hamiltonian $\tilde{H} = FH$ for $\mathcal{N}$ such that (a) and, thus, (b) are satisfied if H is replaced with $\tilde{H}$. As, by assumption, the kernel of the matrix (H^{ab}) is non-trivial, our strong regularity assumption (5.15) guarantees existence and uniqueness of a vertical vector $n_b\, \partial/\partial p_b$ at u which satisfies $H^{ab} n_b = 0$ and $H^b n_b = 1$. If $\mathcal{W}$ is sufficiently small, the function $\tilde{H} = H(H+1) \colon \mathcal{W} \longrightarrow \mathbb{R}$ is, again, a local Hamiltonian for $\mathcal{N}$. It is our goal to prove that the kernel of the matrix $(\tilde{H}^{ab}) = (H^{ab} + 2\, H^a\, H^b)$ is trivial. So let us assume that $\tilde{H}^{ab} Y_b = 0$. We want to demonstrate that this implies $Y_b = 0$. As $H^a\, n_a = 1$, Y_b can be decomposed in the form $Y_b = Z_b + c\, n_b$ where $H^b Z_b = 0$. Then our assumption takes the form $\tilde{H}^{ab} Y_b = H^{ab} Z_b + 2\, c\, H^a = 0$. Owing to our strong regularity assumption this implies that the column vector with components $Z^1, \ldots, Z^n, 2\, c$ is in the kernel of a matrix with non-zero determinant; hence, all these components are zero. $\square$

This proposition shows that, at the level of local Hamiltonians, our strong regularity notion is equivalent to the so-called Condition N introduced by Guckenheimer [53].

For further illustrating strong regularity we introduce the following notation.

Definition 5.2.3. *Let $\mathcal{N}$ be a ray-optical structure on $\mathcal{M}$. For any $q \in \mathcal{M}$, the set*

$$C_q = \left\{ \dot{\lambda}(0) \,\middle|\, \lambda \text{ is a ray with } \lambda(0) = q \right\} \tag{5.17}$$

is called the infinitesimal light cone of $\mathcal{N}$ at q. The set

$$C = \left\{ X \in C_q \,\middle|\, q \in \mathcal{M} \right\} \tag{5.18}$$

is called the bundle of infinitesimal light cones of $\mathcal{N}$.

Here we allow ourselves a slight abuse of language insofar as C is not necessarily a fiber bundle over $\mathcal{M}$. As rays can be arbitrarily reparametrized, $X \in C_q$ implies that $cX \in C_q$ for all $c \in \mathbb{R} \setminus \{0\}$. This justifies calling C_q a "cone". Clearly, a vector $X \in \overset{\circ}{T}_q \mathcal{M}$ is in C_q if and only if it is of the form $X = \mathbb{F}H(u)$ where $u \in \mathcal{N}_q$ and H is a local Hamiltonian for $\mathcal{N}$ which is defined around u. Moreover, for any local Hamiltonian H, the image of the map σ_H defined through (5.16) is a subset of C. Note, however, that even for a global Hamiltonian the image of σ_H does not necessarily cover all of C; the reason is that the image of σ_H need not be invariant under multiplication with negative numbers.

In the case of Example 5.1.1, $C_q = \left\{ X \in \overset{\circ}{T}_q \mathcal{M} \mid g_o(X,X) = 0 \right\}$, i.e., C_q equals the null cone of the Lorentzian metric g_o at q. In particular, C_q is a closed codimension-one submanifold of $\overset{\circ}{T}_q \mathcal{M}$ in this case. The situation is completely different in the case of Example 5.1.2. Here $C_q = \left\{ X \in \overset{\circ}{T}_q \mathcal{M} \mid g_o(X,X) < 0 \right\}$ equals the interior of the null cone of the Lorentzian metric g_o and is, thus, an open subset of $\overset{\circ}{T}_q \mathcal{M}$. For the ray-optical structures of Example 5.1.3 and Example 5.1.4, $C_q = \left\{ cU_q \mid c \in \mathbb{R} \setminus \{0\} \right\}$ is a one-dimensional submanifold of $\overset{\circ}{T}_q \mathcal{M}$ whereas in the case of Example 5.1.5 C_q is all of $\overset{\circ}{T}_q \mathcal{M}$.

These examples show that C has very different features for different ray-optical structures. Moreover, they demonstrate that there is no obvious relation between the geometry of $\mathcal{N}$ and the geometry of C. In the case of Example 5.1.1, which dominates our intuitive ideas of general relativistic ray optics, $\mathcal{N}$ and C are diffeomorphic. In the other cases, however, C looks completely different from $\mathcal{N}$.

The following proposition implies that in the strongly regular case C cannot be diffeomorphic to $\mathcal{N}$.

Proposition 5.2.2. *Let $\mathcal{N}$ be a ray-optical structure on $\mathcal{M}$ and $q \in \mathcal{M}$. If $\mathcal{N}$ is strongly regular at all points $u \in \mathcal{N}_q$, the infinitesimal light cone C_q is an open subset of $\overset{\circ}{T}_q \mathcal{M}$.*

Proof. By Definition 5.2.2, strong regularity implies that the differential of the map σ_H has maximal rank.   □

This observation is exemplified by Example 5.1.2 and Example 5.1.5.

Please recall that strong regularity guarantees that the system of equations (5.10) and (5.11) can be solved for the momentum coordinates $p_a(s)$ and for the stretching factor $k(s)$. It is worthwhile to become clear about the information contained in this system of equations. If a curve $s \longmapsto (x(s), p(s))$ satisfies (5.10) and (5.11) with some $k(s)$ but not necessarily (5.12), it determines at each of its points the same velocity vector as a lifted ray passing through this point. Hence this curve, although not necessarily a lifted ray, describes an object moving at the velocity of light (in the medium for which $\mathcal{N}$ gives the dispersion relation). We now introduce a special name for such curves.

Definition 5.2.4. *Let $\mathcal{N}$ be an arbitrary ray-optical structure on $\mathcal{M}$. A C^{∞} immersion $\xi\colon I \longrightarrow \mathcal{N}$ from a real interval I into $\mathcal{N}$ is called a lifted virtual ray iff*

$$\Omega_{\xi(s)}\big(\dot{\xi}(s), Z_{\xi(s)}\big) = 0 \qquad (5.19)$$

for all $s \in I$ and all $Z_{\xi(s)} \in T_{\xi(s)}\mathcal{N}$ with $(T\tau_{\mathcal{M}}^{*})(Z_{\xi(s)}) = 0$. Then the projected curve $\tau_{\mathcal{M}}^{*} \circ \xi\colon I \longrightarrow \mathcal{M}$ is called a virtual ray.

This notion of "virtual rays" should not be confused with the notion of "virtual images" which is used in elementary optics.

It follows immediately from the definitions that lifted virtual rays and virtual rays can be characterized in the following way.

Proposition 5.2.3. *Let $\mathcal{N}$ be any ray-optical structure $\mathcal{N}$ on $\mathcal{M}$. Then a C^{∞} immersion $\xi\colon I \longrightarrow \mathcal{N}$ is a lifted virtual ray if and only if*

$$(\tau_{\mathcal{M}}^{*} \circ \xi)^{\cdot} = k\,\mathbb{F}H \circ \xi \qquad (5.20)$$

with some function $k\colon I \longrightarrow \mathbb{R} \setminus \{0\}$. Here H is any (local) Hamiltonian for the ray-optical structure $\mathcal{N}$.

A C^{∞} immersion $\lambda\colon I \longrightarrow \mathcal{M}$ is a virtual ray if and only if $\dot{\lambda}(s) \in \mathcal{C}_{\lambda(s)}$ for all $s \in I$. Here $\mathcal{C}_{\lambda(s)}$ denotes the infinitesimal light cone introduced in Definition 5.2.3.

Clearly, a ray is all the more a virtual ray whereas the converse is in general not true. In the case of Example 5.1.1 all g_o-light-like curves in $\mathcal{M}$ are virtual rays but only the g_o-light-like geodesics are rays. In the case of Example 5.1.2 the rays are the g_o-time-like geodesics whereas all g_o-time-like curves are virtual rays. In the case of Example 5.1.3 and 5.1.4 an immersed curve in $\mathcal{M}$ is a ray iff it is a virtual ray iff it is an arbitrarily parametrized integral curve of the vector field U. In the case of Example 5.1.5 all immersed curves in $\mathcal{M}$ are virtual rays whereas only the g_+-geodesics are rays.

If $\mathcal{N}$ is orientable, we can generalize Definition 5.1.3 in the following way.

Definition 5.2.5. *Let $\mathcal{N}$ be a ray-optical structure on $\mathcal{M}$ and $[H]$ be an orientation for $\mathcal{N}$. Then a lifted virtual ray ξ of $\mathcal{N}$ is called positively oriented if (5.20) holds with a positive function k for any $H \in [H]$. Similarly, a virtual ray is called positively oriented if it is the projection of a positively oriented lifted virtual ray.*

In the next proposition we prove that in the strongly hyperregular case there is a global one-to-one correspondence between positively oriented lifted virtual rays and positively oriented virtual rays, i.e., that at each point of a positively oriented virtual ray the momentum covector is uniquely determined.

Proposition 5.2.4. *Let $\mathcal{N}$ be a ray-optical structure on $\mathcal{M}$ and assume that $\mathcal{N}$ is strongly hyperregular. By Definition 5.2.2 this guarantees the existence of a global Hamiltonian H for $\mathcal{N}$ such that the map $\sigma_H \colon \mathcal{N} \times \mathbb{R}^+ \longrightarrow T\mathcal{M}$ defined through (5.16) is a global diffeomorphism onto its image. Choose such a Hamiltonian, thereby defining an orientation $[H]$ for $\mathcal{N}$. Then for every positively oriented virtual ray $\lambda \colon I \longrightarrow \mathcal{M}$ there is a unique positively oriented lifted virtual ray $\xi \colon I \longrightarrow \mathcal{N}$ that projects onto λ, $\tau_{\mathcal{M}}^* \circ \xi = \lambda$.*

Proof. The nontrivial claim is the uniqueness of ξ. So let us assume that ξ_1 and ξ_2 do the job. Since lifted virtual rays have to satisfy (5.20), this implies that $k_1 \, \mathbb{F}H \circ \xi_1 = k_2 \, \mathbb{F}H \circ \xi_2$. Since both ξ_1 and ξ_2 are supposed to be positively oriented, k_1 and k_2 have to be positive such that the last equation can be written in the form $\sigma_H\big(\xi_1(s), k_1(s)\big) = \sigma_H\big(\xi_2(s), k_2(s)\big)$ for all $s \in I$. Since σ_H is a diffeomorphism, this implies $\xi_1 = \xi_2$. $\square$

Example 5.1.5 demonstrates that the restriction to positively oriented lifted virtual rays is, indeed, necessary to get uniqueness.

5.3 Symmetries of ray-optical structures

As the symmetries of a ray-optical structure $\mathcal{N}$ on $\mathcal{M}$ we can view all diffeomorphisms on $\overset{\circ}{T}{}^*\mathcal{M}$ that leave $\mathcal{N}$ invariant. For our purposes it will be reasonable to restrict to those diffeomorphisms on $\overset{\circ}{T}{}^*\mathcal{M}$ that are induced from diffeomorphisms on the base manifold $\mathcal{M}$. (Such diffeomorphisms are called "point transformations" in Hamiltonian mechanics.) To work this out, we have to recall that each diffeomorphism $\psi \colon \mathcal{M} \longrightarrow \mathcal{M}$ induces a cotangent map $T^*\psi \colon T^*\mathcal{M} \longrightarrow T^*\mathcal{M}$ which is again a diffeomorphism, defined by the equation $\big((T^*\psi)(u)\big)(X) = u\big(T\psi(X)\big)$ for all $q \in \mathcal{M}$, $X \in T_q\mathcal{M}$ and $u \in T_{\psi(q)}^*\mathcal{M}$. It is well known and easily verified that $T^*\psi$ leaves the canonical one-form θ and, thus, the canonical two-form Ω invariant, i.e.,

$$(T^*\psi)^*\theta = \theta \quad \text{and} \quad (T^*\psi)^*\Omega = \Omega \,. \tag{5.21}$$

For an invariant proof we refer to Abraham and Marsden [1], Theorem 3.2.12. As an alternative, the proof can be accomplished easily in a natural chart. If ψ is represented, in a local chart, by a map $x \longmapsto x'$, $(T^*\psi)^{-1}$ is represented in the pertaining natural chart by the map $(x, p) \longmapsto (x', p')$ with p' given by (2.71).

After these preparations we are now ready to introduce the following definition.

Definition 5.3.1. *Let $\mathcal{N}$ be a ray-optical structure on $\mathcal{M}$. The symmetry group $G_{\mathcal{N}}$ of $\mathcal{N}$ is, by definition, the set of all diffeomorphism $\psi\colon \mathcal{M} \longrightarrow \mathcal{M}$ such that $T^*\psi$ leaves $\mathcal{N}$ invariant, i.e., such that $(T^*\psi)(u) \in \mathcal{N}$ for all $u \in \mathcal{N}$.*

Clearly, $G_{\mathcal{N}}$ is a group with respect to composition of maps.

For the sake of illustration we take a look at the symmetry groups of our standard examples. In the case of Example 5.1.1, where $\mathcal{N}$ is the set of all light-like covectors of a Lorentzian metric g_o, the symmetry group $G_{\mathcal{N}}$ consists of all diffeomorphisms $\psi\colon \mathcal{M} \longrightarrow \mathcal{M}$ for which the pulled back metric ψ^*g_o has the same cone bundle as g_o. This is the case if and only if ψ^*g_o is conformally equivalent to g_o, i.e., if and only if $\psi^*g_o = e^{2f}g_o$ with some C^∞ function $f\colon \mathcal{M} \longrightarrow \mathbb{R}$. For a proof of this well-known fact we refer, e.g., to Wald [146], p. 445. Hence, in the case of Example 5.1.1 the symmetry group $G_{\mathcal{N}}$ equals the set of all conformal symmetries of the Lorentzian manifold $(\mathcal{M}, g_o)$. In particular, this implies that $G_{\mathcal{N}}$ is a finite-dimensional Lie group. Similarly, in the case of Example 5.1.2 we find that the symmetry group is the group of all isometries of the metric g_o. Again, this is a finite-dimensional Lie group. In the case of Example 5.1.3, on the other hand, the symmetry group $G_{\mathcal{N}}$ consists of all diffeomorphisms $\psi\colon \mathcal{M} \longrightarrow \mathcal{M}$ that map arbitrarily parametrized integral curves of U onto arbitrarily parametrized integral curves of U. In general, this is an infinite-dimensional subgroup of the diffeomorphism group $\mathrm{Diff}(\mathcal{M})$. The same result is found for Example 5.1.4, with the only difference that the diffeomorphisms ψ have to respect the parametrization adapted to U in addition. Finally, in the case of Example 5.1.5, the symmetry group is the group of isometries of the Riemannian metric g_+.

Next we want to show that for any $\psi \in G_{\mathcal{N}}$ the induced cotangent map $T^*\psi$ maps lifted rays onto lifted rays. For later convenience we prove the following more general proposition.

Proposition 5.3.1. *Let $\mathcal{N}$ be a ray-optical structure on $\mathcal{M}$ and $\Psi\colon \overset{\circ}{T}{}^*\mathcal{M} \longrightarrow \overset{\circ}{T}{}^*\mathcal{M}$ be a C^∞ diffeomorphism. Assume that Ψ leaves the canonical two form Ω invariant up to a factor, i.e., $\Psi^*\Omega = f\,\Omega$ with some function $f\colon \overset{\circ}{T}{}^*\mathcal{M} \longrightarrow \mathbb{R}$, and that Ψ is fiber preserving, i.e., $\tau_{\mathcal{M}}^*(\Psi(u_1)) = \tau_{\mathcal{M}}^*(\Psi(u_2))$ whenever $\tau_{\mathcal{M}}^*(u_1) = \tau_{\mathcal{M}}^*(u_2)$. Then the following properties are mutually equivalent.*

(a) Ψ leaves $\mathcal{N}$ invariant, i.e., $\Psi(u) \in \mathcal{N}$ for all $u \in \mathcal{N}$.
(b) Ψ maps each lifted ray of $\mathcal{N}$ onto a lifted ray.
(c) Ψ maps each lifted virtual ray of $\mathcal{N}$ onto a lifted virtual ray.

Proof. First we assume that (a) is satisfied. To prove that then (b) and (c) hold true, we pull back the equation $\Psi^*\Omega = f\,\Omega$ to $\mathcal{N}$. As the diffeomorphism Ψ leaves $\mathcal{N}$ invariant and $\mathcal{N}$ is a closed submanifold of $\overset{\circ}{T}{}^*\mathcal{M}$, Ψ maps $\mathcal{N}$ diffeomorphically onto itself. Hence, $\Psi_{\mathcal{N}}^*\,\Omega_{\mathcal{N}} = f|_{\mathcal{N}}\,\Omega_{\mathcal{N}}$ where $\Psi_{\mathcal{N}}\colon \mathcal{N} \longrightarrow \mathcal{N}$

denotes the restriction of Ψ to $\mathcal{N}$. This implies that, for any C^∞ immersion $\xi \colon I \longrightarrow \mathcal{N}$,

$$\Omega_{\mathcal{N}}\big((\Psi \circ \xi)\dot{}\,, \, \cdot\,\big) = f \circ \xi \; \Omega_{\mathcal{N}}\big(\dot{\xi}, (T\Psi)^{-1}(\cdot)\big)\,. \tag{5.22}$$

Now let us assume that ξ is a lifted ray of $\mathcal{N}$. Then, by (5.19), the right-hand side of (5.22) vanishes, so the left-hand side has to vanish as well, i.e., $\Psi \circ \xi$ has to be a lifted ray as well. This proves the implication "(a)$\Rightarrow$(b)". To prove the implication "(a)$\Rightarrow$(c)", let us assume that ξ is a lifted virtual ray. Then, by Definition 5.2.4, the right-hand side of (5.22) vanishes on all vectors X such that $(T\Psi)^{-1}(X)$ is vertical. Since Ψ is fiber-preserving, the latter condition is equivalent to X being vertical. Thus, the left-hand side of (5.22) has to vanish on all vertical vectors, i.e., $\Psi \circ \xi$ has to be a lifted virtual ray. The proof of the converse implications "(b)$\Rightarrow$(a)" and "(c)$\Rightarrow$(a)" is trivial since a point $u \in \overset{\circ}{T}{}^*M$ is in $\mathcal{N}$ if and only if there is a lifted ray through u if and only if there is a lifted virtual ray through u.   □

If we apply this proposition to the map $\Psi = T^*\psi\big|_{\overset{\circ}{T}{}^*M}$ we see that a diffeomorphism $\psi \colon \mathcal{M} \longrightarrow \mathcal{M}$ is in $G_{\mathcal{N}}$ if and only if its cotangent map $T^*\psi$ maps lifted rays onto lifted rays if and only if $T^*\psi$ maps lifted virtual rays onto lifted virtual rays. This implies, in particular, that any $\psi \in G_{\mathcal{N}}$ maps rays onto rays. Please note, however, that the converse is not true. A diffeomorphism $\psi \colon \mathcal{M} \longrightarrow \mathcal{M}$ that maps rays onto rays need not be in $G_{\mathcal{N}}$. This is in correspondence with our earlier observation that two different ray-optical structures on $\mathcal{M}$ may have the same rays. As an example we may consider a ray-optical structure constructed from a (positive definite) Riemannian metric g_+ as in Example 5.1.5. Then a diffeomorphism $\psi \colon \mathcal{M} \longrightarrow \mathcal{M}$ such that $\psi^* g_+ = c\, g_+$ with a positive constant $c \neq 1$ maps rays onto rays but it is not in the symmetry group $G_{\mathcal{N}}$.

As an alternative, symmetries can be treated in terms of infinitesimal generators. This gives a symmetry algebra rather than a symmetry group. To make this definition precise we need the well-known fact that each vector field K on $\mathcal{M}$ defines a vector field $\bar{K}$ on $T^*\mathcal{M}$ which is called the *canonical lift* of K. Let

$$\Phi \colon \mathcal{V} \subseteq \mathbb{R} \times \mathcal{M} \longrightarrow \mathcal{M}\,, \quad (t,q) \longmapsto \Phi_t(q)\,, \tag{5.23}$$

denote the flow of K, i.e., let $t \longmapsto \Phi_t(q)$ denote the integral curve of K that passes at $t = 0$ through the point q. Then the vector field $\bar{K}$ is defined by the condition that its flow $\bar{\Phi}$ is given by the equation $\bar{\Phi}_t = T^*\Phi_{-t}$. In a natural chart the canonical lift of the vector field $K = K^a(x)\frac{\partial}{\partial x^a}$ takes the form

$$\bar{K} = K^a(x)\frac{\partial}{\partial x^a} - p_b \frac{\partial K^b(x)}{\partial x^a}\frac{\partial}{\partial p_a}\,. \tag{5.24}$$

Comparison with (4.2) shows that $\bar{K}$ is the Hamiltonian vector field of the function $H = \theta(\bar{K}) \colon T^*\mathcal{M} \longrightarrow \mathbb{R}$, i.e., $\bar{K} = X_H$. In terms of a natural chart this function takes the form $H(x,p) = p_a K^a(x)$.

Now the symmetry algebra of a ray-optical structure can be defined in the following way.

Definition 5.3.2. *Let $\mathcal{N}$ be a ray-optical structure on $\mathcal{M}$. The symmetry algebra $\mathcal{G}_{\mathcal{N}}$ of $\mathcal{N}$ is, by definition, the set of all C^{∞} vector fields K on $\mathcal{M}$ such that at all points of $\mathcal{N}$ the canonical lift $\bar{K}$ of K is tangent to $\mathcal{N}$.*

It is easy to check that $\mathcal{G}_{\mathcal{N}}$ is a Lie algebra with respect to the usual Lie bracket of vector fields. Clearly, the one-parameter subgroups of $G_{\mathcal{N}}$ are in one-to-one correspondence with the complete vector fields in $\mathcal{G}_{\mathcal{N}}$.

In analogy to (5.21) the canonical lift $\bar{K}$ of a vector field K satisfies

$$L_{\bar{K}}\theta = 0 \quad \text{and} \quad L_{\bar{K}}\Omega = 0 \tag{5.25}$$

where L denotes the Lie derivative. Hence, by applying (a local version of) Proposition 5.3.1 to the (local) flow of $\bar{K}$ we get the following result.

Proposition 5.3.2. *Let $\mathcal{N}$ be a ray-optical structure on $\mathcal{M}$. Then for a C^{∞} vector field K on $\mathcal{M}$ the following properties are equivalent.*

(a) *K is in $\mathcal{G}_{\mathcal{N}}$.*
(b) *The flow of the canonical lift $\bar{K}$ of K maps lifted rays onto lifted rays.*
(c) *The flow of the canonical lift $\bar{K}$ of K maps lifted virtual rays onto lifted virtual rays.*

In Hamiltonian mechanics it is well-known that symmetries give rise to constants of motion. Similarly, every element of $\mathcal{G}_{\mathcal{N}}$ is associated with a function on $T^*\mathcal{M}$ which is constant along each lifted ray. This is shown in the following proposition.

Proposition 5.3.3. *Let $\mathcal{N}$ be a ray-optical structure on $\mathcal{M}$ and $K \in \mathcal{G}_{\mathcal{N}}$. Then the function $\theta(\bar{K}): T^*\mathcal{M} \longrightarrow \mathbb{R}$ is constant along each lifted ray. Here θ denotes the canonical one-form on $T^*\mathcal{M}$ and $\bar{K}$ denotes the canonical lift of K.*

Proof. We fix a point $u \in \mathcal{N}$ and a local Hamiltonian H for $\mathcal{N}$ around u. Then the definition of the exterior derivative d implies that $(d\theta)(X_H, \bar{K}) = X_H\big(\theta(\bar{K})\big) - \bar{K}\big(\theta(X_H)\big) - \theta\big([X_H, \bar{K}]\big)$ where $[\cdot , \cdot]$ denotes the Lie bracket of vector fields. On the left-hand side we use the definition (4.1) of the Hamiltonian vector field X_H, on the right-hand side we use (5.25). This results in $-dH(\bar{K}) = X_H\big(\theta(\bar{K})\big)$. On $\mathcal{N}$, the left-hand side vanishes since $\bar{K}$ is tangent to $\mathcal{N}$. Hence, the right-hand side has to vanish on $\mathcal{N}$ as well. This implies that the function $\theta(\bar{K})$ is constant along each integral curve of X_H which is contained in $\mathcal{N}$, i.e., along each lifted ray. $\qquad \square$

If $K \in \mathcal{G}_{\mathcal{N}}$ is represented in a local chart as $K = \partial/\partial x^{\mathrm{n}}$, which is possible locally around any point of $\mathcal{M}$ where K does not vanish, the constant of motion $\theta(\bar{K})$ is exactly the corresponding momentum coordinate, $\theta(\bar{K}) = p_{\mathrm{n}}$.

For this reason $\theta(\bar{K})$ is called the *momentum* of the infinitesimal symmetry $K \in \mathcal{G}_{\mathcal{N}}$.

The fact that symmetries imply constants of motion is of particular relevance in view of *dimensional reductions*. Given a ray optical structure $\mathcal{N}$ on $\mathcal{M}$, any subgroup G of the symmetry group $G_{\mathcal{N}}$ can be used to define an equivalence relation on $\mathcal{M}$ by

$$q_1 \sim q_2 \iff \text{there is a } \psi \in G \text{ such that } q_1 = \psi(q_2).$$

If the action of G on $\mathcal{M}$ satisfies some regularity conditions, the quotient space $\hat{\mathcal{M}} = \mathcal{M}/_{\sim}$ can be furnished with a manifold structure such that the natural projection $\mathrm{pr} \colon \mathcal{M} \longrightarrow \hat{\mathcal{M}}$ becomes a submersion. If G is an r-dimensional Lie group, Proposition 5.3.3 gives rise to r constants of motion. Fixing a value for each of them singles out a certain subclass of lifted rays of $\mathcal{N}$. Circumstances permitted, this subclass of lifted rays projects onto a reduced ray-optical structure $\hat{\mathcal{N}}$ on $\hat{\mathcal{M}}$. We shall discuss this reduction formalism in full detail for stationary ray-optical structures on Lorentzian manifolds in Sect. 6.5 below. Relevant background material on the general features of the reduction formalism can be found in Chap. 4 of the book by Abraham and Marsden [1].

The possibility to use symmetries for dimensional reduction is an important motivation for considering ray-optical structures on bare manifolds of unspecified dimension.

We end this section with some remarks on the *isotropy subgroup* $G_{\mathcal{N}}^q$ of $G_{\mathcal{N}}$. This is defined, for each $q \in \mathcal{M}$, by

$$G_{\mathcal{N}}^q = \{ \psi \in G_{\mathcal{N}} \mid \psi(q) = q \} . \tag{5.26}$$

For any $\psi \in G_{\mathcal{N}}^q$, the cotangent map $T^*\psi$ restricted to $T_q^*\mathcal{M}$ gives us a linear automorphism $T_q^*\psi \colon T_q^*\mathcal{M} \longrightarrow T_q^*\mathcal{M}$ that leaves the manifold $\mathcal{N}_q = \mathcal{N} \cap T_q^*\mathcal{M}$ invariant. We introduce the following definition.

Definition 5.3.3. *Let $\mathcal{N}$ be a ray-optical structure on $\mathcal{M}$ and fix a point $q \in \mathcal{M}$. By definition, the* structure group *of $\mathcal{N}$ at q is the set of all linear automorphisms $T_q^*\mathcal{M} \longrightarrow T_q^*\mathcal{M}$ that leave the manifold $\mathcal{N}_q = \mathcal{N} \cap T_q^*\mathcal{M}$ invariant.*

In the case of Example 5.1.2, the structure group at q consists of all Lorentz transformations of the metric $g_o|_q$, whereas in the case of Example 5.1.1 it contains the multiplications with nonzero numbers in addition. In the case of Example 5.1.4, the structure group at q is represented by all invertible matrices of the form

$$(\Lambda_a^b) = \begin{pmatrix} \Lambda_1^1 & \cdots & \Lambda_1^{n-1} & \Lambda_1^n \\ \vdots & & \vdots & \vdots \\ \Lambda_{n-1}^1 & \cdots & \Lambda_{n-1}^{n-1} & \Lambda_{n-1}^n \\ 0 & \cdots & 0 & \Lambda_n^n \end{pmatrix} \tag{5.27}$$

in a basis such that $U^a(q) = \delta_n^a$, whereas in the case of Example 5.1.3 the equation $\Lambda_n^n = 1$ must be satisfied in addition. Finally, in the case of Example 5.1.5, the structure group at q consists of all linear automorphisms that are orthogonal with respect to the metric $g_+|_q$.

As long as $\mathcal{M}$ is a bare manifold, the only distinguished linear automorphisms on $T_q^*\mathcal{M}$ are dilations and inversions, i.e., multiplications with positive or negative numbers. The question of whether or not a ray-optical structure is invariant under dilations and inversions is of particular relevance. Therefore we devote the next subsection to this question.

5.4 Dilation-invariant ray-optical structures

The notion of dilation-invariance for ray-optical structures will give us a mathematically elegant characterization of media which are dispersion-free. For this reason the following definition is of paramount importance in ray optics.

Definition 5.4.1. *A ray-optical structure $\mathcal{N}$ on $\mathcal{M}$ is called* dilation-invariant *at a point $q \in \mathcal{M}$ if $e^t u \in \mathcal{N}_q$ for all $u \in \mathcal{N}_q$ and $t \in \mathbb{R}$. $\mathcal{N}$ is called* reversible *at a point $q \in \mathcal{M}$ if $-u \in \mathcal{N}_q$ for all $u \in \mathcal{N}_q$. $\mathcal{N}$ is called* dilation-invariant *(or* reversible, resp.*) if it is* dilation-invariant *(or* reversible, resp.*) at all points $q \in \mathcal{M}$.*

If $\mathcal{M}$ is to be interpreted as a general-relativistic spacetime, a dilation-invariant ray-optical structure on $\mathcal{M}$ is called *dispersion-free* or *non-dispersive*, otherwise it is called *dispersive*. In Chap. 6 below we shall link up this terminology with the physics textbook definition of non-dispersive media, i.e., we shall use the notions of phase velocity and group velocity for characterizing dilation-invariant ray-optical structures. These notions refer to a time-like vector field; hence, they can only be introduced if there is a Lorentzian metric on $\mathcal{M}$ since otherwise we do not know what is meant by "time-like".

A brief look at our standard examples shows the following. Whereas Examples 5.1.1 and 5.1.3 are dilation-invariant and reversible, Examples 5.1.2, 5.1.4, and 5.1.5 are only reversible but not dilation-invariant.

For each $t \in \mathbb{R}$, the dilation

$$\Phi_t \colon \overset{\circ}{T}{}^*\mathcal{M} \longrightarrow \overset{\circ}{T}{}^*\mathcal{M}, \quad u \longmapsto e^t u \qquad (5.28)$$

is represented in a natural chart by the map $(x,p) \longmapsto (x, e^t p)$. With the help of this representation it is readily verified that Φ_t leaves the canonical one-form $\theta = p_a\, dx^a$ invariant up to a factor,

$$\Phi_t^*\theta = e^t\, \theta. \qquad (5.29)$$

Application of the exterior derivative d yields

$$\Phi_t^* \Omega = e^t\, \Omega \,. \tag{5.30}$$

Applying Proposition 5.3.1 to the map $\Psi = \Phi_t$ proves the following.

Proposition 5.4.1. *For a ray-optical structure $\mathcal{N}$ on $\mathcal{M}$, the following properties are mutually equivalent.*

(a) *$\mathcal{N}$ is dilation-invariant.*
(b) *Each lifted ray $\xi\colon I \longrightarrow \mathcal{N}$ remains a lifted ray if it is multiplied pointwise with a positive number, $\xi(\,\cdot\,) \longmapsto e^t\,\xi(\,\cdot\,)$.*
(c) *Each lifted virtual ray $\xi\colon I \longrightarrow \mathcal{N}$ remains a lifted virtual ray if it is multiplied pointwise with a positive number, $\xi(\,\cdot\,) \longmapsto e^t\,\xi(\,\cdot\,)$.*

Similarly, the inversion $\chi\colon \overset{\circ}{T}{}^*\mathcal{M} \longrightarrow \overset{\circ}{T}{}^*\mathcal{M}$ is represented in each natural chart by the map $(x,p) \longmapsto (x,-p)$. This implies that $\chi^*\theta = -\theta$ and $\chi^*\Omega = -\Omega$. Hence, Proposition 5.3.1 can be applied to the map $\Psi = \chi$ as well, resulting in the following proposition.

Proposition 5.4.2. *For a ray-optical structure $\mathcal{N}$ on $\mathcal{M}$, the following properties are mutually equivalent.*

(a) *$\mathcal{N}$ is reversible.*
(b) *Each lifted ray $\xi\colon I \longrightarrow \mathcal{N}$ remains a lifted ray if it is pointwise inverted, $\xi(\,\cdot\,) \longmapsto -\xi(\,\cdot\,)$.*
(c) *Each lifted virtual ray $\xi\colon I \longrightarrow \mathcal{N}$ remains a lifted virtual ray if it is pointwise inverted, $\xi(\,\cdot\,) \longmapsto -\xi(\,\cdot\,)$.*

The dilation-invariant case can also be characterized in terms of the vector field that generates the one-parameter group of dilations. Since (5.28) satisfies all properties of a global C^∞ flow on $\overset{\circ}{T}{}^*\mathcal{M}$, we can define a C^∞ vector field E on $\overset{\circ}{T}{}^*\mathcal{M}$ by

$$E_u = \tfrac{d}{dt}\left(e^t\,u\right)\big|_{t=0} \tag{5.31}$$

for all $u \in \overset{\circ}{T}{}^*\mathcal{M}$. The integral curves of E are the radial lines in the fibers of the cotangent bundle. This vector field E is known as the *Euler vector field* or *Liouville vector field* on $\overset{\circ}{T}{}^*\mathcal{M}$. In a natural chart E takes the form

$$E = p_a\,\frac{\partial}{\partial p_a}\,. \tag{5.32}$$

(5.29) and (5.30) imply that the Lie derivatives of the canonical one-form and of the canonical two-form with respect to the Euler vector field satisfy

$$L_E\theta = \theta \quad \text{and} \quad L_E\Omega = \Omega\,. \tag{5.33}$$

Moreover , the identities

$$\theta(E) = 0 \quad \text{and} \quad -2\,\Omega(E, \cdot) = \theta \tag{5.34}$$

are easily verified in a natural chart. In terms of the Euler vector field E dilation-invariant ray-optical structures are characterized by the following proposition.

Proposition 5.4.3. *A ray-optical structure $\mathcal{N}$ on $\mathcal{M}$ is dilation-invariant if and only if the Euler vector field E is tangent to $\mathcal{N}$ at all points of $\mathcal{N}$.*

Proof. The "only if" part follows directly from Definition 5.4.1. For the "if" part one has to use the fact that $\mathcal{N}$ is closed in $\overset{\circ}{T}{}^*\mathcal{M}$ in addition.   □

In terms of (local) Hamiltonians dilation-invariance can be characterized in the following way.

Proposition 5.4.4. *A ray-optical structure $\mathcal{N}$ on $\mathcal{M}$ is dilation-invariant if and only if any local Hamiltonian H for $\mathcal{N}$ satisfies the equation $dH(E) = 0$ on $\mathcal{N}$.*

The proof follows immediately from the definitions. By (5.32), the equation $dH(E) = 0$ takes the form

$$p_a \frac{\partial H}{\partial p_a}(x, p) = 0 \tag{5.35}$$

in a natural chart. (5.35) is certainly satisfied on $\mathcal{N}$ if H is a homogeneous function (of any degree) with respect to the momentum coordinates p_a.

If $\mathcal{N}$ is a dilation-invariant ray-optical structure and $u \in \mathcal{N}$, then the whole radial line $\{\, e^t\, u | t \in \mathbb{R} \,\}$ must be in $\mathcal{N}$. This implies that $\mathcal{N}_q = \mathcal{N} \cap T_q^*\mathcal{M}$ necessarily has a non-void intersection with each neighborhood of the origin in $T_q^*\mathcal{M}$. Hence, $\mathcal{N}_q$ cannot be closed in $T_q^*\mathcal{M}$. Typically, the closure $\mathcal{N}_q \cup \{0\}$ of $\mathcal{N}_q$ in $T_q^*\mathcal{M}$ fails to be a smooth manifold at the origin but forms something like a tip or a vertex there like in our Example 5.1.1, see Figure 5.1. For a dilation-invariant ray-optical structure, $\mathcal{N}_q \cup \{0\}$ is a smooth manifold at the origin if and only if it is a hyperplane. This situation is encountered in our pathological Example 5.1.3, see Figure 5.3 (a).

Proposition 5.4.4 can be used to further characterize dilation-invariant ray-optical structures in the following way.

Proposition 5.4.5. *Let $\mathcal{N}$ be a dilation-invariant ray-optical structure on $\mathcal{M}$. Fix a point $u \in \mathcal{N}$ and, on some open neighborhood W of u in $\overset{\circ}{T}{}^*\mathcal{M}$, a natural chart (x, p) and a local Hamiltonian H for $\mathcal{N}$. Then there is a $c \in \mathbb{R}$ such that*

$$\begin{pmatrix} (H^{ab}) & (H^a) \\ (H^b) & 0 \end{pmatrix} \begin{pmatrix} (p_b) \\ c \end{pmatrix} = \begin{pmatrix} 0 \\ 0 \end{pmatrix} \tag{5.36}$$

at u. Here we use the abbreviations (4.7) and the same matrix notation as in (5.15).

Proof. By assumption, H has to satisfy equation (5.35) at all points of $\mathcal{N} \cap \mathcal{W}$. This gives the last row of the matrix equation (5.36). Now consider the set of all vertical vectors $Z_a \frac{\partial}{\partial p_a}$ at the point u. Such a vector is tangent to $\mathcal{N}$ if and only if it satisfies $Z_a \frac{\partial H}{\partial p_a} = 0$. In this case it can be applied to equation (5.35) as a derivation resulting in $p_b Z_a \frac{\partial^2 H}{\partial p_a \, p_b} = 0$ at u. This shows that the equation $p_b \frac{\partial^2 H}{\partial p_a \, p_b} = -c \frac{\partial H}{\partial p_a}$ holds at u with some real number c. As the last row of (5.36) was already verified, this completes the proof. $\qquad \square$

Recalling Definition 5.2.2 of strong regularity, this proposition has the following important consequence.

Corollary 5.4.1. *A dilation-invariant ray-optical structure $\mathcal{N}$ on $\mathcal{M}$ cannot be strongly regular at any point $u \in \mathcal{N}$.*

Proof. Since at least one of the momentum coordinates must be non-zero at u, (5.36) implies that the $(n + 1) \times (n + 1)$ matrix on the left-hand side has zero determinant. On the other hand, non-vanishing of this determinant was the defining property of strong regularity according to Definition 5.2.2. $\qquad \square$

This corollary says that, if $\mathcal{M}$ is a spacetime and $\mathcal{N}$ describes light propagation in a medium on this spacetime, strong regularity can hold only if the medium is dispersive.

Moreover, with the help of Proposition 5.4.5 it is easy to verify that the fiber derivative of a (local) Hamiltonian of a dilation-invariant ray-optical structure maps radial lines $\{ e^t u \mid t \in \mathbb{R} \}$ to radial lines $\{ e^t \mathbb{F}H(u) \mid t \in \mathbb{R} \}$ for $u \in \mathcal{N}$. This implies that, for a dilation invariant ray-optical structure, the infinitesimal light cone C_q is a closed subset of $\overset{\circ}{T}_q\mathcal{M}$ and that it has codimension ≥ 1 if it is a submanifold. (Please recall Definition 5.2.3 and the subsequent discussion.) This general result is exemplified by Example 5.1.1 and Example 5.1.3. On the other hand, transversality of $\mathcal{N}$ to the flow of the Euler vector field is not sufficient for the infinitesimal light cones to be open. This is demonstrated by Example 5.1.4.

The following proposition gives a pointwise characterization of dilation-invariant ray-optical structures that will be of relevance later.

Proposition 5.4.6. *Let $\mathcal{N}$ be a ray-optical structure on $\mathcal{M}$. Then, for any point $u \in \mathcal{N}$, the following properties are mutually equivalent.*

(a) *The Euler vector field E is tangent to $\mathcal{N}$ at the point u.*
(b) *Every characteristic vector field X on $\mathcal{N}$ satisfies $\theta_{\mathcal{N}}(X) = 0$ at u.*
(c) $\theta_{\mathcal{N}} \wedge \Omega_{\mathcal{N}}^{n-1} = 0$ *at u.*

Here $\Omega_{\mathcal{N}}^{n-1}$ means the antisymmetrized tensor product $\Omega_{\mathcal{N}} \wedge \cdots \wedge \Omega_{\mathcal{N}}$ with $(n - 1)$ factors.

Proof. To prove the equivalence of (a) and (b) we recall that locally every characteristic vector field on $\mathcal{N}$ is the Hamiltonian vector field of a local Hamiltonian for $\mathcal{N}$. Thus, (b) holds true if and only if $\theta(X_H) = 0$ at u for any local Hamiltonian H. Owing to the second identity given in (5.34) and the definition (4.1) of the Hamiltonian vector field, this is equivalent to $dH(E) = 0$ at u for any local Hamiltonian H, i.e., it is equivalent to (a).

Now we prove the equivalence of (a) and (c). It is obvious that, at each point of $\overset{\circ}{T}{}^*\mathcal{M}$, $\theta \wedge \Omega^{n-1}$ is a non-zero $(2n - 1)$-form and has, thus, a one-dimensional kernel. From (5.34) we read that this kernel is spanned by the Euler vector field E. Hence the pull-back of $\theta \wedge \Omega^{n-1}$ to $\mathcal{N}$ vanishes exactly at those points where E is tangent to $\mathcal{N}$.   □

According to this proposition, $\theta_{\mathcal{N}} \wedge \Omega_{\mathcal{N}}^{n-1}$ has no zeros if $\mathcal{N}$ is everywhere transverse to the flow of the Euler vector field. In this case $(\mathcal{N}, \theta_{\mathcal{N}})$ is an *exact contact manifold* in the terminology of Abraham and Marsden [1], Definition 5.1.4. In other words, transversality of $\mathcal{N}$ to the flow of the Euler vector field guarantees that $\mathcal{N}$ is orientable and that there is even a canonical volume form, viz. $\theta_{\mathcal{N}} \wedge \Omega_{\mathcal{N}}^{n-1}$, on $\mathcal{N}$. By Proposition 5.1.4, this implies in particular the existence of a global Hamiltonian for such a ray-optical structure.

Proposition 5.4.6 has the following important consequence for lifted virtual rays. (Please recall Definition 5.2.4.)

Proposition 5.4.7. *Let $\mathcal{N}$ be a ray-optical structure on $\mathcal{M}$. A lifted virtual ray $\xi \colon I \longrightarrow \mathcal{N}$ satisfies the equation*

$$\theta_{\xi(s)}\big(\dot{\xi}(s)\big) = 0 \tag{5.37}$$

at the parameter value $s \in I$ if and only if the Euler vector field E is tangent to $\mathcal{N}$ at the point $\xi(s)$.

Proof. By Proposition 5.2.3, the tangent vector of a lifted virtual ray is the sum of a characteristic vector tangent to $\mathcal{N}$ and a vertical vector. Since all vertical vectors are in the kernel of the canonical one-form, now the statement follows from the equivalence of (a) and (b) in Proposition 5.4.6.   □

In Hamiltonian mechanics, integration over the canonical one-form gives the socalled *action functional.* In this terminology, Proposition 5.4.7 says that for a dilation-invariant ray-optical structure the action functional vanishes on all lifted virtual rays. This observation will be of great importance for our discussion of variational principles in Chap. 7 below.

If, on the other hand, the Euler vector field is everywhere transverse to $\mathcal{N}$, Proposition 5.4.7 guarantees that every lifted virtual ray admits a reparametrization ξ such that

$$\theta_{\xi(s)}\big(\dot{\xi}(s)\big) = 1\,. \tag{5.38}$$

This gives a canonical parametrization for each lifted virtual ray ξ which is unique up to an additive constant, $\xi(s) \longmapsto \xi(s + s_o)$. In the case of Example 5.1.5 this distinguished parametrization gives g_+-arc length along each virtual ray $\lambda = \tau_{\mathcal{M}}^* \circ \xi$, i.e., $g_+(\dot{\lambda}, \dot{\lambda}) = 1$. Similarly, in the case of Example 5.1.2 the distinguished parametrization gives g_o-proper time along each virtual ray (= g_o-time-like curve). Finally, in the case of Example 5.1.4 the distinguished parametrization is adapted to U or adapted to $-U$ along each virtual ray (= integral curve of U) $\lambda = \tau_{\mathcal{M}}^* \circ \xi$, i.e., $\dot{\lambda} = \pm U \circ \lambda$.

No such distinguished parametrization exists if $\mathcal{N}$ is dilation-invariant since then every lifted virtual ray satisfies equation (5.37).

5.5 Eikonal equation

From the examples studied in Part I we know that families of rays are associated with families of wave surfaces. In this chapter we want to study the relation of rays and wave surfaces in our general geometrical setting. In particular we want to introduce, for arbitrary ray-optical structures in the sense of Definition 5.1.1, an eikonal equation by which the dynamics of wave surfaces is determined. It is largely a matter of taste whether one considers rays as more fundamental than wave surfaces or vice versa. Our intuitive ideas of light propagation are normally based on the notion of rays, rather than on the notion of wave surfaces. On the other hand, the derivation of ray optics from Maxwell's equations leads to the eikonal equation first and to the ray equation at a later stage, as we have seen in Part I.

Formally the eikonal equation of a ray-optical structure $\mathcal{N}$ on $\mathcal{M}$ can be introduced as the Hamilton-Jacobi equation determined by any (local) Hamiltonian for $\mathcal{N}$. More precisely, we say that a C^∞ function $S: \mathcal{U} \longrightarrow \mathbb{R}$, defined on some open subset $\mathcal{U}$ of $\mathcal{M}$, is a *classical solution of the eikonal equation* of $\mathcal{N}$ iff it satisfies the equation

$$H\big(dS(q)\big) = 0 \quad \text{for all } q \in \mathcal{U} . \tag{5.39}$$

Here the differential of the function S is to be viewed as a local section in the cotangent bundle, i.e., as a map $dS: \mathcal{U} \longrightarrow T^*\mathcal{U} \subseteq T^*\mathcal{M}$, and H denotes any local Hamiltonian for $\mathcal{N}$ whose domain of definition $\mathcal{W} \subseteq \overset{\circ}{T^*}\mathcal{M}$ covers the point $dS(q)$. In a natural chart the eikonal equation (5.39) takes the more familiar form

$$H\big(x, \partial S(x)\big) = 0 . \tag{5.40}$$

(5.39) can be rewritten without any reference to local Hamiltonians as

$$dS(\mathcal{U}) \subset \mathcal{N} \tag{5.41}$$

where $dS(\mathcal{U})$ denotes the image of the section $dS\colon \mathcal{U} \longrightarrow T^*\mathcal{U} \subseteq T^*\mathcal{M}$.
Since $\mathcal{N}$ is a subset of $\overset{\circ}{T}{}^*\mathcal{M}$, (5.41) automatically guarantees that dS has
no zeros. Hence, the function S determines a foliation (or "slicing") of the
open subset $\mathcal{U}$ of $\mathcal{M}$ into smooth hypersurfaces $S = \text{const.}$ which are called
wave surfaces (or *eikonal surfaces*, or *phase surfaces*). The motivation for
this terminology comes, of course, from the approximate-plane-wave method
outlined in Sect. 2.2.

In the case of Example 5.1.1 a wave surface is a g_o-light-like hypersurface
whereas it is a g_o-space-like hypersurface in the case of Example 5.1.2. In
the case of Example 5.1.3 a wave surface is foliated into integral curves of
the vector field U whereas it is transverse to U in the case of Example 5.1.4.
Finally, in the case of Example 5.1.5 a wave surface is a completely arbitrary
hypersurface.

The eikonal equation can be viewed analytically as a partial differential equation for a function S. As an alternative, suggested by (5.41), the
eikonal equation can be viewed geometrically as the problem of finding an
n-dimensional manifold (with certain properties) that is contained in a given
$(2n - 1)$-dimensional manifold. Henceforth we take the geometrical point of
view which is of great advantage for global questions. In other words, we turn
our attention away from the function S and concentrate upon the manifold
$dS(\mathcal{U})$.

For any C^∞ function $S\colon \mathcal{U} \longrightarrow \mathbb{R}$, $dS(\mathcal{U})$ is an n-dimensional C^∞ submanifold of $T^*\mathcal{M}$ and it is transverse to the fibers. Moreover, $dS(\mathcal{U})$ is a
socalled "Lagrangian submanifold" of the symplectic manifold $(T^*\mathcal{M}, \Omega)$.
This notion, which will be at the center of this section, is defined in the
following way.

Definition 5.5.1. *Let* $\mathcal{L} \subseteq T^*\mathcal{M}$ *be an embedded* C^∞ *submanifold of* $T^*\mathcal{M}$
and denote the inclusion map by $j\colon \mathcal{L} \longrightarrow T^*\mathcal{M}$.

(a) $\mathcal{L}$ *is called* isotropic *iff the pull-back with* j *of the canonical two-form* Ω
vanishes, $j^*\Omega = 0$.
(b) $\mathcal{L}$ *is called* Lagrangian *iff* $\mathcal{L}$ *is isotropic and* $\dim(\mathcal{L}) = \dim(\mathcal{M})$.

Definition 5.5.1 admits an obvious generalization for immersed, rather
than embedded, submanifolds of $T^*\mathcal{M}$.

The non-degeneracy of Ω immediately implies that an isotropic submanifold of $T^*\mathcal{M}$ must have dimension $\leq \dim(\mathcal{M})$. Thus, Lagrangian submanifolds are isotropic submanifolds of maximal dimension. The name "Lagrangian submanifold" was introduced by Maslov and Arnold in the 1960s.
It refers to the following characterization of such submanifolds in terms of
the classical *Lagrange brackets*. Consider a k-dimensional embedded subman-
ifold $\mathcal{L}$ of $T^*\mathcal{M}$; let $(u_1, \ldots, u_k)$ be any local chart on the manifold $\mathcal{L}$ and
let $(x^1, \ldots, x^n, p_1, \ldots, p_n)$ be a natural chart (or, more generally, a canonical
chart) on $T^*\mathcal{M}$. Then the classical Lagrange brackets are defined as

$$\left(u_A, u_B\right) = \frac{\partial x^a}{\partial u_A}\frac{\partial p_a}{\partial u_B} - \frac{\partial x^a}{\partial u_B}\frac{\partial p_a}{\partial u_A} \tag{5.42}$$

for $A, B = 1, \ldots, k$. Clearly, the same expression can be written without any reference to a natural (or canonical) chart as

$$\left(u_A, u_B\right) = (j^*\Omega)\left(\frac{\partial}{\partial u_A}, \frac{\partial}{\partial u_B}\right) \tag{5.43}$$

for $A, B = 1, \ldots, k$ where $j^*\Omega$ denotes the pull-back of Ω with the inclusion map $j\colon \mathcal{L} \longrightarrow T^*\mathcal{M}$. From (5.43) it is obvious that the Lagrange brackets vanish for all $A, B = 1, \ldots, k$ if and only if the submanifold $\mathcal{L}$ is isotropic. In other words, Lagrangian submanifolds are submanifolds of maximal dimension for which the Lagrange brackets vanish identically.

Properties of Lagrangian submanifolds are detailed in many articles and textbooks, e.g., in Weinstein [148], Guillemin and Sternberg [55]), Abraham and Marsden [1], and Woodhouse [150]. In the following two propositions we recall some well-known facts which are of particular relevance for us, cf. Figure 5.5.

Proposition 5.5.1. *Let $S\colon \mathcal{U} \longrightarrow \mathbb{R}$ be a C^∞ function defined on an open subset $\mathcal{U}$ of $\mathcal{M}$. Then $\mathcal{L} = dS(\mathcal{U})$ is an embedded Lagrangian submanifold of $T^*\mathcal{M}$ which is everywhere transverse to the fibers.*

Proof. The only non-trivial claim is that $\mathcal{L}$ is isotropic. To prove this, we recall that the canonical one-form θ on $T^*\mathcal{M}$ satisfies $\beta^*\theta = \beta$ where β is any local section in $T^*\mathcal{M}$. Hence, $(dS)^*\theta = dS$. Now we apply the exterior derivative d to this equation. Upon using the identity $dd = 0$ and the fact that d commutes with the pull-back operation, this results in $(dS)^*\Omega = 0$. As the inclusion map $j\colon \mathcal{L} \longrightarrow T^*\mathcal{M}$ can be written in the form $j = dS \circ \tau^*_{\mathcal{M}}$, this implies $j^*\Omega = (\tau^*_{\mathcal{M}})^*(dS)^*\Omega = 0$.   □

Proposition 5.5.2. *Let $\mathcal{L}$ be an n-dimensional embedded Lagrangian C^∞ submanifold of $T^*\mathcal{M}$ which is everywhere transverse to the fibers. Then $\mathcal{L}$ can be represented, locally around each of its points, in the form $\mathcal{L} = dS(\mathcal{U})$ where $S\colon \mathcal{U} \longrightarrow \mathbb{R}$ is a C^∞ function defined on an open subset $\mathcal{U}$ of $\mathcal{M}$. Moreover, if $\mathcal{L}$ is simply connected, $\mathcal{L}$ can be globally represented in this way. S is then called a* generating function *for $\mathcal{L}$.*

Proof. Since $\mathcal{L}$ is an n-dimensional C^∞ submanifold of $T^*\mathcal{M}$ transverse to the fibers, it is the image of a local section in $T^*\mathcal{M}$. Thus, there is a (necessarily open) subset $\mathcal{U}$ of $\mathcal{M}$ and a one-form $\beta\colon \mathcal{U} \longrightarrow T^*U \subseteq T^*\mathcal{M}$ such that $\mathcal{L} = \beta(\mathcal{U})$. Since $\mathcal{L}$ is Lagrangian, $\beta^*\Omega = 0$. On the other hand, the defining property of the canonical one-form θ on $T^*\mathcal{M}$ guarantees that $\beta^*\theta = \beta$ and, thus, $\beta^*\Omega = -d\beta$. Comparison of these two results gives $d\beta = 0$. If $\mathcal{U}$ is simply connected, this implies that β is of the form $\beta = dS$ with some function $S\colon \mathcal{U} \longrightarrow \mathbb{R}$ which is unique up to an additive constant. (S can be

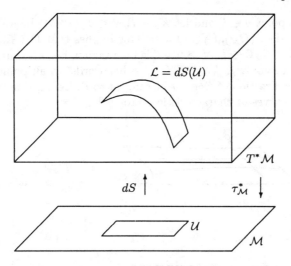

Fig. 5.5. A Lagrangian submanifold $\mathcal{L}$ of $T^*\mathcal{M}$ which is everywhere transverse to the fibers is locally generated by a function S on $\mathcal{M}$, i.e., $\mathcal{L} = dS(\mathcal{U})$.

defined by fixing a point q' in $\mathcal{U}$ and setting $S(q) = \int_{q'}^{q} \beta$ where the integral is to be performed along any path from q' to q. If $\mathcal{U}$ is simply connected, the equation $d\beta = 0$ guarantees that the result is independent of the path chosen. This follows from the well known Stokes theorem, see, e.g., Abraham and Marsden [1], p. 138). Since $\mathcal{U}$ is simply connected if and only if $\mathcal{L}$ is simply connected, this proves the second claim. The first claim follows from the fact that each point in $\mathcal{U}$ has a simply connected neighborhood. □

These two propositions have the following consequences, which are illustrated in Figure 5.5. Any classical solution $S: \mathcal{U} \longrightarrow \mathbb{R}$ of the eikonal equation determines a Lagrangian submanifold $\mathcal{L} = dS(\mathcal{U})$ of $T^*\mathcal{M}$ which is transverse to the fibers and completely contained in $\mathcal{N}$. Conversely, to any Lagrangian submanifold $\mathcal{L}$ of $T^*\mathcal{M}$ which is transverse to the fibers and completely contained in $\mathcal{N}$ we can find, on each simply connected subset $\mathcal{U}$ of $\tau^*_\mathcal{M}(\mathcal{L})$, a classical solution S of the eikonal equation such that $\mathcal{L} \cap (\tau^*_\mathcal{M})^{-1}(\mathcal{U}) = dS(\mathcal{U})$; this solution S is unique up to an additive constant. These observations suggest the following definition.

Definition 5.5.2. *Let $\mathcal{N}$ be a ray-optical structure on $\mathcal{M}$. A generalized solution of the eikonal equation of $\mathcal{N}$ is a Lagrangian C^∞ submanifold $\mathcal{L}$ of $T^*\mathcal{M}$ which is completely contained in $\mathcal{N}$.*

The following proposition guaranteees that any generalized solution of the eikonal equation determines an $(n-1)$-parameter family of lifted rays.

Proposition 5.5.3. *Let $\mathcal{N}$ be a ray-optical structure on $\mathcal{M}$ and let $\mathcal{L}$ be an embedded Lagrangian C^∞ submanifold of $T^*\mathcal{M}$ which is completely contained in $\mathcal{N}$. Then $\mathcal{L}$ is foliated into lifted rays.*

Proof. Fix a point $u \in \mathcal{L}$ and let $X_u \in T_u\mathcal{N} \subset T_u(T^*\mathcal{M})$ be a characteristic vector for $\mathcal{N}$, i.e., $(\Omega_\mathcal{N})_u(X_u, \cdot) = 0$. This implies that $\Omega_u(X_u, Z_u) = 0$ for all vectors $Z_u \in T_u\mathcal{L} \subset T_u\mathcal{N}$. Since $\mathcal{L}$ is Lagrangian, i.e., maximally isotropic, this can be true only if $X_u \in T_u\mathcal{L}$. In other words, at all points $u \in \mathcal{L}$ the characteristic direction of $\mathcal{N}$ must be tangent to $\mathcal{L}$. Hence, $\mathcal{L}$ must be foliated into integral curves of characteristic vector fields.   □

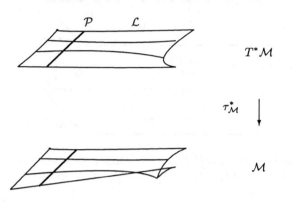

Fig. 5.6. A generalized solution $\mathcal{L}$ to the eikonal equation can be constructed, according to Proposition 5.5.4, by applying the characteristic flow to an appropriately chosen isotropic submanifold $\mathcal{P}$.

If we combine this observation with Proposition 5.5.1, we find that each classical solution $S : \mathcal{U} \longrightarrow \mathbb{R}$ of the eikonal equation is associated with a congruence of rays on $\mathcal{U}$. Those rays are the projections to $\mathcal{M}$ of the lifted rays into which $\mathcal{L} = dS(\mathcal{U})$ is foliated. (In terms of a local Hamiltonian and a natural chart this construction is already known to us from Sect. 2.4.) In other words, a classical solution $S : \mathcal{U} \longrightarrow \mathbb{R}$ of the eikonal equation determines, on its domain of definition $\mathcal{U} \subseteq \mathcal{M}$, not only a "slicing" into wave surfaces $S = \text{const.}$ but also a "threading" into rays. Later in this section we shall inquire whether rays and wave surfaces are transverse to each other.

The following proposition gives a construction method for generalized solutions of the eikonal equation, please cf. Figure 5.6.

Proposition 5.5.4. *Let $\mathcal{N}$ be a ray-optical structure on $\mathcal{M}$. Fix an $(n-1)$-dimensional embedded C^∞ submanifold $\mathcal{P}$ of $T^*\mathcal{M}$ such that $\mathcal{P}$ is completely contained in $\mathcal{N}$, isotropic, and non-characteristic. By the latter condition we mean that, at all points $u \in \mathcal{P} \subset \mathcal{N}$, the characteristic direction of $\mathcal{N}$ is non-tangent to $\mathcal{P}$. Let $\mathcal{L}$ be the set of all points in $T^*\mathcal{M}$ that can be connected to a point of $\mathcal{P}$ by a lifted ray. Then $\mathcal{L}$ is a generalized solution of the eikonal equation of $\mathcal{N}$. (In general, $\mathcal{L}$ is only an immersed but not an embedded submanifold of $T^*\mathcal{M}$. However, if we restrict to an appropriate neighborhood of $\mathcal{P}$ this construction always gives an embedded submanifold.)*

Proof. $\mathcal{L}$ is defined as the image of $\mathcal{P}$ under the flow of a characteristic vector field. Since $\mathcal{P}$ is $(n-1)$-dimensional and non-characteristic, $\mathcal{L}$ must be an n-dimensional immersed submanifold of $\mathcal{N}$. What remains to be shown is that $\mathcal{L}$ is isotropic, i.e., that the pull-back of Ω to $\mathcal{L}$ vanishes. If $\Phi: (s, u) \longmapsto \Phi_s(u)$ denotes the flow of a characteristic vector field on $\mathcal{N}$, the image of $\mathcal{P}$ under Φ_s is an isotropic submanifold (for each $s \in \mathbb{R}$ for which this image is non-empty). This follows from the fact that the Lie derivative of Ω with respect to a Hamiltonian vector field on $T^*\mathcal{M}$ vanishes. Hence, at each point of $\mathcal{L}$ the tangent space to $\mathcal{L}$ is spanned by the characteristic direction and by the tangent space to an isotropic submanifold. This proves that $\mathcal{L}$ must be isotropic. $\Box$

At the level of (local) Hamiltonians this is a standard result, cf., e.g., Abraham and Marsden [1], Lemma 5.3.29.

The construction of Proposition 5.5.4 can be carried through, in particular, for the special choice $\mathcal{P} = \mathcal{N}_q = \mathcal{N} \cap T_q^*\mathcal{M}$ where q is any point in $\mathcal{M}$. Since $\mathcal{N}_q$ is completely contained in the fiber $T_q^*\mathcal{M}$, it is, indeed, non-characteristic and isotropic. Hence the image of $\mathcal{P}$ under the characteristic flow gives a generalized solution $\mathcal{L}$ of the eikonal equation. Clearly, this $\mathcal{L}$ cannot be transverse to the fibers at q. The projection of $\mathcal{L}$ to $\mathcal{M}$ gives the set of all points in $\mathcal{M}$ that can be joined to q by a ray.

On the other hand, it is also possible to choose the initial surface $\mathcal{P}$ in such a way that the projection $\tau_{\mathcal{M}}^*$ maps $\mathcal{P}$ diffeomorphically onto an $(n-1)$-dimensional submanifold $\tau_{\mathcal{M}}^*(\mathcal{P})$ of $\mathcal{M}$. In this case the resulting generalized solution $\mathcal{L}$ of the eikonal equation is transverse to the fibers, and thus of the form $dS(\mathcal{U})$, near $\mathcal{P}$. It is foliated into lifted rays which, if projected to $\mathcal{M}$, give a congruence of rays that intersect $\tau_{\mathcal{M}}^*(\mathcal{P})$ transversely. Farther away from $\mathcal{P}$, however, $\mathcal{L}$ need not be transverse to the fibers and neighboring rays may intersect each other, see Figure 5.6. This shows that it is necessary to consider generalized solutions, rather than just classical solutions, of the eikonal equation if one wants to treat global questions.

Proposition 5.5.4 has the following interesting consequence.

Proposition 5.5.5. *Let $\xi: I \longrightarrow \mathcal{N}$ be a lifted ray of a ray-optical structure $\mathcal{N}$ on $\mathcal{M}$ and $s \in I$. Then there is an $\varepsilon > 0$ and a classical solution $S: \mathcal{U} \longrightarrow \mathbb{R}$ of the eikonal equation for $\mathcal{N}$ such that $\xi(s') \in dS(\mathcal{U})$ for all $s' \in \,]s - \varepsilon, s + \varepsilon[$.*

Proof. Construct a generalized solution $\mathcal{L}$ of the eikonal equation according to Proposition 5.5.4, with an initial manifold $\mathcal{P}$ that passes through the point $\xi(s)$ and is transverse to the fibers at that point. Then $\mathcal{L}$ must be transverse to the fibers on some neighborhood of $\lambda(s)$, i.e., it can be written as the image of a differential dS on that neighborhood. As, by construction, ξ is contained in $\mathcal{L}$, this concludes the proof. $\Box$

Quite generally, Proposition 5.5.4 gives a generalized solution of the eikonal equation in the form of an $(n-1)$-parameter family of lifted rays, parametrized by the points of $\mathcal{P}$. It is crucial to realize that only very special $(n-1)$-parameter families of lifted rays give rise to a generalized solution of the eikonal equation. This special property is in the condition of $\mathcal{P}$ being isotropic which guarantees that $\mathcal{L}$ is Lagrangian. This condition on the $(n-1)$-parameter family of lifted rays can be viewed as an integrability condition. The following proposition is helpful to clarify the geometric meaning of this integrability condition.

Proposition 5.5.6. *Let $\mathcal{L}$ be an embedded C^∞ submanifold of $T^*\mathcal{M}$, denote the inclusion map by $j\colon \mathcal{L} \longrightarrow T^*\mathcal{M}$ and let $\theta_{\mathcal{L}} = j^*\theta$ and $\Omega_L = j^*\Omega$. Then the following two statements are equivalent.*

(a) $\Omega_{\mathcal{L}} = 0$, *i.e., $\mathcal{L}$ is an isotropic submanifold of $T^*\mathcal{M}$.*
(b) *On every simply connected open subset $\bar{\mathcal{U}}$ of $\mathcal{L}$ there is a C^∞ function $\bar{S}\colon \bar{\mathcal{U}} \longrightarrow \mathbb{R}$, unique up to an additive constant, such that $d\bar{S} = \theta_{\mathcal{L}}|_{\bar{\mathcal{U}}}$.*

Proof. Since $\Omega = -d\theta$ and the exterior derivate d commutes with the pull-back operation, $\Omega_{\mathcal{L}} = 0$ is equivalent to $d\theta_{\mathcal{L}} = 0$. On a simply connected subset this equation is satisfied if and only if $\theta_{\mathcal{L}}$ is the differential of a function, please cf. the proof of Proposition 5.5.2. $\qquad\square$

As a consequence, an n-dimensional submanifold $\mathcal{L}$ of $\mathcal{N}$ is a generalized solution of the eikonal equation if and only if the kernel distribution of the one-form $\theta_{\mathcal{L}}$ is locally integrable.

If we supplement the hypotheses of Proposition 5.5.6 with the assumption that $\theta_{\mathcal{L}}$ has no zeros, the isotropy condition $\Omega_{\mathcal{L}} = 0$ guarantees that $\mathcal{L}$ is locally foliated into hypersurfaces $\bar{S} = $ const. If we specialize from the isotropic to the Lagrangian case, the situation that $\theta_{\mathcal{L}}$ has no zeros can be characterized with the help of the Euler vector field (5.31) in the following way.

Proposition 5.5.7. *Let $\mathcal{L}$ be an embedded Lagrangian submanifold of $T^*\mathcal{M}$ and $u \in \mathcal{L}$. Then the pull-back to $\mathcal{L}$ of the canonical one-form θ has a zero at u if and only if the Euler vector field E is tangent to $\mathcal{L}$ at u.*

Proof. Let us assume that $\theta_u(X_u) = 0$ for all $X_u \in T_u\mathcal{L} \subset T_u(T^*\mathcal{M})$. By (5.34) this is equivalent to $\Omega_u(E_u, X_u) = 0$ for all $X_u \in T_u\mathcal{L} \subset T_u(T^*\mathcal{M})$. As $\mathcal{L}$ is Lagrangian (i.e., maximally isotropic), this is equivalent to $E_u \in T_u\mathcal{L} \subset T_u(T^*\mathcal{M})$. $\qquad\square$

If a Lagrangian submanifold of $T^*\mathcal{M}$ is invariant under the flow of the Euler vector field, it is called *conic* (cf. Guckenheimer [52]) or *homogeneous* (cf. Guillemin and Sternberg [55]). Conic Lagrangian submanifolds are of relevance as generalized solutions of the eikonal equation for dilation-invariant ray-optical structures. They are necessarily non-transverse to the fibers, i.e., they cannot be associated with classical solutions of the eikonal equation. By Proposition 5.5.7, the pull-back of the canonical one-form θ to a conic

Lagrangian submanifold vanishes identically, i.e., its kernel distribution does not give a foliation into smooth hypersurfaces.

Let us consider, on the other hand, a Lagrangian submanifold $\mathcal{L}$ of $T^*\mathcal{M}$ such that the Euler vector field is nowhere tangent to $\mathcal{L}$. In this case Proposition 5.5.7 guarantees that the pull-back of θ to $\mathcal{L}$ has no zeros. As a consequence of Proposition 5.5.6, the kernel distribution of this one-form defines a foliation of $\mathcal{L}$ into smooth hypersurfaces. We introduce the following terminology.

Definition 5.5.3. *Let $\mathcal{L}$ be a generalized solution of the eikonal equation of a ray-optical structure $\mathcal{N}$ on $\mathcal{M}$. Assume that the Euler vector field is nowhere tangent to $\mathcal{L}$ such that the pull-back $\theta_{\mathcal{L}}$ to $\mathcal{N}$ of the canonical one-form θ has no zeros. Then an integral manifold of the kernel distribution of $\theta_{\mathcal{L}}$ is called a lifted wave surface of $\mathcal{L}$.*

On each simply connected open subset of $\mathcal{L}$, a lifted wave surface of $\mathcal{L}$ can be represented as a surface $\bar{S} = $ const. where $\bar{S}$ satisfies $d\bar{S} = \theta_{\mathcal{L}}$. If, in the situation of Definition 5.5.3, $\mathcal{L}$ is transverse to the fibers of $T^*\mathcal{M}$, the projection $\tau_{\mathcal{M}}^*$ maps each lifted wave surface onto a smooth hypersurface in $\mathcal{M}$ which, in agreement with our earlier terminology, is called a *wave surface* associated with $\mathcal{L}$. If $\mathcal{L}$ is not transverse to the fibers, the image of a lifted wave surface under the projection $\tau_{\mathcal{M}}^*$ need not be a smooth submanifold of $\mathcal{M}$ and could be called a *generalized wave surface*.

We have, thus, generalized our earlier observation that a classical solution $S \colon \mathcal{U} \longrightarrow \mathbb{R}$ of the eikonal equation is associated with a "slicing" of $\mathcal{U}$ into wave surfaces and a "threading" of $\mathcal{U}$ into rays. A generalized solution $\mathcal{L}$ of the eikonal equation is associated with a "slicing" of $\mathcal{L}$ into lifted wave surfaces and a "threading" of $\mathcal{L}$ into lifted rays, provided that the Euler vector field is nowhere tangent to $\mathcal{L}$. The question of whether lifted rays are transverse to lifted wave surfaces is answered in the following proposition.

Proposition 5.5.8. *Let $\mathcal{L}$ be a generalized solution of the eikonal equation of a ray-optical structure $\mathcal{N}$ on $\mathcal{M}$. Assume that the Euler vector field is nowhere tangent to $\mathcal{L}$, i.e., that $\mathcal{L}$ is foliated not only into lifted rays but also into lifted wave surfaces. Then for any point $u \in \mathcal{L} \subset \mathcal{N}$ the following two statements are equivalent.*

(a) The lifted ray through u is tangent to the lifted wave surface through u.
(b) The Euler vector field E is tangent to $\mathcal{N}$ at u.

Here we speak of "the" lifted ray through u in the sense that this lifted ray is unique up to reparametrization and extension.

Proof. Let $X_u \in T_u\mathcal{N} \subset T_u(T^*\mathcal{M})$ be tangent to the lifted ray through u, i.e., let X_u be a vector that spans the characteristic direction at u. By Definition 5.5.3, (a) is satisfied if and only if $\theta_u(X_u) = 0$. By Proposition 5.4.6, this is equivalent to (b). $\qquad\qquad\square$

Let us apply this proposition to a ray-optical structure $\mathcal{N}$ which is everywhere transverse to the flow of the Euler vector field, such as given by Example 5.1.2, 5.1.4 or 5.1.5. Then the Euler vector field cannot be tangent to a generalized solution of the eikonal equation, i.e., it is automatically guaranteed that every generalized solution $\mathcal{L}$ of the eikonal equation is foliated into lifted wave surfaces. By Proposition 5.5.8, those lifted wave surfaces are always transverse to the lifted rays into which $\mathcal{L}$ is foliated. Any real valued (local) function $\bar{S}$ on $\mathcal{L}$ with $d\bar{S} = \theta_{\mathcal{L}}$ gives a (local) parametrization on each of those lifted rays. This distinguished parametrization, which is unique (globally along the lifted ray) up to an additive constant, was already mentioned in Sect. 5.4, see (5.38).

The situation is completely different for a dilation-invariant ray-optical structure $\mathcal{N}$, such as given by Example 5.1.1 or 5.1.3. For a generalized solution $\mathcal{L}$ of the eikonal equation, the Euler vector field E may or may not be tangent to $\mathcal{L}$ at any of its points. Only in the case that E is nowhere tangent to $\mathcal{L}$ is $\mathcal{L}$ foliated into lifted wave surfaces. By Proposition 5.5.8, those lifted wave surfaces are then foliated into lifted rays. In other words, any real valued function $\bar{S}$ on $\mathcal{L}$ with $d\bar{S} = \theta_{\mathcal{L}}$ is constant along each of those lifted rays.

For an arbitrary ray-optical structure $\mathcal{N}$ on $\mathcal{M}$ these considerations apply to the maximal open subset of $\mathcal{N}$ on which E is non-tangent to $\mathcal{N}$ and to the maximal open subset of $\mathcal{N}$ on which E is tangent to $\mathcal{N}$. A full discussion requires an appropriate matching procedure in addition. This is rather cumbersome and we abstain from working out an example.

5.6 Caustics

In the last section we have seen that generalized solutions $\mathcal{L}$ of the eikonal equation are foliated into lifted rays. If projected to $\mathcal{M}$ those lifted rays give a congruence of rays as long as $\mathcal{L}$ is transverse to the fibers of $T^*\mathcal{M}$. At points where $\mathcal{L}$ fails to be transverse to the fibers neighboring rays start intersecting, see Figure 5.6. In optical terminology, this indicates the formation of a "caustic". Therefore we introduce the following mathematical definition.

Definition 5.6.1. *Let $\mathcal{L} \subset T^*\mathcal{M}$ be an embedded Lagrangian C^∞ submanifold of $T^*\mathcal{M}$ and denote the restriction to $\mathcal{L}$ of the cotangent bundle projection $\tau_{\mathcal{M}}^*$ by $\kappa = \tau_{\mathcal{M}}^*|_{\mathcal{L}} : \mathcal{L} \longrightarrow \mathcal{M}$. Then $u \in \mathcal{L}$ is called a* critical point *of $\mathcal{L}$ iff the tangent map $T_u\kappa : T_u\mathcal{L} \longrightarrow T_{\kappa(u)}\mathcal{M}$ is not surjective. The set*

$$\mathrm{Caust}_{\mathcal{L}} = \{ \kappa(u) \in \mathcal{M} \mid u \text{ is a critical point of } \mathcal{L} \} \qquad (5.44)$$

is called the caustic *of $\mathcal{L}$.*

Clearly, the critical points of $\mathcal{L}$ are exactly those points where $\mathcal{L}$ is not transverse to the fibers of $T^*\mathcal{M}$. In other words, $\mathcal{L}$ is everywhere transverse to the fibers if and only if $\mathrm{Caust}_{\mathcal{L}} = \emptyset$.

In any case, Caust$_\mathcal{L}$ is a set of measure zero in $\mathcal{M}$. This is an immediate consequence of the well-known *Sard theorem* which is proven, e.g., in Abraham and Robbin [2], p. 37. In general, Caust$_\mathcal{L}$ features cusps, edges and vertices, i.e., Caust$_\mathcal{L}$ is not a submanifold of $\mathcal{M}$. Thus, the geometry of caustics can be very complicated even locally. Quite generally, the variety of cusps, edges and vertices possible is so vast that a complete classification is not feasible. However, V. Arnold was able to locally classify all caustic types which are stable in a certain sense. For the details of this highly technical work we refer to Arnold, Gusein-Zade and Varchenko [9]. Arnold's formalism was applied to general relativity, e.g., by Friedrich and Stewart [45], by Petters [114], by Hasse, Kriele and Perlick [57], and by Low [87].

Here we want, of course, to apply Definition 5.6.1 to the case that $\mathcal{L}$ is a generalized solution of the eikonal equation of a ray-optical structure $\mathcal{N}$ on $\mathcal{M}$. Then $\mathcal{L}$ is foliated into lifted rays which can be projected to $\mathcal{M}$ to give a family of rays. In this situation, Caust$_\mathcal{L}$ is the set of all points in $\mathcal{M}$ where infinitesimally neighboring rays intersect each other. To put this rigorously we introduce the following notation (see Figure 5.7).

Definition 5.6.2. *Let $\mathcal{N}$ be a ray-optical structure on $\mathcal{M}$.*

(a) *A C^∞ vector field Z on $\mathcal{N}$ is called a* field of connecting vectors *iff, for every characteristic vector field X on $\mathcal{N}$, the Lie bracket $[Z, X]$ is, again, characteristic.*

(b) *Let $\xi: I \longrightarrow \mathcal{N}$ be a lifted ray of $\mathcal{N}$ and let $\bar{J}: I \longrightarrow T\mathcal{N}$ be a C^∞ map with $\bar{J}(s) \in T_{\xi(s)}\mathcal{N}$ for all $s \in I$. $\bar{J}$ is called a* lifted Jacobi field *along ξ iff it can be represented, locally around any parameter value $s \in I$, in the form $\bar{J} = Z \circ \xi$ where Z is a field of connecting vectors. Two lifted Jacobi fields along ξ are called* equivalent *iff they differ by a multiple of the tangent field of ξ. The respective equivalence classes are called* lifted Jacobi classes. *A lifted Jacobi field is called* trivial *if it is equivalent to the zero vector field, i.e., if it is parallel to the tangent field of ξ.*

(c) *If $\bar{J}$ is a lifted Jacobi field along ξ, $J = T\tau^*_\mathcal{M} \circ \bar{J}$ is called a* Jacobi field *along the ray $\lambda = \tau^*_\mathcal{M} \circ \xi$. Two Jacobi fields along λ are called* equivalent *if they differ by a multiple of the tangent field of λ. The respective equivalence classes are called* Jacobi classes. *A Jacobi field is called* trivial *if it is equivalent to the zero vector field, i.e., if it is parallel to the tangent field of λ.*

For a ray-optical structure $\mathcal{N}$ whose rays are geodesics, such as in our Examples 5.1.1, 5.1.2 and 5.1.5, Definition 5.6.2 (c) reproduces the standard textbook definition of Jacobi fields. (Note, however, that those standard textbooks usually assume their geodesics to be affinely parametrized whereas our rays are arbitrarily parametrized.)

If $\bar{J}$ is a lifted Jacobi field along a lifted ray ξ, the "arrow-head" of $\bar{J}$ can be thought as tracing a neighboring lifted ray which is infinitesimally close to ξ. All members of a lifted Jacobi class trace the same neighboring lifted ray.

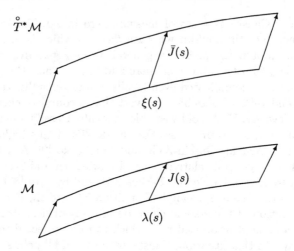

Fig. 5.7. A lifted Jacobi field $\bar{J}$ connects a lifted ray ξ with a neighboring lifted ray; a Jacobi field J connects a ray λ with a neighboring ray.

To construct a lifted Jacobi field along a lifted ray $\xi \colon [s_1, s_2] \longrightarrow \mathcal{N}$ we consider a variation of ξ, i.e., a C^∞ map $\eta \colon \ ] - \varepsilon_o, \varepsilon_o[\times [s_1, s_2] \longrightarrow \mathcal{N}$ such that $\eta(\varepsilon, \cdot)$ is a lifted ray for all $\varepsilon \in \ ] - \varepsilon_o, \varepsilon_o[$ and $\eta(0, \cdot) = \xi$. Then differentiation with respect to the variational parameter ε at $\varepsilon = 0$ gives a lifted Jacobi field along ξ, $\bar{J}(s) = \eta(\cdot, s)^{\cdot}|_{\varepsilon=0}$. In a natural chart, denoting the derivative with respect to the variational parameter by δ such that

$$\bar{J} = \delta x^a \frac{\partial}{\partial x^a} + \delta p_a \frac{\partial}{\partial p_a} \,, \tag{5.45}$$

a lifted Jacobi field is determined by the set of equations

$$\delta \Big(H(x, p) \Big) = 0 \,, \tag{5.46}$$

$$\delta \Big(\dot{x}^a - k \frac{\partial H}{\partial p_a}(x, p) \Big) = 0 \,, \tag{5.47}$$

$$\delta \Big(\dot{p}_a + k \frac{\partial H}{\partial x^a}(x, p) \Big) = 0 \,, \tag{5.48}$$

where H is any (local) Hamiltonian for the ray-optical structure considered. (5.46), (5.47) and (5.48) are, of course, just the conditions that the ray equations (5.10), (5.11) and (5.12) are to be preserved. The δ-derivatives in (5.46), (5.47) and (5.48) can be evaluated with the help of the usual product and chain rules. As derivatives with respect to the variational parameter and with respect to the curve parameter commute, (5.47) and (5.48) give us a system of first order linear differential equations for δp_a and δx^a. Any solution of this system, with any δk, that satisfies (5.46) gives us a lifted Jacobi field. It is easy to verify the following fact. To any initial values $\delta x^a(s_1)$, $\delta p_a(s_1)$

that satisfy (5.46) there is a solution δx^a, δp_a of the full system (5.46), (5.47) and (5.48), and it is unique up to adding a multiple of the tangent field. This freedom corresponds to the freedom of choosing δk at will.

This observation proves the following result which can be verified directly from Definition 5.6.2 as well, without refering to the coordinate representation (5.46), (5.47) and (5.48).

Proposition 5.6.1. *The set of all lifted Jacobi fields along a lifted ray ξ is an infinite dimensional real vector space. The set of lifted Jacobi classes along ξ is a $(2n-2)$-dimensional real vector space, corresponding to the $(2n-2)$ directions transverse to the characteristic direction in $\mathcal{N}$.*

As a consequence, the set of Jacobi fields along a ray λ is an infinite dimensional real vector space. The set of Jacobi classes along λ is a finite dimensional real vector space of dimension $\leq (2n-2)$. Since there is a ray through each point of $\mathcal{M}$, the dimension cannot be smaller than $(n-1)$. This minimal value is realized, e.g., in Examples 5.1.3 and 5.1.4. We shall now demonstrate that the maximal value is realized in strongly regular ray-optical structures. (Please recall Definition 5.2.2.) The proof will be based on the observation that in the strongly regular case Jacobi classes are determined by a second order linear differential equation that admits an existence and uniqueness theorem; so the dimension of the space of Jacobi classes can be found out by counting the allowed values for the inditial data.

Proposition 5.6.2. *Let $\mathcal{N}$ be a strongly regular ray-optical structure on $\mathcal{M}$ and $\lambda\colon [s_1, s_2] \longrightarrow \mathcal{M}$ be a ray of $\mathcal{N}$. Choose an arbitrary affine connection ∇ on $\mathcal{M}$. Then for any two vectors X and Y in $T_{\lambda(s_1)}\mathcal{M}$ there is a Jacobi field J along λ with $J(s_1) = X$ and $\nabla_{\dot{\lambda}}(s_1)J = Y$. This Jacobi field is unique up to transformations $J \longmapsto J + w\dot{\lambda}$ with $w(s_1) = 0$. As a consequence, for a strongly regular ray-optical structure the vector space of Jacobi classes along any ray λ has dimension $(2n-2)$, corresponding to the $(n-1)$ components of X and the $(n-1)$ components of Y transverse to the tangent vector $\dot{\lambda}(s_1)$.*

Proof. Let $\xi : [s_1, s_2] \longrightarrow \mathcal{N}$ be a lifted ray that projects onto λ. First we give the proof of the proposition under the additional assumption that ξ can be covered by the domain of a Hamiltonian and by a natural chart. Then our assumption of strong regularity guarantees that, in the natural chart chosen, condition (5.15) holds along ξ; we can thus introduce the inverse matrix by

$$\begin{pmatrix} (G_{ca}) & (G_c) \\ (G_a) & G \end{pmatrix} \begin{pmatrix} (H^{ab}) & (H^a) \\ (H^b) & 0 \end{pmatrix} = \begin{pmatrix} 1 & 0 \\ 0 & 1 \end{pmatrix} \qquad (5.49)$$

where the components G_{ca}, G_a and G are to be viewed as functions of the curve parameter s. Please recall that, in terms of their coordinate representation (5.45), lifted Jacobi fields along ξ are determined by (5.46), (5.47), and (5.48). It is our goal to eliminate δk and δp_a from these equations and to get a second order differential equation for δx^a alone, i.e., to get an equation for Jacobi fields rather than for lifted Jacobi fields. To that end we observe that (5.46) and (5.47) can be written as a matrix equation in the following way.

$$\begin{pmatrix} (H^{ab}) & (H^a) \\ (H^b) & 0 \end{pmatrix} \begin{pmatrix} \delta p_b \\ \frac{\delta k}{k} \end{pmatrix} = \begin{pmatrix} \frac{1}{k}\delta \dot{x}^a - \frac{\partial^2 H}{\partial p_a \partial x^b} \delta x^b \\ -\frac{\partial H}{\partial x^a} \delta x^a \end{pmatrix}.$$ (5.50)

With the help of (5.49), (5.50) can be solved for δp_b and δk,

$$\begin{pmatrix} \delta p_c \\ \frac{\delta k}{k} \end{pmatrix} = \begin{pmatrix} (G_{ca}) & (G_c) \\ (G_a) & G \end{pmatrix} \begin{pmatrix} \frac{1}{k}\delta \dot{x}^a - \frac{\partial^2 H}{\partial x^b} \delta x^b \\ -\frac{\partial H}{\partial x^b} \delta x^b \end{pmatrix}.$$ (5.51)

With the help of (5.50), we may eliminate δp_c and δk from (5.48) which gives an equation of the form

$$G_{ab}\, \delta \ddot{x}^b + B_{ab}\, \delta \dot{x}^b + C_{ab}\, \delta x^b = 0 \,.$$ (5.52)

Here G_{ab} has the same meaning as in (5.49) and (5.50) whereas B_{ab} and C_{ab} are some functions of s the special form of which will be of no interest in the following. By construction, $J = \delta x^b \frac{\partial}{\partial x^b}$ is a Jacobi field if and only if the δx^b satisfy (5.52). From (5.49) we read that

$$G_{ca}H^a = 0 \,.$$ (5.53)

Thus, for each parameter value s the matrix $(G_{ca}(s))$ has a non-trivial kernel which is spanned by the tangent vector of the ray, so (5.52) cannot be solved for the second derivatives. This reflects the fact that initial values $\delta x^a(s_1)$ and $\delta \dot{x}^a(s_1)$ do not fix a solution δx^a of (5.52) uniquely but leave the freedom of adding multiples of the tangent field. At each parameter value s we may introduce the $(n-1)$-dimensional vector space

$$L(s) = \left\{ (z^a) \in \mathbb{R}^n \mid G_a(s)\, z^a = 0 \right\}$$ (5.54)

which is transverse to the tangent vector $\dot{x}^a(s) = k(s)\, H^a(s)$ since, by (5.49),

$$G_a H^a = 1 \,.$$ (5.55)

From (5.49) we read that for all (z^a) in $L(s)$ the equation

$$H^{ba}(s)G_{ac}(s)z^c = z^b$$ (5.56)

holds true, i.e., that on $L(s)$ the matrix $(G_{ac}(s))$ is invertible, with $(H^{ba}(s))$ being its inverse. If we restrict to Jacobi fields with

$$(\delta x^a(s)) \in L(s) \quad \text{for all } s,$$ (5.57)

then (5.52) gives us a second order differential equation for δx^a that admits an existence and uniqueness theorem. In order to prove this it is convenient to choose the coordinates in such a way that $H^a = \delta^a_n$ and $G_a = \delta^n_a$ along λ, which is possible owing to (5.55). With this choice of coordinates (5.57) implies that $(\delta\dot{x}^a(s)) \in L(s)$ and $(\delta\ddot{x}^a(s)) \in L(s)$ for all s. As a consequence, multiplying (5.52) with H^{ca} results in

$$\delta\ddot{x}^c + H^{ca} B_{ab} \delta\dot{x}^b + H^{ca} C_{ab} \delta x^b = 0 \,. \tag{5.58}$$

Giving initial values $J(s_1) = X$ and $\nabla_{\dot{\lambda}(s_1)} J = Y$ is equivalent to giving initial values $\delta x^a(s_1)$ and $\delta\dot{x}^a(s_1)$. By adding appriopriate multiples of $\dot{x}^a(s_1)$ we get initial values in $L(s_1)$ which determine a unique solution δx^b of (5.58). Then $\delta x^b + f\dot{x}^b$, with $f(s_1)$ chosen appropriately, is a Jacobi field that satisfies the original initial conditions. Except for its value at the initial point, f can be chosen at will. This completes the proof under the assumption that λ can be covered by a chart of the desired form.

In the general case we divide the domain of λ into subintervals such that the restriction of λ to each subinterval can be covered by a local chart in which the equations $H^a = \delta^a_n$ and $G_a = \delta^n_a$ hold along λ. Then we get the desired Jacobi fields by solving (5.58) piecewise, where on each subinterval the initial values are determined by the end values on the preceding subinterval. $\Box$

Quite generally, Jacobi fields and lifted Jacobi fields can be used to characterize caustics in the following way. Let $\mathcal{L}$ be a generalized solution of the eikonal equation of a ray-optical structure $\mathcal{N}$ on $\mathcal{M}$ and let $\xi \colon [s_1, s_2] \longrightarrow \mathcal{L} \subset \mathcal{N}$ be a lifted ray through the point $u = \xi(s_2) \in \mathcal{L}$. Then u is a critical point of $\mathcal{L}$, in the sense of Definition 5.6.1, if and only if there is a non-zero vertical vector $Z_u \in T_u\mathcal{L}$. By the above argument, the existence of such a vector Z_u is equivalent to the existence of a non-trivial lifted Jacobi field $\bar{J}$ along ξ such that $\bar{J}$ is everywhere tangent to $\mathcal{L}$ and $\bar{J}(s_2)$ is vertical. (Please note that $\bar{J}$ is everywhere tangent to $\mathcal{L}$ if $\bar{J}(s_2)$ is tangent to $\mathcal{L}$, owing to the Lagrange property of $\mathcal{L}$.) Verticality of $\bar{J}(s_2)$ indicates an intersection of the ray $\tau^*_{\mathcal{M}} \circ \xi$ with the "infinitesimally neighboring ray" $J = T\tau^*_{\mathcal{M}} \circ \bar{J}$. In this sense, $\text{Caust}_{\mathcal{L}}$ is the set of all points where infinitesimally neighboring members of the family of rays determined by $\mathcal{L}$ have an intersection. Note that, in general, J may be zero on a whole interval, i.e., the two neighboring rays may coincide on a whole interval. To exclude this unwanted situation one introduces the following definition.

Definition 5.6.3. *Let* $\lambda \colon I \longrightarrow \mathcal{N}$ *be a ray of a ray-optical structure* $\mathcal{N}$ *on* $\mathcal{M}$ *and fix two different parameter values* $s_1, s_2 \in I$. *Let* $\text{Jac}(\lambda, s_1, s_2)$ *denote the vector space of Jacobi classes* $[J]$ *along* λ *such that there is a* $J \in [J]$ *with* $J(s_1) = 0$ *and* $J(s_2) = 0$. *Then the point* $\lambda(s_2)$ *is called* conjugate *to* $\lambda(s_1)$ *along* λ *iff the dimension of* $\text{Jac}(\lambda, s_1, s_2)$ *is non-zero and this dimension is called the* multiplicity *of the conjugate point.*

For the ray-optical structures of Example 5.1.5, Definition 5.6.3 coincides with the standard textbook definition of conjugate points in Riemannian geometry. For the ray-optical structures of Example 5.1.1 (and 5.1.2, respectively) it coincides with the definition of light-like (and time-like, respectively) conjugate points in Lorentzian geometry, cf. Beem, Ehrlich, and Easley [11]. The ray-optical structures of Example 5.1.3 and 5.1.4 do not admit any conjugate points.

Later we shall come back to the notion of Jacobi fields and of conjugate points. In particular, we shall explicitly evaluate the differential equation for Jacobi fields of isotropic ray-optical structures on Lorentzian manifolds in Sect. 6.4, and we shall use the notion of conjugate points and its multiplicity to develop a Morse theory for rays of strongly (hyper-)regular ray-optical structures in Sect. 7.5. In the latter context, the following observation will be important.

Proposition 5.6.3. *Let $\lambda : I \longmapsto \mathcal{N}$ be a ray of a strongly regular ray-optical structure $\mathcal{N}$ and fix a parameter value $s_1 \in I$. Then the following holds true.*

(a) *If $\lambda(s)$ is conjugate to $\lambda(s_1)$ along λ, its multiplicity cannot be bigger than $(n-1)$.*

(b) *There is an $\varepsilon > 0$ such that for $0 < |s - s_1| < \varepsilon$ the point $\lambda(s)$ cannot be conjugate to $\lambda(s_1)$ along λ.*

(c) *Let ξ be a lifted ray that projects onto λ and assume that, with respect to any local Hamiltonian and any natural chart, the matrix in (5.15) is not only non-degenerate but even positive definite at all points of ξ. If $\lambda(s_2)$ is conjugate to $\lambda(s_1)$, then there is an $\varepsilon > 0$ such that for $0 < |s - s_2| < \varepsilon$ the point $\lambda(s)$ cannot be conjugate to $\lambda(s_1)$ along λ.*

Proof. By Proposition 5.6.2, the vector space of Jacobi classes that vanish at a particular point has dimension $(n-1)$. Hence, the vector space of Jacobi classes that vanish at two points cannot be bigger than $(n-1)$. This proves (a).

To prove (b), we consider the set of all Jacobi fields J that vanish at s_1. By Proposition 5.6.2, the derivatives $\nabla_{\dot{\lambda}(s_1)} J$ of those Jacobi fields span an $(n-1)$-dimensional vector space transverse to $\dot{\lambda}(s_1)$, where ∇ denotes any affine connection on $\mathcal{M}$. By Taylor's theorem, this implies that for $0 < |s - s_1| < \varepsilon$ the values $J(s)$ of those Jacobi fields span an $(n-1)$-dimensional vector space transverse to $\dot{\lambda}(s)$. This proves (b).

We now turn to the proof of (c) which is more difficult. If the point $\lambda(s_2)$ is conjugate to $\lambda(s_1)$ along λ, part (a) implies that the multiplicity m of this conjugate point has to satisfy the inequality $m \leq n - 1$. In this situation we can find Jacobi fields $J_1, \ldots, J_{n-1}$ along λ such that

– the vector fields $J_1, \ldots, J_{n-1}, \dot{\lambda}$ are linearly independent over $\mathbb{R}$;

– $J_1(s_1) = \cdots = J_{n-1}(s_1) = 0$;

– $J_1(s_2) = \cdots = J_m(s_2) = 0$;

– the vectors $J_{m+1}(s_2), \ldots, J_{n-1}(s_2), \dot{\lambda}(s_2)$ are linearly independent.

A vector field along λ that vanishes at s_1 is a Jacobi field if and only if it differs from a linear combination of $J_1, \ldots, J_{n-1}$ with constant coefficients by a multiple of $\dot{\lambda}$. Here we made use of Proposition 5.6.2.

Now we give the proof of (c) under the additional assumption that the lifted ray $\xi|_{[s_1,s_2]}$ is contained in the domain of a natural chart (x, p) and of a Hamiltonian H for $\mathcal{N}$. For $A = 1, \ldots, (n-1)$, our Jacobi fields are then represented in the form

$$J_A = \delta_A x^a \, \frac{\partial}{\partial x^a} \tag{5.59}$$

with

$$\delta_A x^a(s_1) = 0 \quad \text{for } A = 1, \ldots, (n-1) \, ; \tag{5.60}$$

$$\delta_I x^a(s_2) = 0 \quad \text{for } I = 1, \ldots, m \, ; \tag{5.61}$$

$$\left(\delta_{m+1} x^a(s_2)\right), \ldots, \left(\delta_{n-1} x^a(s_2)\right), \left(\dot{x}^a(s_2)\right) \text{ are linearly independent} . \tag{5.62}$$

We are still free to change the Jacobi fields by a transformation of the form $J_A \longmapsto J_A + f_A \dot{\lambda}$ with functions f_A that satisfy $f_A(s_1) = 0$ for all indices $A = 1, \ldots, (n-1)$ and $f_I(s_2) = 0$ for all indices $I = 1, \ldots, m$. As shown in the proof of Proposition 5.6.2, we may use this freedom in such a way that, in an appropriately chosen chart, the second order differential equation (5.58) is satisfied by $\delta x^a = \delta_A x^a$ for all $A = 1, \ldots, (n-1)$. Then (5.61) implies that

$$\left(\delta_1 \dot{x}^a(s_2)\right), \ldots, \left(\delta_m \dot{x}^a(s_2)\right) \text{ are linearly independent.} \tag{5.63}$$

Otherwise there would be a non-zero solution $\delta x^a(s) = c^1 \delta_1 x^a(s) + \cdots + c^m \delta_m x^a(s)$ of the linear differential equation (5.58) with $\delta x^a(s_2) = 0$ and $\delta \dot{x}^a(s_2) = 0$, which is impossible.

In the following we have to use the positive-definiteness assumption of (c). This assumption implies that, along ξ, we have (5.49) at our disposal, with both matrices on the left-hand side positive definite. (Here we make use of the elementary fact that the inverse of a positive definite matrix is, again, positive definite.) For $A = 1, \ldots, (n-1)$, the $\delta_A x^a$ are the components of a Jacobi field. Hence, inserting $\delta x^a = \delta_A x^a$ into (5.51) determines $\delta p_b = \delta_A p_b$ and $\delta k = \delta_A k$ in such a way that (5.46), (5.47), and (5.48) are satisfied, i.e., such that

$$\bar{J}_A = \delta_A x^a \, \frac{\partial}{\partial x^a} + \delta_A p_a \, \frac{\partial}{\partial p_a} \tag{5.64}$$

is a lifted Jacobi field along ξ for each $A = 1, \ldots, (n-1)$. With the help of (5.47) and (5.48) it is readily verified that

$$\left(\delta_J p_a \, \delta_I x^a - \delta_I p_a \, \delta_J x^a\right)^{\cdot} = 0 \tag{5.65}$$

for any two indices I and J between 1 and $(n-1)$. By (5.60), this implies that

$$\delta_J p_a \, \delta_I x^a - \delta_I p_a \, \delta_J x^a = 0 \,. \tag{5.66}$$

Evaluating this equation at the parameter value s_2, and using (5.61), we find

$$(\delta_I p_a \delta_J x^a)(s_2) = 0 \quad \text{for } 1 \le I \le m < J \le (n-1) \,. \tag{5.67}$$

As $\delta_I p_a$ is determined by (5.52), with the index I affixed to all variational derivatives, (5.67) can be rewritten in matrix form as

$$\Big((\delta_J x^b(s_2)) \quad 0 \Big) \begin{pmatrix} (G_{ba}(s_2)) & (G_b(s_2)) \\ (G_a(s_2)) & G(s_2) \end{pmatrix} \begin{pmatrix} (\delta_I \dot{x}^a(s_2)) \\ 0 \end{pmatrix} = 0 \tag{5.68}$$

for $1 \le I \le m < J \le (n-1)$. Moreover, (5.53) and (5.55) imply that

$$\Big((\dot{x}^b(s_2)) \quad 0 \Big) \begin{pmatrix} (G_{ba}(s_2)) & (G_b(s_2)) \\ (G_a(s_2)) & G(s_2) \end{pmatrix} \begin{pmatrix} (\delta_I \dot{x}^a(s_2)) \\ 0 \end{pmatrix} = 0 \,. \tag{5.69}$$

(5.68) and (5.69) demonstrate that, with respect to a positive definite matrix, the space spanned by

$$\big(\delta_1 \dot{x}^a(s_2) \big), \ldots, \big(\delta_m \dot{x}^a(s_2) \big) \tag{5.70}$$

is orthogonal to the space spanned by

$$\big(\delta_{m+1} x^a(s_2) \big), \ldots, \big(\delta_{n-1} x^a(s_2) \big), \big(\dot{x}^a(s_2) \big) \,. \tag{5.71}$$

By (5.62) and (5.63), this implies that the vectors

$$\big(\delta_1 \dot{x}^a(s_2) \big), \ldots, \big(\delta_m \dot{x}^a(s_2) \big), \big(\delta_{m+1} x^a(s_2) \big), \ldots, \big(\delta_{n-1} x^a(s_2) \big), \big(\dot{x}^a(s_2) \big) \tag{5.72}$$

are linearly independent.

Now we define

$$\delta_I y^a(s) = \begin{cases} \dfrac{\delta_I x^a(s)}{s - s_2} & \text{if } s \ne s_2 \\ \delta_I \dot{x}^a(s) & \text{if } s = s_2 \end{cases} \quad \text{for } I = 1, \ldots, m \,, \tag{5.73}$$

$$\delta_J y^a(s) = \delta_J x^a(s) \quad \text{for } J = m+1, \ldots, (n-1) \,.$$

The Bernoulli-l'Hôpital rule guarantees that not only the $\delta_J y^a$ but also the $\delta_I y^a$ are continuous functions of the parameter s. We have just proven that $\big(\delta_1 y^a(s) \big), \ldots, \big(\delta_{n-1} y^a(s) \big), \big(\dot{x}^a(s) \big)$ are linearly independent for $s = s_2$. By continuity, the same must be true for $s \ne s_2$ as long as $|s - s_2|$ is sufficiently small. As a consequence, we can read from (5.73) that the vectors

$(\delta_1 x^a(s)), \ldots, (\delta_{n-1} x^a(s)), (\dot{x}^a(s))$ are linearly independent for $s \neq s_2$ and $|s - s_2|$ sufficiently small, i.e., that at those points the Jacobi fields span an $(n-1)$-dimensional space transverse to the tangent vector. This completes the proof for the case that ξ can be covered by the domain of an appropriate natural chart and a Hamiltonian.

In the general case one has to use a patching of several charts and to evaluate all relevant differential equations piecewise. $\square$

It is important to realize that part (c) of this proposition is not true without the positive-definiteness assumption. A counter-example can be found in an article by Helfer [62]. What Helfer constructs is a space-like geodesic in a Lorentzian manifold along which a whole interval is conjugate to some point. This example can be translated into our terminology by considering a Hamiltonian of the form $H(x,p) = \frac{1}{2}\left(g^{ab}(x)p_a p_b - 1\right)$, with g^{ab} the contravariant components of a Lorentzian metric. Read in this way, Helfer's construction gives a ray of a strongly regular ray-optical structure along which a whole interval is conjugate to some point. It should be noted that Helfer's metric is of class C^∞ but not analytic. As a matter of fact, it is not difficult to verify that for analytic Hamiltonians part (c) of Proposition 5.6.3 is true even without the positive-definiteness assumption. In other words, in the analytic category it is true that conjugate points are isolated along every ray of a strongly ray-optical structure. To prove this, it suffices to observe that, in the notation used in the proof of Proposition 5.6.3, the determinant

$$D(s) = \det \begin{pmatrix} \delta_1 x^1(s) & \cdots & \delta_{n-1} x^n(s) & \dot{x}^1(s) \\ \vdots & & \vdots & \vdots \\ \delta_1 x^n(s) & \cdots & \delta_{n-1} x^n(s) & \dot{x}^n(s) \end{pmatrix} \tag{5.74}$$

must be analytic if the Hamiltonian is analytic. Hence, if D does not vanish identically, then its zeros must be isolated. We shall come back to Proposition 5.6.3 in Sect. 7.5 below.

We end this section with a short remark on the characterization of caustics in terms of (lifted) wave surfaces, rather than in terms of (lifted) rays. To that end we consider a generalized solution $\mathcal{L}$ of the eikonal equation of a ray-optical structure $\mathcal{N}$ on $\mathcal{M}$ and we assume that the Euler vector field E is nowhere tangent to $\mathcal{L}$. By Definition 5.5.3, only in this case is the notion of lifted wave surfaces defined. By Definition 5.6.1, $u \in \mathcal{L}$ is a critical point of $\mathcal{L}$ if and only if there is a non-zero vertical vector $Z_u \in T_u\mathcal{L}$. Since all vertical vectors are in the kernel of the canonical one-form θ, this vector Z_u must be tangent to the lifted wave surface $\bar{\mathcal{S}}$ through u. Hence, the existence of such a vector Z_u indicates that the restriction to $\bar{\mathcal{S}}$ of the projection $\tau^*_{\mathcal{M}}$ cannot be an immersion at u, i.e., that $\tau^*_{\mathcal{M}}(\bar{\mathcal{S}})$ is not a codimension-one submanifold of $\mathcal{M}$ near $\tau^*_{\mathcal{M}}(u)$. In other words, $\mathrm{Caust}_{\mathcal{L}}$ is the set of all points where the generalized wave surfaces associated with $\mathcal{L}$ fail to be codimension-one submanifolds.

6. Ray-optical structures on Lorentzian manifolds

In Chap. 5 we have established the notion of a ray-optical structure on a bare manifold $\mathcal{M}$. From now on we shall assume that there is a Lorentzian metric g given on $\mathcal{M}$. This metric is to be interpreted as a spacetime metric in the sense of general relativity, although for the mathematical formalism it will not be necessary to specialize to the case $\dim(\mathcal{M}) = 4$. We shall assume, however, that $\dim(\mathcal{M}) > 2$ to exclude some pathologies.

6.1 The vacuum ray-optical structure

The metric g determines a distinguished ray-optical structure

$$\mathcal{N}^g = \{\, u \in T^*\mathcal{M} \mid g^{\#}(u, u) = 0 \,\} \tag{6.1}$$

on $\mathcal{M}$, just by the construction of Example 5.1.1 with $g_o = g$. We shall refer to $\mathcal{N}^g$ as to the *vacuum ray-optical structure* on $(\mathcal{M}, g)$. The rays of $\mathcal{N}^g$ are the g-light-like geodesics which are to be interpreted as vacuum light rays according to general relativity. It is important to realize that any conformally equivalent metric $\tilde{g}$ (i.e., any metric of the form $\tilde{g} = e^{2f}\, g$ with some C^{∞} function $f\colon \mathcal{M} \longrightarrow \mathbb{R}$) determines the same vacuum ray-optical structure as g. Up to conformal equivalence, $\mathcal{N}^g$ determines g uniquely. Hence, the causal structure of the Lorentzian manifold $(\mathcal{M}, g)$ is completely coded in the ray-optical structure $\mathcal{N}^g$.

At each point $q \in \mathcal{M}$, $\mathcal{N}^g \cap T_q^*\mathcal{M}$ consists of two connected components. (Here our assumption $\dim(\mathcal{M}) > 2$ is essential.) Thus, $\mathcal{N}^g$ has either one or two connected components. The Lorentzian manifold $(\mathcal{M}, g)$ is called *time-orientable* iff $\mathcal{N}^g$ has two connected components, cf., e.g., Sachs and Wu [126], p. 24, or Wald [146], p. 189. In this case, each connected component of $\mathcal{N}^g$ may be viewed as a ray-optical structure in its own right. One of them gives rays with *future-pointing momenta* whereas the other one gives rays with *past-pointing momenta*. If $(\mathcal{M}, g)$ is not time-orientable, i.e., if $\mathcal{N}^g$ has only one connected component, no such distinction can be made in a globally consistent way.

Ray-optical structures $\mathcal{N}$ on $\mathcal{M}$ which are different from $\mathcal{N}^g$ are to be interpreted as giving light propagation in a medium. If $\mathcal{N}$ is dilation-invariant in the sense of Definition 5.4.1, the medium is called *non-dispersive*, otherwise it is called *dispersive*. The following proposition characterizes the vacuum ray-optical structure $\mathcal{N}^g$ in comparison to other ray-optical structures on $\mathcal{M}$.

Proposition 6.1.1. *Let $\mathcal{N}$ be a ray-optical structure on $\mathcal{M}$. Assume that $\mathcal{N}$ is regular at all points $u \in \mathcal{N}$ in the sense of* Definition 5.2.1 *and that all the rays of $\mathcal{N}$ are g-light-like. Then $\mathcal{N} \subseteq \mathcal{N}^g$. (That is to say, if $(\mathcal{M}, g)$ is not time-orientable, $\mathcal{N} = \mathcal{N}^g$; if $(\mathcal{M}, g)$ is time-orientable, either $\mathcal{N} = \mathcal{N}^g$ or $\mathcal{N}$ is one of the two connected components of $\mathcal{N}^g$.)*

Proof. Fix a point $q \in \mathcal{M}$. Since all the rays of the ray-optical structure $\mathcal{N}$ are g-light-like, $\mathcal{N}_q = \mathcal{N} \cap T_q^* \mathcal{M}$ is a light-like hypersurface of the Minkowski space $(T_q^* \mathcal{M}, g_q^{\#})$. By elementary Minkowski geometry, this implies that $\mathcal{N}_q$ is ruled by light-like straight lines. Since $\mathcal{N}_q$ is closed in $\overset{\circ}{T}_q^* \mathcal{M}$, any such line either runs from infinity to infinity or it runs from the origin to infinity. We shall prove that the first case is impossible. By contradiction, let us assume that there is a light-like straight line L in $\mathcal{N}_q$ that runs from infinity to infinity. This means that all light-like straight lines in $\mathcal{N}_q$ which are infinitesimally close to L also have to run from infinity to infinity, without intersecting L. We decompose the motion of those neighboring lines in the familiar way into rotation, shear and expansion. Since the lines are surface forming, the rotation vanishes. Since the neighboring lines have no intersection with L, shear and expansion also vanish. (Non-vanishing shear gives an intersection with L of those neighboring lines that lie in the principal shear directions. In the case of vanishing shear, non-vanishing expansion gives an intersection with L of all neighboring lines.) Hence, all the neighboring lines have to be parallel to L. It is easy to verify that this conclusion contradicts our assumption that $\mathcal{N}$ is everywhere regular. Thus, we have proven that $\mathcal{N}_q$ is ruled by light-like straight lines running from the origin to infinity. As a consequence, $\mathcal{N}_q$ must be a subset of $\mathcal{N}_q^g = \mathcal{N}^g \cap T_q^* \mathcal{M}$.   □

This proposition can be rephrased in the following way. As long as regularity is not violated, the velocity of light in a medium is necessarily different from the vacuum velocity of light, at least for some rays. To demonstrate that the regularity assumption is, indeed, necessary one can consider Example 5.1.3 with a light-like vector field U.

For an arbitrary ray-optical structure on our Lorentzian manifold the rays can be time-like, light-like or space-like; they can even change their causal character from point to point. If we assume that rays can be used to transmit signals (and this, after all, is a basic idea of ray optics), the rules of general relativity prohibit space-like rays. We use the following terminology.

Definition 6.1.1. *A ray-optical structure $\mathcal{N}$ on $\mathcal{M}$ is called* causal *with respect to g if all rays of $\mathcal{N}$ are everywhere g-time-like or g-light-like.*

The ray-optical structures of Examples 5.1.1 and 5.1.2 are causal with respect to g iff the metric g_o is "narrower" than g, i.e., iff the inequality $g_o(X, X) \leq 0$ implies the inequality $g(X, X) \leq 0$. The ray-optical structures of Examples 5.1.3 and 5.1.4 are causal with respect to g iff the vector field U satisfies the inequality $g(U, U) \leq 0$.

The question of whether or not all physically reasonable ray-optical structures on a spacetime have to be causal is a bit subtle. If ray optics is viewed as an approximation scheme to wave optics, the energy of a wave field does not propagate exactly along rays; this is true only in an approximate sense. (We have discussed this issue in Sect. 2.7, at least for a special class of media.) On the basis of this observation, non-causal rays are not necessarily to be discarded altogether as unphysical. In the next section we shall introduce the notions of phase velocity and group velocity for a ray-optical structure $\mathcal{N}$, and we shall see that $\mathcal{N}$ is causal iff the group velocity does not exceed the vacuum velocity of light. It is well known that there are physically relevant optical media in which the group velocity exceeds the vacuum velocity of light. For a comprehensive discussion of this issue we refer to Brillouin [22].

6.2 Observer fields, frequency, and redshift

Several basic concepts of ray optics which are familiar from elementary textbooks depend on the notion of frequency. For a ray-optical structure on our Lorentzian manifold $(\mathcal{M}, g)$ this notion can be introduced after a time-like vector field has been chosen.

A time-like C^∞ vector field, given on some open subset $\mathcal{U}$ of $\mathcal{M}$ will be called a (local) *observer field* henceforth. In the case $\mathcal{U} = \mathcal{M}$ we speak of a *global observer field*. This terminology refers to the fact that the integral curves of such a vector field can be interpreted as the worldlines of observers. A global observer field exists if and only if $(\mathcal{M}, g)$ is time-orientable, see, e.g., Wald [146], Lemma 8.1.1. In the following we assume that we have an observer field V given on some open subset $\mathcal{U}$ of $\mathcal{M}$ which satisfies the normalization condition $g(V, V) = -1$. This normalization condition means that the integral curves of V are parametrized by proper time.

At each point $q \in \mathcal{U} \subseteq \mathcal{M}$, we write V_q for the value at q of the time-like vector field V. We decompose the tangent space $T_q\mathcal{M}$ into the one-dimensional time-like subspace spanned by V_q and its $(n-1)$-dimensional orthocomplement

$$H_q\mathcal{M} = \{ Y \in T_q\mathcal{M} \mid g_q(V_q, Y) = 0 \} . \tag{6.2}$$

Similarly, we decompose the cotangent space $T_q^*\mathcal{M}$ into the one-dimensional subspace spanned by the covector $g_q(V_q, \cdot)$ and its $(n-1)$-dimensional orthocomplement

$$H_q^*\mathcal{M} = \{ u \in T_q^*\mathcal{M} \mid u(V_q) = 0 \} . \tag{6.3}$$

As suggested by our notation, $H_q^* \mathcal{M}$ can be identified with the dual space of $H_q \mathcal{M}$.

Now let $\mathcal{N}$ be a ray-optical structure on $\mathcal{M}$. To each point $u \in \mathcal{N}$ we assign the following quantities, denoting the footpoint of u by $q = \tau_{\mathcal{M}}^*(u)$. The *frequency*

$$\omega(u) = -u(V_q) \in \mathbb{R} ; \tag{6.4}$$

the *spatial wave covector*

$$k(u) = u - \omega(u)\, g(V_q, \cdot) \in H_q^* \mathcal{M} ; \tag{6.5}$$

the *phase velocity*

$$w(u) = \frac{\omega(u)}{\|k(u)\|^2}\, k(u) \in H_q^* \mathcal{M} ; \tag{6.6}$$

the *ray velocity* or *group velocity*

$$v(u) = \frac{-(\mathbb{F}H)(u)}{g_q\left(V_q, (\mathbb{F}H)(u)\right)} - V_q \in H_q \mathcal{M} . \tag{6.7}$$

Here $\| \cdot \|$ denotes the norm induced on $H_q^* \mathcal{M}$ by our Lorentzian metric and $\mathbb{F}H$ denotes the fiber derivative of a local Hamiltonian H for $\mathcal{N}$.

In a natural chart these definitions take the following form.

$$\omega(x,p) = -p_a\, V^a(x) , \tag{6.8}$$

$$k_a(x,p) = p_a - \omega(x,p)\, g_{ab}(x)\, V^b(x) , \tag{6.9}$$

$$w_a(x,p) = \frac{\omega(x,p)}{g^{bc}(x)\, k_b(x,p)\, k_c(x,p)}\, k_a(x,p) , \tag{6.10}$$

$$v^a(x,p) = \frac{-\frac{\partial H}{\partial p_a}(x,p)}{g_{cd}(x)\, V^c(x)\frac{\partial H}{\partial p_d}(x,p)} - V^a(x) . \tag{6.11}$$

If we evaluate (6.8) and (6.9) along a classical solution $S: \mathcal{U} \longrightarrow \mathbb{R}$ of the eikonal equation of $\mathcal{N}$, we reproduce the frequency function (2.38) and the spatial wave covector field (2.39), respectively. (The parameter α can be absorbed by a redefinition of the eikonal function S.) This justifies the terminology. The name "phase velocity" for $w(u)$ refers, of course, to the same situation. At each point $u \in dS(\mathcal{U})$, the covector $w(u)$ determines the spatial velocity of the wave surfaces (="phase surfaces") with respect to the observer field V. Geometrically, the norm of $w(u)$ is a measure for the pseudo-Euclidean angle γ between the covector $g_q(V_q, \cdot)$ and the covector u according to the formula $\sinh^2\gamma = \left(1 - \|w(u)\|^2\right)^{-1}$, as can be read from equation (6.6).

The ray velocity (6.7), on the other hand, admits an obvious physical interpretation if evaluated along a lifted ray. The direction of the vector $v(u)$

determines the spatial direction in which the ray is moving, viewed with the eyes of an observer traveling along an integral curve of V; the norm of $v(u)$ is a measure for the pseudo-Euclidean angle between the tangent vector to the ray and the tangent vector to the observer's worldline, i.e., for the relative velocity of the ray with respect to the observer field chosen. This justifies the term "ray velocity". To verify that, moreover, (6.7) is equivalent to the familiar textbook definition of the "group velocity", we proceed in the following way. First we have to assume that $g_q(V_q, \mathbb{F}H(u)) \neq 0$ to make sure that $v(u)$, as defined by (6.7), is non-singular. This condition is satisfied if and only if at the point u the straight line parallel to $g_q(V_q, \cdot)$ in $T_q^* \mathcal{M}$ is transverse to $\mathcal{N}_q = \mathcal{N} \cap T_q^* \mathcal{M}$. (For causal ray-optical structures this transversality condition is automatically satisfied.) Clearly, this is the case if and only if the manifold $\mathcal{N}_q \subset T_q^* \mathcal{M} \cong H_q^* \mathcal{M} \times \mathbb{R}$ is the graph of a function $f \colon H_q^* \mathcal{M} \longrightarrow \mathbb{R}$ locally around u. (Globally, however, $\mathcal{N}_q$ need not be the graph of a single-valued function. In typical cases, such as in our Examples 5.1.1, 5.1.2, and 5.1.4, $\mathcal{N}_q$ has several "branches".) Then a quick calculation shows that $v(u)$, as defined by equation (6.7), is equal to the differential $(df)_u \in (H_q^* \mathcal{M})^* \cong H_q \mathcal{M}$. In other words, to calculate $v(u)$ we have to write the frequency as a function of the spatial wave covector by means of the dispersion relation and we have to calculate the gradient of this function. This is exactly the usual textbook definition of the group velocity.

By (6.6), the phase velocity has a zero at points $u \in \mathcal{N}$ where the frequency vanishes, and it has a singularity at points $u \in \mathcal{N}$ where the spatial wave covector vanishes. Either case is to be viewed as a pathological behavior and indicates a "bad" choice of observer. (If no other choice is possible, the ray-optical structure is to be viewed as "bad".) Similarly, we can read from (6.7) that the ray velocity has a zero at points $u \in \mathcal{N}$ where $\mathbb{F}H(u)$ is parallel to our observer field and that it has a singularity at points $u \in \mathcal{N}$ where $\mathbb{F}H(u)$ is orthogonal to our observer field. The first case indicates a "bad" choice of observer, whereas the second case cannot happen if our ray-optical structure is causal in the sense of Definition 6.1.1. As a matter of fact, a ray-optical structure is causal if and only if $\|v(u)\| \leq 1$ for all $u \in \mathcal{N}$. Here $\| \cdot \|$ denotes the norm induced by our Lorentzian metric on the vector space $H_q \mathcal{M}$ at the point $q = \tau_{\mathcal{M}}^*(u)$. In other words, a ray-optical structure is causal if and only if the ray velocity ($=$ group velocity) is bounded by the vacuum velocity of light.

Note that basically the phase velocity is a spatial covector whereas the ray velocity is a spatial vector. We can, of course, use the metric g to identify vectors and covectors. In particular, we can use the metric to make the spatial wave covector into a vector, i.e., we can introduce, in terms of a natural chart, the quantity

$$k^a(x, p) = g^{ab}(x) \, k_b(x, p) \,. \tag{6.12}$$

This vector is sometimes called the *vector of normal slowness*, following Sir William R. Hamilton. Here, "normal" refers to the fact that, along any classical solution of the eikonal equation, this vector is orthogonal to all spatial vectors which are tangent to the wave surfaces; "slowness" refers to the fact that the phase velocity decreases if the length of this vector increases.

The properties of a specific ray-optical structure are nicely visualized by indicating, for a fixed frequency, the phase velocity and the group velocity for each spatial direction. To make this idea precise we fix a ray-optical structure $\mathcal{N}$, a normalized observer field V, a point $q \in \mathcal{M}$, and a real number $\omega_o \in \mathbb{R}$. Then we define the sets

$$\mathfrak{F}^* = \{\, w(u) \,|\, u \in \mathcal{N}_q,\, \omega(u) = \omega_o \,\} \subset H_q^*\mathcal{M} \subset T_q^*\mathcal{M}, \qquad (6.13)$$

$$\mathfrak{F} = \{\, v(u) \,|\, u \in \mathcal{N}_q,\, \omega(u) = \omega_o \,\} \subset H_q\mathcal{M} \subset T_q\mathcal{M}. \qquad (6.14)$$

$\mathfrak{F}^*$ is usually called the *figuratrix* of the medium whereas $\mathfrak{F}$ is called the *indicatrix*. Both names originate from variational calculus.

In general, figuratrix and indicatrix depend on the frequency value ω_o, i.e., if we switch from ω_o to $c\omega_o$ with some real number c, figuratrix and indicatrix undergo a deformation. If we restrict ourselves to the case $c > 0$ (i.e., if we fix the sign of the frequency), there is no such deformation, for any observer field and for any point $q \in \mathcal{M}$, if and only if our ray-optical structure is dilation invariant. In other words, the dilation invariant case is characterized by the property that phase velocity and group velocity are independent of the frequency, as long as the sign of the frequency is fixed. This is the defining property of a non-dispersive medium according to standard textbooks on optics. We have thus justified our earlier claim that for a ray-optical structure on a Lorentzian manifold the attributes "dilation invariant" and "non-dispersive" are synonymous.

As an example, we consider a Hamiltonian of the form

$$H(x,p) = \tfrac{1}{2}\bigl(g^{ab}(x)p_a p_b + h(x)\bigr) \qquad (6.15)$$

in natural coordinates, which comprises Example 5.1.1 ($h(x) = 0$ and $g_o^{ab}(x) = g^{ab}(x)$) and Example 5.1.2 ($h(x) \neq 0$ and $g_o^{ab}(x) = g^{ab}(x)/h(x)$). Then the figuratrix at the point with coordinates x is a sphere of radius $\omega_o^2/\bigl(\omega_o^2 - h(x)\bigr)$ around the origin in $H_q^*\mathcal{M}$, whereas the indicatrix is a sphere of radius $\bigl(\omega_o^2 - h(x)\bigr)/\omega_o^2$ around the origin in $H_q\mathcal{M}$.

For a pathological ray-optical structure of the type given in Example 5.1.3, on the other hand, the figuratrix is a sphere through the origin and the indicatrix is a single point.

Now we turn to the question of how the frequency changes along a ray. In other words, we want to discuss the general relativistic redshift (or blueshift) for light propagation in media. Along any curve in $\mathcal{N}$, given in terms of a natural chart as a map $s \longmapsto (x(s), p(s))$, differentiation of the frequency function (6.8) yields

$$\tfrac{d}{ds}\omega\bigl(x(s), p(s)\bigr) = -\dot{p}_a(s)\, V^a\bigl(x(s)\bigr) - p_a(s)\,\frac{\partial V^a\bigl(x(s)\bigr)}{\partial x^b}\,\dot{x}^b(s). \qquad (6.16)$$

If we introduce the canonical lift $\bar{V}$ of the observer field V according to (5.24), equation (6.16) can be rewritten in coordinate-free form as

$$\tfrac{d}{ds}\omega\big(x(s),p(s)\big) = \Omega_{\xi(s)}\big(\dot{\xi}(s), \bar{V}_{\xi(s)}\big) \tag{6.17}$$

for any C^∞ curve $\xi\colon I \longrightarrow \mathcal{N}$. Since $\mathcal{N}$ is transverse to the fibers of $T^*\mathcal{M}$, $\bar{V}$ can be decomposed, at each point of $\mathcal{N}$, into a vector tangent to $\mathcal{N}$ and a vector tangent to the fiber. (This decomposition is, of course, not unique.) If ξ is a lifted ray, only the component tangent to the fiber gives a contribution to (6.17). Therefore, it is justified to say that a C^∞ curve $\xi\colon I \longrightarrow \mathcal{N}$ satisfies the *redshift law* of lifted rays with respect to the observer field V if

$$\Omega_{\xi(s)}\big(\dot{\xi}(s), Q(s)\big) = 0 \tag{6.18}$$

for all C^∞ maps $Q\colon I \longrightarrow T\mathcal{N}$ with $Q(s) \in T_{\xi(s)}\mathcal{N}$ and $T\tau^*_{\mathcal{M}}(Q(s)) = V_{\tau^*_{\mathcal{M}}(\xi(s))}$ for $s \in I$. The redshift law for lifted rays can be expressed more conveniently if we use a Hamiltonian H for $\mathcal{N}$. Then a lifted ray $\xi\colon I \longrightarrow \mathcal{N}$ with a parametrization adapted to H (i.e., such that (5.9) holds with $k = 1$) satisfies, by (6.17), the redshift law

$$\tfrac{d}{ds}\omega\big(x(s),p(s)\big) = dH(\bar{V})\big|_{\xi(s)} . \tag{6.19}$$

To illustrate the redshift law with an example, we consider a ray-optical structure that is generated by a Hamiltonian of the form

$$H(x,p) = \tfrac{1}{2}\, g_o^{ab}(x)\, p_a\, p_b + c \tag{6.20}$$

where g_o^{ab} are the contravariant components of a Lorentzian metric g_o and c is a real constant. Our Examples 5.1.1 and 5.1.2 are of this form. A lifted ray of such a ray-optical structure gives a geodesic of the metric g_o if projected to $\mathcal{M}$. The parametrization is adapted to H if this geodesic is affinely parametrized and if, in addition,

$$p_a(s) = (g_o)_{ab}\big(x(s)\big)\, \dot{x}^b(s) . \tag{6.21}$$

In this situation, the frequency with respect to a normalized observer field V is given by

$$\omega\big(x(s),p(s)\big) = -(g_o)_{ab}\big(x(s)\big)\, \dot{x}^b(s)\, V^a\big(x(s)\big) \tag{6.22}$$

along the lifted ray.

If we switch to coordinate-free notation, denoting the lifted ray by $\xi\colon I \to \mathcal{N}$ and its projection to $\mathcal{M}$ by $\lambda = \tau^*_{\mathcal{M}} \circ \xi$, (6.22) implies

$$\frac{\omega\big(\xi(s_2)\big)}{\omega\big(\xi(s_1)\big)} = \frac{(g_o)_{\lambda(s_2)}\big(\dot{\lambda}(s_2), V_{\lambda(s_2)}\big)}{(g_o)_{\lambda(s_1)}\big(\dot{\lambda}(s_1), V_{\lambda(s_1)}\big)} \tag{6.23}$$

for any two parameter values s_1, $s_2 \in I$. Please note that ξ enters into the right-hand side of (6.23) only in terms of its projection λ. As (6.23) remains true after an affine reparametrization, it gives the redshift along any ray which is parametrized affinely (as a geodesic of g_o).

The redshift formula (6.23) applies, in particular, to the vacuum ray-optical structure, where we have to read $g_o = g$. In this special context equation (6.23) is well known. It was found by Kermack, McCrea and Whittacker [71] and rediscovered by Schrödinger [129]. A particularly clear derivation was given by Brill [20]. As an alternative, the vacuum redshift formula can also be expressed in terms of acceleration, expansion and shear of the observer field V. Details can be found in articles by Ehlers [37], by Hasse and Perlick [58] and by Perlick [107].

Now we turn back to arbitrary ray-optical structures. The following proposition characterizes the redshift-free case.

Proposition 6.2.1. *Consider a ray-optical structure $\mathcal{N}$ and a global observer field V with $g(V,V) = -1$ on $\mathcal{M}$. Then the following two properties are equivalent.*

(a) *The frequency with respect to V is constant along each lifted ray ξ of $\mathcal{N}$.*
(b) *$V \in \mathcal{G}_{\mathcal{N}}$. (Please recall Definition 5.3.2 of the symmetry algebra $\mathcal{G}_{\mathcal{N}}$.)*

Proof. The general redshift formula (6.17) implies that (a) is true if and only if the function $\Omega(X, \bar{V})$ vanishes identically on $\mathcal{N}$ whenever X is a characteristic vector field on $\mathcal{N}$. Since, at each point of $\mathcal{N}$, the kernel of $\Omega(X, \cdot)$ coincides with the tangent space to $\mathcal{N}$, this condition is satisfied if and only if $\bar{V}$ is tangent to $\mathcal{N}$ at all points of $\mathcal{N}$. By Definition 5.3.2, this is equivalent to (b). □

For an arbitrary ray-optical structure $\mathcal{N}$ on $\mathcal{M}$ the symmetry algebra $\mathcal{G}_{\mathcal{N}}$ need not contain a time-like vector field normalized to $g(V,V) = -1$. Thus, only in very special cases is it possible to find a normalized observer field V such that the frequency is constant along all lifted rays.

Proposition 6.2.1 can be specialized to our standard examples for which we have analyzed the symmetries in Sect. 5.3. This gives the following results. For the ray-optical structures of Example 5.1.1 the frequency with respect to V is constant along each lifted ray iff V is a conformal Killing vector field of the metric g_o, i.e., iff the Lie derivative $L_V g_o$ is a multiple of g_o. (In the vacuum case $g_o = g$, the normalization condition $g(V,V) = -1$ then requires V to be a Killing vector field, i.e., $L_V g = 0$.) For the ray-optical structures of Example 5.1.2 the frequency with respect to V is constant along each lifted ray iff V is a Killing vector field of the metric g_o, i.e., iff $L_V g_o = 0$. For the ray-optical structures of Example 5.1.3 the frequency with respect to V is constant along each lifted ray iff the Lie bracket $[V, U]$ is a multiple of U. Finally, for the ray-optical structures of Example 5.1.4 the frequency with respect to V is constant along each lifted ray iff $[V, U] = 0$.

It is also interesting to consider normalized observer fields which are not in $\mathcal{G}_{\mathcal{N}}$ but can be rescaled to give an element of $\mathcal{G}_{\mathcal{N}}$. In this case the redshift does not vanish but it admits a representation in terms of a "redshift potential". This situation, which is of particular interest in cosmology, is characterized by the following proposition.

Proposition 6.2.2. *Consider a ray-optical structure $\mathcal{N}$ and a global observer field V with $g(V, V) = -1$ on $\mathcal{M}$. Then for any C^∞ function $f : \mathcal{M} \longrightarrow \mathbb{R}$ the following two properties are equivalent.*

(a) *f is a redshift potential in the following sense. If $\xi : I \longrightarrow \mathcal{N}$ is a lifted ray of $\mathcal{N}$ with projection $\lambda = \tau_{\mathcal{M}}^* \circ \xi$, the frequency ω with respect to V satisfies*

$$e^{f\left(\lambda(s)\right)} \omega\big(\xi(s)\big) = \text{const.} \tag{6.24}$$

(b) *$e^f V \in \mathcal{G}_{\mathcal{N}}$.*

Proof. We write $W = e^f V$. Then the general redshift formula (6.17) implies

$$\frac{d}{ds}\left(e^{f(\lambda(s))} \omega\big(\xi(s)\big)\right) = \Omega_{\xi(s)}\big(\dot{\xi}(s), \bar{W}_{\xi(s)}\big) \tag{6.25}$$

as can be easily checked with the help of the coordinate expression (5.24) for the canonical lift of a vector field. The right-hand side of (6.25) vanishes for all lifted rays if and only if $\bar{W}$ is tangent to $\mathcal{N}$ at all points of $\mathcal{N}$, i.e., if and only if $W \in \mathcal{G}_{\mathcal{N}}$. (This argument is analogous to the proof of Proposition 6.2.1). $\square$

If the frequency has no zeros, (6.24) implies

$$\ln \frac{\omega\big(\xi(s_2)\big)}{\omega\big(\xi(s_1)\big)} = f\big(\lambda(s_1)\big) - f\big(\lambda(s_2)\big) \tag{6.26}$$

for any two parameter values s_1 and s_2. It is this expression to which the name "redshift potential" refers. By Proposition 6.2.2, a ray-optical structure admits a redshift potential, for an appropriately chosen observer field, if and only if there is a time-like vector field in the symmetry algebra $\mathcal{G}_{\mathcal{N}}$. A ray-optical structure with this property is called "stationary", a notion we are going to discuss in full detail in Sect. 6.5 below. For the vacuum ray-optical structure the notion of a "redshift potential" (or "redshift function") was investigated in papers by Dautcourt [30] and by Hasse and Perlick [58]. Please recall that for the vacuum ray-optical structure $\mathcal{N} = \mathcal{N}^g$ the condition $W \in \mathcal{G}_{\mathcal{N}}$ means that W is a conformal Killing vector field of the metric g.

6.3 Isotropic ray-optical structures

With the Lorentzian metric g given on $\mathcal{M}$ we can define, for each point $q \in \mathcal{M}$, the set of *Lorentz transformations* on the cotangent space $T_q^*\mathcal{M}$ by

$$\mathrm{Lor}(q) = \Big\{ \Lambda \colon T_q^*\mathcal{M} \longrightarrow T_q^*\mathcal{M} \Big| \qquad (6.27)$$

Λ is a linear automorphism with $g_q^\#(\Lambda(\cdot), \Lambda(\cdot)) = g_q^\#(\cdot, \cdot) \Big\}$.

The Lorentz transformations $\mathrm{Lor}(q)$ foliate the punctured cotangent space $\overset{\circ}{T}_q^*\mathcal{M}$ into orbits. Here, a subset Q of $\overset{\circ}{T}_q^*\mathcal{M}$ is called an *orbit* iff it is of the form $Q = \big\{ \Lambda(u) \, \big| \, \Lambda \in \mathrm{Lor}(q) \big\}$ for some $u \in \overset{\circ}{T}_q^*\mathcal{M}$. The geometry of the orbits is sketched in Figure 6.1.

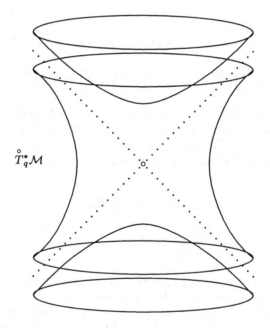

$\overset{\circ}{T}_q^*\mathcal{M}$

Fig. 6.1. The orbits of $\mathrm{Lor}(q)$ are the g-light-like cone, a family of g-space-like two-shell hyperboloids and a family of g-time-like one-shell hyperboloids.

Please recall that we have defined the structure group of a ray-optical structure in Definition 5.3.3. The next definition characterizes the situation that, at a point $q \in \mathcal{M}$, the set of Lorentz transformations (6.27) is completely contained in the structure group of $\mathcal{N}$.

Definition 6.3.1. *A ray-optical structure $\mathcal{N}$ on $(\mathcal{M}, g)$ is called* Lorentz invariant *at a point $q \in \mathcal{M}$ iff $\Lambda(u) \in \mathcal{N}_q$ for all $u \in \mathcal{N}_q$ and $\Lambda \in \mathrm{Lor}(q)$. $\mathcal{N}$ is called* Lorentz invariant *iff it is Lorentz invariant at all points $q \in \mathcal{M}$.*

If a ray-optical structure $\mathcal{N}$ is Lorentz invariant at a point q, $\mathcal{N}_q$ must be an orbit of $\mathrm{Lor}(q)$ or a union of several orbits which are separated by open neighborhoods in $\overset{\circ}{T}{}^*_q\mathcal{M}$. If $\mathcal{N}$ is causal, only space-like and light-like orbits come into question, i.e., the one-shell hyperboloids of Figure 6.1 are excluded.

The following proposition is rather trivial.

Proposition 6.3.1. *Let $\mathcal{N}$ be a ray-optical structure on $(\mathcal{M}, g)$ which is dilation invariant and Lorentz invariant at all points $q \in \mathcal{M}$. Then $\mathcal{N}$ is the vacuum ray-optical structure, $\mathcal{N} = \mathcal{N}^g$.*

Proof. By assumption, $\mathcal{N}_q$ is an orbit of $\mathrm{Lor}(q)$ or a union of several orbits, and $\mathcal{N}_q$ is dilation invariant. Clearly, the only codimension-one submanifold $\mathcal{N}_q \subset \overset{\circ}{T}{}^*_q\mathcal{M}$ with these properties is the double cone $\mathcal{N}_q = \mathcal{N}^g_q$. $\qquad\square$

This proposition implies that a non-dispersive medium necessarily has to break Lorentz invariance.

The class of Lorentz invariant ray-optical structures is rather small. We get much larger classes if we require invariance not under the full Lorentz group but only under certain subgroups. If we fix a vector $U_q \in T_q\mathcal{M}$ with $g_q(U_q, U_q) = -1$, we can consider the subgroup of *spatial rotations*

$$\mathrm{Rot}(U_q) = \big\{\, \Lambda \in \mathrm{Lor}(q) \mid \Lambda\big(g_q(U_q, \,\cdot\,)\big) = g_q(U_q, \,\cdot\,)\,\big\} \qquad (6.28)$$

with respect to U_q. This gives rise to the following definition.

Definition 6.3.2. *A ray-optical structure $\mathcal{N}$ on $(\mathcal{M}, g)$ is called* isotropic *at a point $q \in \mathcal{M}$ with respect to a normalized time-like vector $U_q \in T_q\mathcal{M}$ iff $\Lambda(u) \in \mathcal{N}_q$ for all $u \in \mathcal{N}_q$ and $\Lambda \in \mathrm{Rot}(U_q)$. $\mathcal{N}$ is called* isotropic *with respect to a global observer field U with $g(U, U) = -1$ iff $\mathcal{N}$ is isotropic at all points $q \in \mathcal{M}$ with respect to the vector $U_q = U(q)$ ($= U$ evaluated at q).*

Instead of "isotropic" one might use more precise attributes such as "spatially isotropic" or "invariant under spatial rotations". For the sake of brevity, however, we stick with the terminology of Definition 6.3.2.

If a ray-optical structure is Lorentz invariant at q, then it is in particular isotropic at q with respect to all normalized time-like vectors $U_q \in T_q\mathcal{M}$. In addition, we already know some examples of isotropic ray-optical structures that need not be Lorentz invariant. In Example 2.5.1 of Part I we have derived the Hamiltonian

$$H(x, p) = \frac{1}{2}\left(\frac{g^{ab}(x) + U^a(x)\,U^b(x)}{n(x)^2} - U^a(x)\,U^b(x)\right) p_a\, p_b \qquad (6.29)$$

for light propagation in a linear and isotropic electromagnetic medium, cf. eq. (2.87). Clearly, this Hamiltonian generates a ray-optical structure which is isotropic with respect to the rest system U of the medium. We are now going to prove that this example is much more general than it seems to be. As a matter of fact, every isotropic ray-optical structure is locally generated by a Hamiltonian of the form (6.29) provided that it is dilation invariant. If it is not dilation invariant, the only modification is in the fact that the index of refraction must be allowed to depend on the frequency, i.e., we have to write $n\big(x, -U^a(x)\,p_a\big)$ instead of $n(x)$.

It needs a little bit of preparation to prove this fact. Let us assume that the ray-optical structure $\mathcal{N}$ on $(\mathcal{M}, g)$ is isotropic with respect to an observer field U. For notational convenience we introduce a natural chart (x, p) around a point $u \in \mathcal{N}$. We have to assume that neither the frequency $\omega(x, p)$, given by (6.8) with $V^a = U^a$, nor the spatial wave covector $k_a(x, p)$, given by (6.9) with $V^a = U^a$, has a zero at u. Then there is a neighborhood $\mathcal{W} \subseteq T^*\mathcal{M}$ of u on which frequency and spatial wave covector are different from zero. If this neighborhood $\mathcal{W}$ has been chosen appropriately, our isotropy assumption guarantees that the norm $\sqrt{g^{ab}(x)\,k_a(x, p)\,k_b(x, p)}$ of the spatial wave covector must be a function of x and $\omega(x, p)$ on $\mathcal{N} \cap \mathcal{W}$, as is nicely illustrated by the orbit structure of Figure 6.1. Hence there is a strictly positive real valued function n, defined on some subset of $\mathcal{M} \times \mathbb{R}$, such that

$$n\big(x, \omega(x, p)\big)|\omega(x, p)| = \sqrt{g^{ab}(x)\,k_a(x, p)\,k_b(x, p)} \tag{6.30}$$

on $\mathcal{N} \cap \mathcal{W}$. This construction locally assigns a frequency-dependent index of refraction n to any isotropic ray-optical structure. Comparison with (6.10) shows that n is reciprocal to the norm of the phase velocity. Since we use units making the vacuum velocity of light equal to 1, this can be rephrased as saying that $1/n$ gives the phase velocity in units of the vacuum velocity of light. It is to be emphasized that, as long as there are no additional assumptions on our isotropic ray-optical structure $\mathcal{N}$, the index of refraction is a local concept. Here, "local" refers in particular to the necessity of restricting the fibers of $T^*\mathcal{M}$. Globally, n might be a "multi-valued function" corresponding to various "branches" of $\mathcal{N}$.

On an appropriate neighborhood $\mathcal{W} \subseteq T^*\mathcal{M}$, at least, the index of refraction is well-defined and can be used to introduce a local Hamiltonian

$$H(x, p) = \frac{1}{2}\left(\frac{g^{ab}(x) + U^a(x)\,U^b(x)}{n\big(x, -U^c(x)p_c\big)^2} - U^a(x)\,U^b(x)\right) p_a\,p_b\,. \tag{6.31}$$

It is an immediate consequence of (6.30) that H vanishes on $\mathcal{N} \cap \mathcal{W}$. Moreover, our assumption that the spatial wave covector has no zeros implies that $(\partial H/\partial p_a)$ has no zeros on $\mathcal{N} \cap \mathcal{W}$. Hence, (6.31) gives, indeed, a local Hamiltonian for our ray-optical structure $\mathcal{N}$. If n is frequency-independent, i.e., constant with respect to its second argument, the rays determined by the Hamiltonian (6.31) are the light-like geodesics of a Lorentzian metric. In the frequency-dependent case they are associated with a sort of "generalized Lorentzian metric" investigated in detail by Miron and Kawaguchi [97].

To sum up, we have proven that any isotropic ray-optical structure is locally generated by a Hamiltonian of the form (6.31) near any point of $\mathcal{N}$ where frequency and spatial wave covector are non-vanishing. Here, frequency and spatial wave covector are meant with respect to the normalized observer field U distinguished by the isotropy assumption. Hence, a ray-optical structure on $(\mathcal{M}, g)$ which is isotropic with respect to some given normalized observer field U is unambiguously characterized by an index of refraction $n \colon \mathcal{M} \times \mathbb{R} \longrightarrow \mathbb{R}^+$, on a neighborhood $\mathcal{W} \subseteq \overset{\circ}{T}{}^* \mathcal{M}$ near any point where the ray-optical structure is well-behaved. It is easy to check that such a ray-optical structure is

(a) causal iff

$$\left(n(x, \omega) + \frac{\omega}{n(x, \omega)^2} \frac{\partial n}{\partial \omega}(x, \omega) \right)^2 \geq 1 \, ; \qquad (6.32)$$

(b) dilation-invariant iff

$$\frac{\partial n}{\partial \omega}(x, \omega) = 0 \, ; \qquad (6.33)$$

(c) Lorentz invariant iff n is of the form

$$n(x, \omega)^2 = 1 - \frac{h(x)}{\omega^2} \, . \qquad (6.34)$$

From (b) and (c) we can read an alternative proof of Proposition 6.3.1. (c) is just a different way of saying that a Lorentz invariant ray-optical structure is locally generated by a Hamiltonian of the form

$$H(x, p) = \tfrac{1}{2} \left(g^{ab}(x) \, p_a \, p_b + h(x) \right) . \qquad (6.35)$$

Please note that the Hamiltonian (3.46) for light propagation in a non-magnetized plasma is of this form. In this case the function $h(x)$ is given by the plasma frequency (3.51), $h(x) = \omega_p(x)^2 = \frac{e^2}{m} \overset{\circ}{n}(x)$.

6.4 Light bundles in isotropic media

In this section we investigate the dynamics of infinitesimal bundles of light rays in an isotropic non-dispersive medium, thereby generalizing several standard results of ordinary optics. For our purposes it will be necessary to assume that the medium is irrotational and that it is globally characterized by a single-valued index of refraction which has no zeros and no singularities. According to the results of the preceding section, this implies that the rays are exactly the light-like geodesics of a Lorentzian metric g_o, called the "optical metric" henceforth. The contravariant components g_o^{ab} of the optical metric are determined by writing (6.29) in the form $H(x,p) = \frac{1}{2}g_o^{ab}(x)p_ap_b$, i.e., g_o is related to the spacetime metric g by

$$g_o(X,Y) = n^2 g(X,Y) + (n^2 - 1) g(U,X) g(U,Y) \qquad (6.36)$$

for all vector fields X and Y on $\mathcal{M}$. Here, U is a given observer field with $g(U,U) = -1$ on $\mathcal{M}$ which is supposed to be hypersurface-orthogonal and $n \colon \mathcal{M} \longrightarrow \mathbb{R}^+$ is a given C^∞ function. In the following we refer to U as to the "rest system" of the medium and to n as to the "index of refraction". These assumptions include, of course, vacuum light propagation as the special case $n = 1$.

Now let us fix a ray $\lambda \colon I \longrightarrow \mathcal{M}$, i.e., a light-like g_o-geodesic, where I denotes a real interval. For the sake of simplicity we choose an affine parametrization such that the tangent field $K = \dot{\lambda} \colon I \longrightarrow T\mathcal{M}$ of λ satisfies the equations

$$g_o(K,K) = 0 \quad \text{and} \quad (\nabla_o)_K K = 0 \qquad (6.37)$$

where ∇_o denotes the Levi-Civita connection of the metric g_o. Infinitesimally neighboring rays are mathematically modeled by Jacobi fields along λ. (Please recall that Jacobi fields are defined, for arbitrary ray-optical structures, by Definition 5.6.2.) In the case at hand, a Jacobi field along λ is a C^∞ map $J \colon I \longrightarrow T\mathcal{M}$ with $\tau_{\mathcal{M}} \circ J = \lambda$ that satisfies the following two conditions.

$$(\nabla_o)_K (\nabla_o)_K J - R_o(K,J,K) \quad \text{is a multiple of } K, \qquad (6.38)$$
$$g_o(K,J) = \text{const.}, \qquad (6.39)$$

where R_o denotes the curvature tensor of the connection ∇_o. (6.38) assures that "the arrow-head of J is tracing a neighboring g_o-geodesic" and (6.39) assures that this neighboring geodesic is again g_o-light-like.

For analyzing the motion of such Jacobi fields it is convenient to refer to an appropriate basis of vector fields along λ. We introduce the following definition which makes sense for arbitrary curves λ in our Lorentzian manifold $(\mathcal{M}, g)$.

Definition 6.4.1. *Let* $\lambda \colon I \longrightarrow \mathcal{M}$ *be a* C^∞ *curve and denote its tangent field by* K. *Then* $(E_1, \ldots, E_{n-2})$ *is called a* Sachs bein *along* λ *iff for* $A, B = 1, \ldots, n-2$

(a) E_A *is a* C^∞ *vector field along* λ, *i.e.,* $E_A \colon I \longrightarrow T\mathcal{M}$ *with* $\tau_{\mathcal{M}} \circ E_A = \lambda$;
(b) $g(E_A, E_B) = \delta_{AB}$ *and* $g(K, E_A) = 0$;
(c) $g(E_A, \nabla_K E_B) = 0$.

A Sachs bein is called adapted *to an observer field* V *iff the vector* $E_A(s)$ *is* g-*orthogonal to the vector* $V_{\lambda(s)}$ *for all* $A = 1, \ldots, n-2$ *and all* $s \in I$.

Whenever refering to a Sachs bein we use the summation convention for capital indices $A, B, C, \ldots$ running from 1 to n − 2.

It is easy to check that, for an arbitrary curve λ and an arbitrary observer field V on $\mathcal{M}$, there is a Sachs bein along λ which is adapted to V and that it is unique up to transformations of the form

$$E_A(s) \longmapsto \mathcal{O}^B{}_A \, E_B(s) \tag{6.40}$$

where $(\mathcal{O}^B{}_A)$ is a constant orthogonal matrix, i.e., $\mathcal{O}^D{}_A \mathcal{O}^C{}_B \, \delta_{DC} = \delta_{AB}$. In the literature, the name "Sachs bein" is usually restricted to the case that λ is a light-like geodesic with respect to g. In this situation, which was considered in the original paper by Sachs [125], condition (b) of Definition 6.4.1 assures that $K, E_1, \ldots, E_{n-2}$ span the g-orthocomplement of K and condition (c) requires that, apart from the freedom of adding multiples of K, each E_A is ∇-parallel along λ.

Here, however, we are considering the case that λ is a light-like geodesic of the optical metric rather than of the spacetime metric. In the following we fix a Sachs bein along λ that is adapted to the distinguished observer field U, i.e., that satisfies in addition to (6.42) and (6.43) the condition

$$g(U, E_A) = g_o(U, E_A) = 0 \,. \tag{6.41}$$

With the help of (6.36) and (6.41), conditions (b) and (c) of Definition 6.4.1 can be rewritten in terms of the optical metric in the following way.

$$g_o(E_A, E_B) = n^2 \, \delta_{AB} \quad \text{and} \quad g_o(K, E_A) = 0 \,, \tag{6.42}$$

$$(\nabla_o)_K \left(\tfrac{1}{n} E_A \right) \text{ is a multiple of } K \,. \tag{6.43}$$

Whereas (6.42) is obvious, it needs a bit of work to verify (6.43). One has to calculate the difference tensor $\nabla - \nabla_o$ from (6.36) and to use the assumption of U being hypersurface-orthogonal (= irrotational).

In this situation every vector field J along λ can be represented as a linear combination

$$J(s) = J^A(s) \, E_A(s) + v(s) \, U_{\lambda(s)} + w(s) \, K(s) \,, \tag{6.44}$$

with scalar coefficients $J^A(s)$, $v(s)$ and $w(s)$. Jacobi fields are determined by inserting (6.44) into (6.38) and (6.39). This gives conditions on the coefficients J^A and v but not on w because a Jacobi field remains a Jacobi field if an arbitrary multiple of the tangent field is added. If $v = 0$, J is g-orthogonal to the observer field U up to the irrelevant term proportional to the tangent field, i.e., the connecting vector from λ to the neighboring ray is purely spatial with respect to the distinguished observer field U. (In the vacuum case $n = 1$ this is true for all observer fields simultaneously.) Hence, the dynamics of infinitesimally thin bundles of light rays in our isotropic medium is given by inserting (6.44) with $v = 0$ into (6.38) and (6.39). As (6.39) is automatically satisfied with const. $= 0$, we only have to care about (6.38). This gives the system of second order linear differential equations

$$n(s)\frac{d^2}{ds^2}\left(n(s)J^A(s)\right) = S^A{}_B(s)\,J^B(s) \qquad (6.45)$$

for the coefficients J^A, where

$$S^A{}_B = \delta^{AC}\,g_o\big(E_C, R_o(K, E_B, K)\big)\,. \qquad (6.46)$$

Here and in the following we write $n(s)$ for $n\big(\lambda(s)\big)$. Owing to the symmetries of the curvature tensor, $S^A{}_B$ satisfies the identity

$$S^A{}_B\,\delta^{BC} = S^C{}_B\,\delta^{BA}\,. \qquad (6.47)$$

If the Sachs bein is changed according to (6.40), $S^A{}_B$ undergoes the transformation

$$S^A{}_B(s) \longmapsto \delta^{AC}\,\mathcal{O}^D{}_C\,\mathcal{O}^F{}_B\,S^G{}_F(s)\,\delta_{DG}\,. \qquad (6.48)$$

Now let us assume that we have a matrix valued function $s \longmapsto L(s) = \big(L^A{}_B(s)\big)$ that satisfies the matrix analogue of the differential equation (6.45), i.e.,

$$n(s)\frac{d^2}{ds^2}\left(n(s)L^A{}_C(s)\right) = S^A{}_B(s)\,L^B{}_C(s)\,. \qquad (6.49)$$

Then any $(c^1, \ldots, c^{n-2}) \in \mathbb{R}^{n-2}$ determines a solution

$$J^A(s) = L^A{}_B(s)\,c^B \qquad (6.50)$$

of (6.45) and, upon inserting into (6.44) with $v = 0$ and w arbitrary, a solution J of (6.38) and (6.39) with const. $= 0$. In other words, such a matrix-valued function L determines an $(n-2)$-parameter family of infinitesimally neighboring rays around the central ray λ. If $\det\big(L(s)\big) \neq 0$, those neighboring rays fill the space (not the spacetime!) around λ completely. For that reason, we call a solution $L = (L^A{}_B)$ of (6.49) that satisfies $\det\big(L(s)\big) \neq 0$, with the possible exception of some isolated parameter values s, an *infinitesimal bundle* of rays around λ. More precisely, L should be called the *representation* of such a bundle with respect to the Sachs bein chosen. If we change the Sachs bein, we have, of course, to change L according to

$$L^A{}_B(s) \longmapsto \delta^{AC}\,\mathcal{O}^D{}_C\,\mathcal{O}^F{}_B\,L^G{}_F(s)\,\delta_{DG}\,. \qquad (6.51)$$

Since n has no zeros, (6.49) can be solved for $\frac{d^2}{ds^2}L^A{}_C(s)$. Thus, arbitrary initial values for L and for its first derivative determine a unique solution.

Invertibility of $L(s)$ for almost all parameter values s implies that the equation

$$D^A{}_B(s)\,L^B{}_C(s) = \dot{L}^A{}_C(s) \qquad (6.52)$$

defines a matrix $D(s) = (D^A{}_B(s))$ for almost all parameter values s. (At those isolated values s where $\det(L(s)) = 0$, some components of D become infinite.) Hence, the J^A from (6.50) satisfy

$$\dot{J}^A(s) = D^A{}_B(s)\, J^B(s) \tag{6.53}$$

for almost all s. (6.53) demonstrates that the matrix $D(s)$ measures the motion of the infinitesimal bundle around λ with respect to the Sachs bein chosen. If we decompose $D(s)$ into a symmetrical and an antisymmetrical part according to

$$D^A{}_B(s) = \theta^A{}_B(s) + \omega^A{}_B(s)\,, \tag{6.54}$$

$$\theta^A{}_B(s)\, \delta^{BC} - \theta^C{}_B(s)\, \delta^{BA} = 0\,, \tag{6.55}$$

$$\omega^A{}_B(s)\, \delta^{BC} + \omega^C{}_B(s)\, \delta^{BA} = 0\,, \tag{6.56}$$

the symmetrical part $\theta^A{}_B(s)$ gives the *deformation* and the antisymmetrical part $\omega^A{}_B(s)$ gives the *rotation* of the infinitesimal bundle with respect to the Sachs bein. The symmetrical part can be further decomposed according to

$$\theta^A{}_B(s) = \sigma^A{}_B(s) + \tfrac{\theta(s)}{(n-2)}\, \delta^A_B\,, \tag{6.57}$$

where $\theta(s) = \theta^A{}_A(s)$ gives the *expansion* and $\sigma^A{}_B(s)$, which is defined through (6.57), gives the *shear* of the infinitesimal bundle with respect to the Sachs bein.

It is easy to derive propagation equations for these quantities. If we calculate the derivative of (6.52), we find that the second order differential equation (6.49) for $L^A{}_B$ implies the first order differential equation

$$\dot{D}^A{}_B(s) + \tfrac{2\dot{n}(s)}{n(s)}\, D^A{}_B(s) =$$
$$\tfrac{1}{n(s)^2} S^A{}_B(s) - D^A{}_C(s) D^C{}_B(s) - \frac{\ddot{n}(s)}{n(s)} \delta^A_B \tag{6.58}$$

for $D^A{}_B$. Symmetrization respectively antisymmetrization results in

$$\dot{\theta}^A{}_B(s) = \tfrac{1}{n(s)^2} S^A{}_B(s) - \theta^A{}_C(s)\, \theta^C{}_B(s) -$$
$$\omega^A{}_C(s)\, \omega^C{}_B - \tfrac{2\dot{n}(s)}{n(s)}\, \theta^A{}_B(s) - \tfrac{\ddot{n}(s)}{n(s)} \delta^A_B\,, \tag{6.59}$$

$$\dot{\omega}^A{}_B(s) = \omega^A{}_C(s)\, \theta^C{}_B(s) - \theta^A{}_C(s)\, \omega^C{}_B(s) - \tfrac{2\dot{n}(s)}{n(s)}\, \omega^A{}_B(s)\,. \tag{6.60}$$

For the vacuum case $n = 1$, the propagation equations (6.59) and (6.60) are well-known and can be found in many textbooks on general relativity, see, e.g., Wald [146], p. 222. In particular, the trace of (6.59) gives the well-known *focusing equation* for vacuum light rays in the case $n = 1$. Please note that in the generalization considered here $S^A{}_B(s)$ involves the curvature tensor of the optical metric according to (6.46).

Now we use the propagation equations (6.59) and (6.60) to prove three theorems on light propagation in a general-relativistic medium of the kind under consideration and thus, in particular, in vacuum. All three have famous counter-parts in ordinary optics.

Theorem 6.4.1. *In an isotropic, dispersion-free and non-rotating medium the following holds true. If for an infinitesimal bundle of rays the rotation vanishes for one parameter value s, then it vanishes for all s.*

Proof. This is an immediate consequence of the fact that the rotation satisfies the homogeneous differential equation (6.60). □

This theorem can be viewed as a general relativistic analogue of the *Malus theorem* of ordinary optics. In its most elementary version, found by Malus in 1808, this classical theorem can be formulated in the following way (cf., e.g., Born and Wolf [16]). A family of straight lines that starts surface-orthogonal remains surface-orthogonal (a) after reflexion at an arbitrarily curved surface according to the usual reflexion law and (b) after refraction at an arbitrarily curved surface according to Snell's law. In other words, a surface-orthogonal bundle of light rays remains surface-orthogonal after passing through any system of mirrors and lenses. Theorem 6.4.1 gives a similar statement for light rays in an isotropic, non-dispersive, and non-rotating medium on a general-relativistic spacetime. The analogy comes from the fact that a two-parameter family of straight lines in ordinary Euclidean 3-space is surface orthogonal iff, around any member of this family, the infinitesimally neighboring members are irrotational. In this case rotation is to be measured with respect to ordinary Euclidean parallel transport.

The other two theorems refer to infinitesimal bundles of rays which are homocentric, i.e., to the case that $L(s)$ is the zero matrix for one particular parameter value s.

Theorem 6.4.2. *In an isotropic, non-dispersive and non-rotating medium the following holds true. If an infinitesimal bundle of rays is homocentric, then its rotation vanishes.*

Proof. Let $L = (L^A{}_B)$ be a solution of the differential equation (6.49) which can be rewritten in matrix notation as

$$n(s)\frac{d^2}{ds^2}\big(n(s)L(s)\big) = S(s)\,L(s) \tag{6.61}$$

with $S = (S^A{}_B)$. Owing to (6.47), transposition of (6.61) results in

$$n(s)\frac{d^2}{ds^2}\big(n(s)L^T(s)\big) = L^T(s)\,S(s)\,, \tag{6.62}$$

where $(\cdot)^T$ denotes the transpose of a matrix. (6.61) and (6.62) together imply that the matrix

$$C(s) = n(s)^2 \left(\dot{L}^T(s) \, L(s) - L^T(s) \, \dot{L}(s) \right) \tag{6.63}$$

has vanishing derivative, $\dot{C}(s) = 0$. Hence, $C(s)$ is a constant matrix. If L is a homocentric infinitesimal bundle, $L(s_o)$ is the zero matrix for some specific parameter value s_o. So $C(s) = C(s_o)$ must be the zero matrix for all s. With $C = 0$, we multiply the matrix equation (6.63) from the left by $(L^{-1})^T(s)$ and from the right by $L^{-1}(s)$. This can be done for almost all parameter values s. By (6.52), the resulting equation reads $D^T(s) - D(s) = 0$, i.e., the antisymmetrical part of $D(s)$ vanishes. □

This theorem is analogous to the elementary fact that, in ordinary optics, homocentric bundles are surface-orthogonal.

We now turn to the last of our three theorems which is of particular relevance for cosmology.

Theorem 6.4.3. (Reciprocity theorem) *In an isotropic, dispersion-free and non-rotating medium, the following holds true. If L_1 and L_2 are two infinitesimal bundles around the same central ray λ and if both L_1 and L_2 are homocentric, with $L_1(s_1) = 0$ and $L_2(s_2) = 0$, then*

$$\frac{\left| \det\!\big(n(s_2) \, L_1(s_2) \big) \right|}{\left| \det\!\big(n(s_1) \dot{L}_1(s_1) \big) \right|} = \frac{\left| \det\!\big(n(s_1) L_2(s_1) \big) \right|}{\left| \det\!\big(n(s_2) \dot{L}_2(s_2) \big) \right|} . \tag{6.64}$$

Proof. We use the same matrix notation as in the proof of Theorem 6.4.2. By assumption, the differential equation (6.61) and its transposed version (6.62) are satisfied by $L = L_1$ and $L = L_2$. This implies that the matrix

$$C_{12}(s) = n(s)^2 \left(\dot{L}_1^T(s) \, L_2(s) - L_1^T(s) \, \dot{L}_2(s) \right) \tag{6.65}$$

has vanishing derivative, $\dot{C}_{12}(s) = 0$. Hence, $C_{12}(s)$ is a constant matrix. In particular, the equation

$$C_{12}(s_1) = C_{12}(s_2) \tag{6.66}$$

has to hold. Owing to our hypothesis $L_1(s_1) = 0$ and $L_2(s_2) = 0$, (6.66) simplifies to

$$n(s_1)^2 \, \dot{L}_1^T(s_1) \, L_2(s_1) = -n(s_2)^2 \, L_1^T(s_2) \, \dot{L}_2(s_2) . \tag{6.67}$$

(6.64) is an obvious consequence of (6.67). □

Please note that the denominators in (6.64) are different from zero since L_1 and L_2 are infinitesimal bundles and n is strictly positive.

To give a physical interpretation to this theorem we now assume that our underlying Lorentzian manifold is 4-dimensional and that, with respect to the time orientation defined by the distinguished observer field U, $\lambda(s_2)$ is in the future of $\lambda(s_1)$. We restrict our consideration to the section of λ between these two points. We consider the totality of all Jacobi fields, defined via eq. (6.50) with $L = L_1$, with $\delta_{AB} c^A c^B \leq 1$. This can be interpreted as a pencil of light rays issuing from the point $\lambda(s_1)$. Similarly, the analogous construction carried through with $L = L_2$ gives a pencil of light rays received at the point $\lambda(s_2)$. Now the numerators in (6.64), up to a factor π and up to the square of the index of refraction, give the cross-sectional areas of these two pencils at the points $\lambda(s_1)$ and $\lambda(s_2)$, respectively. It is important to realize that these quantities have an invariant geometrical meaning, independent of the Sachs bein and of the g_o-affine parametrization chosen for λ. On the other hand, the denominators in (6.64), again up to the square of the index of refraction, are measuring the opening angle of the respective pencil at its focal point. These quantities are, again, independent of the choice of the Sachs bein; however, they do depend on the affine parametrization chosen for λ. If we switch to another affine parametrization by a transformation $s \longmapsto as + b$ with real constants $a \neq 0$ and b, on both sides of (6.64) the denominator is getting a factor $|a|^{-1}$. As for a ray (i.e., for a g_o-light-like geodesic) the choice of a particular affine parametrization is a matter of arbitrariness, this argument shows that the denominators of (6.64) are "unphysical" in the sense that they cannot be measured. Therefore we introduce the quantities

$$d_{\text{lum}} = \sqrt{\frac{|\det(L_1(s_2))|}{|\det(\dot{L}_1(s_1))|}} \; \left|g_{\lambda(s_1)}(K(s_1), U(s_1))\right|, \qquad (6.68)$$

$$d_{\text{ang}} = \sqrt{\frac{|\det(L_2(s_1))|}{|\det(\dot{L}_2(s_2))|}} \; \left|g_{\lambda(s_2)}(K(s_2), U(s_2))\right|, \qquad (6.69)$$

where $U(s)$ denotes the distinguished observer field at the point $\lambda(s)$. d_{lum} and d_{ang} are invariant with respect to changing the affine parametrization of λ since the tangent field K in the numerator is stretched with the same factor as the derivative operator (overdot) in the denominator. Here it is essential that we restrict to the case $\dim(\mathcal{M}) = 4$ such that L_1 and L_2 are (2×2)-matrices. d_{lum}^2 relates the cross-sectional area of the L_1-pencil at $\lambda(s_2)$ to its opening angle at $\lambda(s_1)$ where the latter is now measured as a solid angle in the local rest space of the observer $U(s_1)$, see Figure 6.2. In cosmology d_{lum} is known as the *corrected luminosity distance* from $\lambda(s_1)$ to $\lambda(s_2)$, whereas the analogously defined quantity d_{ang} is known as the *angular diameter distance* from $\lambda(s_1)$ to $\lambda(s_2)$. Now the reciprocity law (6.64) can be rewritten in the form

$$d_{\text{lum}} = d_{\text{ang}} \frac{n(s_1)^2 \left|g_{\lambda(s_1)}(K(s_1), U(s_1))\right|}{n(s_2)^2 \left|g_{\lambda(s_2)}(K(s_2), U(s_2))\right|}. \qquad (6.70)$$

Please note that, by (6.23), the factor

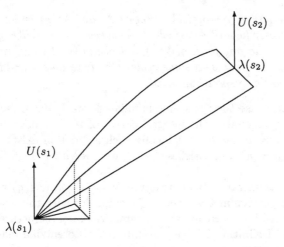

Fig. 6.2. This illustration shows a pencil with central ray λ, issuing from a light source at $\lambda(s_1)$ to an observer at $\lambda(s_2)$. The corrected luminosity distance relates the cross-sectional area of the pencil at $\lambda(s_2)$ to its opening angle at $\lambda(s_1)$, measured in the rest space of the observer $U(s_1)$. The angular diameter distance is defined in an analogous manner for a pencil around λ with focus at $\lambda(s_2)$.

$$\frac{g_{\lambda(s_1)}\left(K(s_1), U(s_1)\right)}{g_{\lambda(s_2)}\left(K(s_2), U(s_2)\right)} = \frac{(g_o)_{\lambda(s_1)}\left(K(s_1), U(s_1)\right)}{(g_o)_{\lambda(s_2)}\left(K(s_2), U(s_2)\right)} \qquad (6.71)$$

gives the redshift under which $U(s_1)$ is seen by $U(s_2)$.

In the vacuum case $n = 1$, (6.70) gives a remarkable relation between corrected luminosity distance, angular diameter distance and redshift. (The observers $U(s_1)$ and $U(s_2)$ are arbitrary in the vacuum case.) As to the literature on this subject we refer to Etherington [41] who discovered the law (6.70) for the case $n = 1$; to Ellis [40] whose proof of the reciprocity law for the case $n = 1$ served as a model for our proof of Theorem 6.4.3; and to Schneider, Ehlers and Falco [128], Sect. 3.5, who give a detailed discussion of the reciprocity theorem for vacuum light rays and of its relevance for cosmology. It should also be mentioned that the name "reciprocity theorem" goes back to Straubel [135] who introduced the analogue of Theorem 6.4.3 in ordinary optics. This classical reciprocity theorem is closely related to the socalled *sine condition* for stigmatic imaging. The latter can be traced back well into the 19th century to Clausius, Helmholtz and Abbé. It is discussed, e.g., in the standard textbook by Born and Wolf [16], p. 166.

6.5 Stationary ray-optical structures

Recalling Definition 5.3.2 of the symmetry algebra $\mathcal{G}_\mathcal{N}$, stationarity of a ray-optical structure $\mathcal{N}$ can be introduced in the following way.

Definition 6.5.1. *A ray-optical structure $\mathcal{N}$ on $(\mathcal{M}, g)$ is called* stationary *iff there is a vector field $W \in \mathcal{G}_\mathcal{N}$ which is everywhere time-like, $g(W, W) < 0$. If, in addition, the one-form $g(W, \cdot)$ is globally (or locally, respectively) of the form $g(W, \cdot) = h\, dt$ with some scalar C^∞ functions h and t, $\mathcal{N}$ is called globally (or locally, respectively)* static.

In the static case the spacetime is foliated into hypersurfaces $t = $ const. which are g-orthogonal to the integral curves of W. It follows from the well-known Frobenius Theorem that locally such a foliation exists if and only if the one-form $\kappa = g(W, \cdot)$ satisfies the equation $\kappa \wedge dk = 0$, cf. Sachs and Wu [126], p. 53 .

For the vacuum ray-optical structure $\mathcal{N} = \mathcal{N}^g$ on $(\mathcal{M}, g)$ we know from Sect. 5.3 that the symmetry algebra $\mathcal{G}_\mathcal{N}$ coincides with the set of conformal Killing vector fields of g. Hence, the vacuum ray-optical structure is stationary in the sense of Definition 6.5.1 if and only if the Lorentzian manifold $(\mathcal{M}, g)$ is conformally stationary.

In terms of local Hamiltonians, stationary ray-optical structures can be characterized in the following way.

Proposition 6.5.1. *Let $\mathcal{N}$ be a stationary ray-optical structure on $(\mathcal{M}, g)$ and fix a point $u \in \mathcal{N}$. Then there is a local Hamiltonian H for $\mathcal{N}$, defined on a neighborhood of u, such that $dH(\bar{W}) = 0$. Here $\bar{W}$ denotes the canonical lift (5.24) of the time-like vector field $W \in \mathcal{G}_\mathcal{N}$.*

Proof. Since the vector field W is time-like, $\bar{W}(u) \neq 0$. Hence, we can choose a codimension-one C^∞ submanifold $\mathcal{P}$ of $T^*\mathcal{M}$ through u that is transverse to the flow of $\bar{W}$. As the assumption $W \in \mathcal{G}_\mathcal{N}$ implies that $\bar{W}(u)$ is tangent to $\mathcal{N}$, it is then automatically guaranteed that $\mathcal{P}$ is transverse to $\mathcal{N}$ at u. This transversality property implies that $\mathcal{N} \cap \mathcal{P}$ is a codimension-one C^∞ submanifold of $\mathcal{P}$. If $\mathcal{P}$ is small enough, this guarantees the existence of a C^∞ function $h \colon \mathcal{P} \longrightarrow \mathbb{R}$ such that the differential dh has no zeros and $\mathcal{N} \cap \mathcal{P} = \{ w \in \mathcal{P} \mid h(w) = 0 \}$. Then the conditions $H|_\mathcal{P} = h$ and $dH(\bar{W}) = 0$ define a real valued function H on a neighborhood of u. By construction, H is a local Hamiltonian for $\mathcal{N}$. $\square$

In a natural chart, induced by a chart $(x^1, \ldots, x^n)$ on $\mathcal{M}$ with $W = \partial/\partial x^n$, the equation $dH(\bar{W}) = 0$ means that H is independent of the coordinate x^n.

For a stationary ray-optical structure, the time-like vector field $W \in \mathcal{G}_\mathcal{N}$ defines a normalized observer field $V = e^{-f} W$, where

$$ f = \tfrac{1}{2} \ln\left(- g(W, W) \right) . \tag{6.72} $$

By Proposition 6.2.2, f is a redshift potential for this observer field. This observation is closely related to the fact that, by Proposition 5.3.3, the momentum $\theta(\bar{W}) \colon T^*\mathcal{M} \longrightarrow \mathbb{R}$ is constant along each lifted ray. The existence of this constant of motion is crucial for the dimensional reduction of stationary ray-optical structures. We are now going to discuss this reduction formalism in detail.

The goal is to study, for a stationary ray-optical structure $\mathcal{N}$ on our n-dimensional spacetime $(\mathcal{M}, g)$, the dynamics of rays in terms of their projections onto an $(n - 1)$-dimensional space $\hat{\mathcal{M}}$. In the first step we have to construct the space $\hat{\mathcal{M}}$. The obvious idea is to introduce $\hat{\mathcal{M}}$ as a quotient space of $\mathcal{M}$, calling two points of $\mathcal{M}$ equivalent iff they are connected by an integral curve of the time-like vector field $W \in \mathcal{G}_{\mathcal{N}}$ and to hope that $\hat{\mathcal{M}}$ is a smooth Hausdorff manifold. This is, indeed, always true locally, i.e., if we restrict to a sufficiently small neighborhood of an arbitrary point in $\mathcal{M}$. Globally, however, the topological space $\hat{\mathcal{M}}$ may violate the Hausdorff axiom and need not admit a smooth manifold structure such that the natural projection $\pi : \mathcal{M} \longrightarrow \hat{\mathcal{M}}$ becomes a submersion. E.g., for a time-like vector field W on Minkowski space with one point removed the quotient space necessarily violates the Hausdorff property. Also, it is easy to verify that the quotient space cannot be a smooth manifold if an integral curve of W is almost closed, coming back into any neighborhood of some point infinitely often without being periodic.

It is, thus, necessary, to introduce additional assumptions to make sure that the quotient space $\hat{\mathcal{M}}$ is a smooth Hausdorff manifold. To put this rigorously we introduce the following terminology, cf. Figure 6.3.

Definition 6.5.2. *Let W be a time-like C^∞ vector field on $(\mathcal{M}, g)$. A C^∞ function $t \colon \mathcal{M} \to \mathbb{R}$ is called a* global timing function *for W iff*

(a) *$dt(W) = 1$ and*
(b) *for any t_1 and t_2 in $\mathbb{R}$ the flow of W maps the hypersurface $t = t_1$ diffeomorphically onto the hypersurface $t = t_2$.*

It would be misleading to call t a "time function", rather than a "timing function", since the hypersurfaces $t = $ const. need not be space-like with respect to the Lorentzian metric g.

If t is a global timing function for W, the above-mentioned quotient space $\hat{\mathcal{M}}$ can be identified with any of the hypersurfaces $t = $ const.; this identification makes $\hat{\mathcal{M}}$ into an $(n - 1)$-dimensional C^∞ manifold such that the natural projection $\pi \colon \mathcal{M} \to \hat{\mathcal{M}}$ becomes a submersion. Then the map $(\pi, t) \colon \mathcal{M} \to \hat{\mathcal{M}} \times \mathbb{R}$ is a global diffeomorphism.

The above-mentioned counter-examples demonstrate that, for an arbitrary time-like vector field W, a global timing function need not exist. It is interesting to note the following result. If W is a time-like vector field on a Lorentzian manifold that has no closed integral curves, then the Hausdorff property of the quotient space $\hat{\mathcal{M}}$ guarantees the existence of a global timing function for an appropriate reparametrization of W. The proof can be taken over from Harris [56], Theorem 2.

If $t \colon \mathcal{M} \to \mathbb{R}$ is a global timing function for W, a second function $t' \colon \mathcal{M} \to \mathbb{R}$ is, again, a global timing function for W if and only if t' is of the form

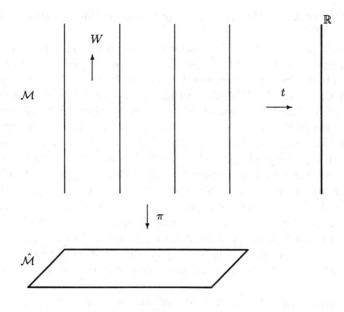

Fig. 6.3. A global timing function t for a time-like vector field W allows to write spacetime $\mathcal{M}$ as a product of space $\hat{\mathcal{M}}$ and time $\mathbb{R}$.

$$t' = t + \hat{h} \circ \pi \tag{6.73}$$

where $\hat{h} \colon \hat{\mathcal{M}} \to \mathbb{R}$ is any C^∞ function. In bundle theoretical language, two different global timing functions for W define two different global trivializations for the fiber bundle $\pi \colon \mathcal{M} \to \hat{\mathcal{M}}$. If t can be chosen in such a way that the hypersurfaces $t = \text{const.}$ are g-orthogonal to W, this additional condition fixes the timing function uniquely up to an additive constant. However, since it is our goal to study stationary ray-optical structures, and not only globally static ones, we have to deal with situations where such a choice is not possible.

Now let us assume that we have a global timing function $t \colon \mathcal{M} \to \mathbb{R}$ for W. Then the global diffeomorphism $(\pi, t) \colon \mathcal{M} \to \hat{\mathcal{M}} \times \mathbb{R}$ induces a splitting of the cotangent spaces $T_q^* \mathcal{M} \cong T_{\pi(q)}^* \hat{\mathcal{M}} \oplus T_{t(q)}^* \mathbb{R}$ for all points $q \in \mathcal{M}$. Projecting onto the first factor gives a *reduction map*

$$\text{red} \colon T^* \mathcal{M} \longrightarrow T^* \hat{\mathcal{M}} . \tag{6.74}$$

If we change the timing function according to (6.73), the reduction map undergoes the transformation

$$\text{red}(u) \longmapsto \text{red}'(u) = \text{red}(u) + (d\hat{h})_{\pi(q)} \tag{6.75}$$

for $u \in T_q^* \mathcal{M}$. In local coordinates, the reduction map is most easily expressed if we choose coordinates $(x^1, \ldots, x^n)$ on $\mathcal{M}$ with $t = x^n$ and $W = \partial/\partial x^n$. Then we can use $(x^1, \ldots, x^{n-1})$ as coordinates on $\hat{\mathcal{M}}$ and the reduction map takes the form

$$(x^1, \ldots, x^n, p_1, \ldots, p_n) \longmapsto (x^1, \ldots, x^{n-1}, p_1, \ldots, p_{n-1}). \tag{6.76}$$

Please note that in such a coordinate system the momentum coordinate p_n coincides with the function $\theta(\bar{W})$ which is a constant of motion according to Proposition 5.3.3.

We are now ready to formulate a reduction theorem for stationary ray-optical structures. Roughly speaking, this theorem says that a stationary ray-optical structure $\mathcal{N}$ on our n-dimensional spacetime $(\mathcal{M}, g)$ induces a one-parameter family $\hat{\mathcal{N}}_{\omega_o}$ of ray-optical structures on the $(n-1)$-dimensional quotient space $\hat{\mathcal{M}}$, where the family parameter $\omega_o \in \mathbb{R}$ is given by the value of the conserved momentum, $\omega_o = -\theta(\bar{W})$. For the construction of the reduced ray-optical structure $\hat{\mathcal{N}}_{\omega_o}$ it is necessary to choose a global timing function t for the time-like vector field $W \in \mathcal{G}_\mathcal{N}$, i.e., in situations where such a t does not exist the reduction does not work globally. Moreover, the theorem requires two transversality assumptions. Even locally for the reduction process to give a reduced ray-optical structure near a point $u \in \mathcal{N}$ it is necessary that (i) the covector u is not a multiple of the differential dt at the point $q = \tau_\mathcal{M}^*(u)$; and (ii) the fiber derivative $\mathbb{F}H(u)$ is not a multiple of W at the point $q = \tau_\mathcal{M}^*(u)$, where H is any local Hamiltonian for $\mathcal{N}$. Note that $\mathbb{F}H(u)$ is a multiple of W at those points where the ray velocity with respect to the normalized observer field $V = e^{-f}W$ has a zero, see (6.7). Hence, the second transversality assumption just excludes all points where the ray-optical structure has a pathological behavior.

The precise formulation of the reduction theorem reads as follows.

Theorem 6.5.1. (**Reduction theorem for stationary ray-optical structures**) *Let $\mathcal{N}$ be a stationary ray-optical structure on $(\mathcal{M}, g)$ and let $t: \mathcal{M} \to \mathbb{R}$ be a global timing function for the time-like vector field $W \in \mathcal{G}_\mathcal{N}$. As outlined above, this induces a global diffeomorphism $(\pi, t): \mathcal{M} \to \hat{\mathcal{M}} \times \mathbb{R}$ and a reduction map* red: $T^*\mathcal{M} \to T^*\hat{\mathcal{M}}$. *Now fix a value $\omega_o \in \mathbb{R}$ such that the set $\mathcal{Q}_{\omega_o} = \{ u \in \mathcal{N} \mid \theta(\bar{W}) = -\omega_o \}$ is non-empty. Assume that for all points $u \in \mathcal{Q}_{\omega_o}$ (i) the covector u is not a multiple of the differential dt at $q = \tau_\mathcal{M}^*(u)$; and (ii) the fiber derivative $\mathbb{F}H(u)$ is not a multiple of W at $q = \tau_\mathcal{M}^*(u)$, where H is any local Hamiltonian for $\mathcal{N}$. Then $\hat{\mathcal{N}}_{\omega_o} = \mathrm{red}(\mathcal{Q}_{\omega_o})$ is a ray-optical structure on $\hat{\mathcal{M}}$. A C^∞ curve $\hat{\xi}: I \to \hat{\mathcal{N}}_{\omega_o}$ is a lifted ray (or a lifted virtual ray, respectively) of $\hat{\mathcal{N}}_{\omega_o}$ if and only if it can be written in the form $\hat{\xi} = \mathrm{red} \circ \xi$ where $\xi: I \to \mathcal{Q}_{\omega_o} \subset \mathcal{N}$ is a lifted ray (or a lifted virtual ray, respectively) of $\mathcal{N}$.*

Proof. Recall that $\bar{W}$ is the Hamiltonian vector field of the function $\theta(\bar{W})$,

$$d\big(\theta(\bar{W})\big) = \Omega(\bar{W}, \cdot)\,. \tag{6.77}$$

This equation shows that the differential $d\big(\theta(\bar{W})\big)$ has no zeros. Hence, the set $\mathcal{K}_{\omega_o} = \{\, u \in T^*\mathcal{M} \mid \big(\theta(\bar{W})\big)(u) = -\omega_o \,\}$ is a codimension-one submanifold of $T^*\mathcal{M}$. On $\mathcal{K}_{\omega_o}$, the canonical two-form Ω induces a two-form Ω_{ω_o} with a one-dimensional kernel. At each point of $\mathcal{K}_{\omega_o}$, this kernel is spanned by $\bar{W}$, as can be read from (6.77). Let us call two points of $\mathcal{K}_{\omega_o}$ equivalent if they can be connected by an integral curve of $\bar{W}$. Then the quotient space $\mathcal{K}_{\omega_o}/\!\sim$ carries a Hausdorff manifold structure such that the natural projection $\mathcal{K}_{\omega_o} \longrightarrow \mathcal{K}_{\omega_o}/\!\sim$ becomes a submersion. This follows from the fact that W admits a global timing function. Moreover, the two-form Ω_{ω_o} induces a symplectic structure on $\mathcal{K}_{\omega_o}/\!\sim$. It is worthwhile to reconsider this construction in terms of a natural chart induced by coordinates $(x^1, \dots, x^n)$ on $\mathcal{M}$ with $t = x^n$ and $W = \partial/\partial x^n$. Then $\mathcal{K}_{\omega_o}$ is given by the equation $p_n = -\omega_o$, i.e., $\mathcal{K}_{\omega_o}$ is parametrized by the coordinates $(x^1, \dots, x^n, p_1, \dots, p_{n-1})$. Forming the quotient $\mathcal{K}_{\omega_o}/\!\sim$ comes up to factoring out the coordinate x^n. This shows that $\mathcal{K}_{\omega_o}/\!\sim$ can be identified, as a symplectic manifold, with the cotangent bundle $T^*\hat{\mathcal{M}}$, and that the natural projection $\mathcal{K}_{\omega_o} \longrightarrow \mathcal{K}_{\omega_o}/\!\sim$ can be identified with the restriction of the reduction map (6.74) to $\mathcal{K}_{\omega_o}$.

Now we consider the set $\mathcal{Q}_{\omega_o} = \mathcal{N} \cap \mathcal{K}_{\omega_o}$. For all points $u \in \mathcal{Q}_{\omega_o}$, $\mathbb{F}H(u)$ is linearly independent of $W_{\tau_{\mathcal{M}}^*(u)}$ by assumption. Thus, the characteristic direction of $\mathcal{K}_{\omega_o}$ (i.e., the direction spanned by $\bar{W}$) and the characteristic direction of $\mathcal{N}$ (i.e., the lifted ray direction) do not coincide. This implies that $\mathcal{N}$ and $\mathcal{K}_{\omega_o}$ have a transverse intersection at all points $u \in \mathcal{Q}_{\omega_o}$. Thus, $\mathcal{Q}_{\omega_o}$ is a closed codimension-one submanifold of $\mathcal{K}_{\omega_o}$, i.e., a manifold of dimension $(2n-2)$. Since $W \in \mathcal{G}_{\mathcal{N}}$, $\mathcal{Q}_{\omega_o}$ is invariant under the flow of $\bar{W}$. This implies that the set $\hat{\mathcal{N}}_{\omega_o} = \mathrm{red}(\mathcal{Q}_{\omega_o})$ is a closed submanifold of $T^*\hat{\mathcal{M}} \cong \mathcal{K}_{\omega_o}/\!\sim$ of dimension $(2n-3)$. Since we assume that $u \in \mathcal{Q}_{\omega_o}$ is never a multiple of $dt_{\tau_{\mathcal{M}}^*(u)}$, $\hat{\mathcal{N}}_{\omega_o}$ does not meet the zero section in $T^*\mathcal{M}$; since we assume that for $u \in \mathcal{Q}_{\omega_o}$ the fiber derivative $\mathbb{F}H(u)$ is never a multiple of $W_{\tau_{\mathcal{M}}^*(u)}$, $\hat{\mathcal{N}}_{\omega_o}$ is everywhere transverse to the fibers of $T^*\mathcal{M}$. This proves that $\hat{\mathcal{N}}_{\omega_o}$ is, indeed, a ray-optical structure on $\hat{\mathcal{M}}$. To prove the rest of the proposition, we observe that $\mathcal{Q}_{\omega_o}$ is foliated into the two-surfaces spanned by lifted rays and by integral curves of $\bar{W}$. If we denote the pull-back of the canonical two-form Ω to $\mathcal{Q}_{\omega_o}$ by Ω_{ω_o}, these two-surfaces can be characterized as the integral manifolds of the kernel of Ω_{ω_o}. Hence, red maps any such two-surface onto the image of a lifted ray of $\hat{\mathcal{N}}_{\omega_o}$. This proves that for each lifted ray $\hat{\xi} \colon I \longrightarrow \hat{\mathcal{N}}_{\omega_o}$ there is a one-parameter family of lifted rays $\xi \colon I \longrightarrow \mathcal{Q}_{\omega_o}$ of $\mathcal{N}$ such that $\hat{\xi} = \mathrm{red} \circ \xi$. Since the map red is fiber preserving, this result remains true if "lifted ray" is replaced with "lifted virtual ray". $\square$

Please note that at a point $u \in \mathcal{Q}_{\omega_o} \cap T_q^*\mathcal{M}$ the frequency with respect to the normalized observer field $V = e^{-f}W$ is equal to $e^{-f(q)}\,\omega_o$. If $\mathcal{N}$ is reversible one can restrict to values $\omega_o \geq 0$ without loss of generality.

Locally around a point $u \in \mathcal{N}$ the reduction can be carried through in the following way. It is convenient to introduce coordinates $(x^1, \ldots, x^n)$ on $\mathcal{M}$ such that $t = x^n$ and $W = \partial/\partial x^n$. By Proposition 6.5.1, we can choose a local Hamiltonian H for $\mathcal{N}$ around u which is independent of x^n. It is easy to verify that a local Hamiltonian $\hat{H}$ for the reduced ray-optical structure $\hat{\mathcal{N}}_{\omega_o}$ is given simply by setting p_n equal to $-\omega_o$, i.e.,

$$\hat{H}\left(x^1, \ldots, x^{n-1}, p_1, \ldots, p_{n-1}\right) = H\left(x^1, \ldots, x^{n-1}, p_1, \ldots, p_{n-1}, -\omega_o\right). \quad (6.78)$$

It is important to realize that the reduced ray-optical structure $\hat{\mathcal{N}}_{\omega_o}$ depends on the choice of the global timing function for W in the following way. If, in the situation of Theorem 6.5.1, the global timing function is changed according to (6.73), the reduced ray-optical structure changes according to

$$\hat{\mathcal{N}}_{\omega_o} \longrightarrow \hat{\mathcal{N}}'_{\omega_o} = \hat{\mathcal{N}}_{\omega_o} + d\hat{h}(\hat{\mathcal{M}}). \quad (6.79)$$

$\hat{\mathcal{N}}'_{\omega_o}$ is again a ray-optical structure on $\hat{\mathcal{M}}$ provided that the transversality condition (i) of Theorem 6.5.1 is satisfied for the new global timing function as well as for the old one. The proof of (6.79) follows immediately from the transformation behavior (6.75) of the reduction map.

In this situation, Theorem 6.5.1 gives a natural one-to-one relation between lifted rays of $\hat{\mathcal{N}}_{\omega_o}$ and lifted rays of $\hat{\mathcal{N}}'_{\omega_o}$. This relation is defined by associating a lifted ray $\hat{\xi}$ of $\hat{\mathcal{N}}_{\omega_o}$ with a lifted ray $\hat{\xi}'$ of $\hat{\mathcal{N}}'_{\omega_o}$ iff they are representable in the form $\hat{\xi} = \text{red} \circ \xi$ and $\hat{\xi}' = \text{red}' \circ \xi$, with the same lifted ray $\xi \colon I \longrightarrow \mathcal{Q}_{\omega_o}$ of $\mathcal{N}$. By (6.75), $\hat{\xi}$ and $\hat{\xi}'$ are then related by

$$\hat{\xi}'(s) = \hat{\xi}(s) + d\hat{h}\big|_{\tau_{\hat{\mathcal{M}}} \cdot (\hat{\xi}(s))} \quad (6.80)$$

for $s \in I$. There is an analoguous one-to-one relation for lifted virtual rays.

In terms of a natural chart on $T^* \hat{\mathcal{M}}$, (6.80) takes the form

$$x'^\rho(s) = x^\rho(s), \quad (6.81)$$

$$p'_\rho(s) = p_\rho(s) + \frac{\partial \hat{h}}{\partial x^\rho}\left(x^1(s), \ldots, x^{n-1}(s)\right) \quad (6.82)$$

where $\rho = 1, \ldots, n-1$. This observation implies that the rays of $\hat{\mathcal{N}}_{\omega_o}$ and $\hat{\mathcal{N}}'_{\omega_o}$ coincide although the lifted rays do not. Similarly, the virtual rays of $\hat{\mathcal{N}}_{\omega_o}$ and $\hat{\mathcal{N}}'_{\omega_o}$ coincide although the lifted virtual rays do not. From (6.80) or (6.82) we can also read the transformation behavior of wave surfaces. The function $\hat{S} \colon \hat{\mathcal{M}} \longrightarrow \mathbb{R}$ is a classical solution of the eikonal equation of $\hat{\mathcal{N}}_{\omega_o}$ if and only if the function $\hat{S} + \hat{h}$ is a classical solution of the eikonal equation of $\hat{\mathcal{N}}'_{\omega_o}$. Clearly, both solutions are associated with the same family of rays. There is a far-reaching formal analogy between this situation and the dynamics of charged particles moving in a magnetostatic field. The change of the global timing function corresponds to a *gauge transformation* of the magnetostatic potential. In both cases, the canonical momentum coordinates undergo a transformation of the form (6.82). In view of this analogy one might say that the rays of $\hat{N}_{\omega_o}$ are "gauge invariant" whereas the wave surfaces are not. This "gauge freedom" can be removed if our stationary ray-optical structure is globally static. Then we can "fix the gauge" by choosing the hypersurfaces $t = $ const. g-orthogonal to the integral curves of W. If, however, our stationary ray-optical structure is not globally static, then there is no distinguished choice for the global timing function and we have to live with the "gauge freedom".

Having clarified the dependence of the reduced ray-optical structure $\hat{N}_{\omega_o}$ on the choice of the global timing function, we are now going to investigate its dependence on the parameter ω_o which fixes the frequency of the rays. If the assumptions of Theorem 6.5.1 are satisfied for two real numbers ω_o and ω'_o, the reduced ray-optical structures $\hat{N}_{\omega_o}$ and $\hat{N}_{\omega'_o}$ are in general completely different. If, however, the stationary ray-optical structure N is dilation-invariant in the sense of Definition 5.4.1 (i.e., if the medium under consideration is non-dispersive), then $\hat{N}_{\omega_o}$ and $\hat{N}_{\omega'_o}$ are related by the following proposition.

Proposition 6.5.2. *Assume that all the assumptions of* Theorem 6.5.1 *are satisfied and that, in addition,* N *is dilation-invariant. Then the assumptions of* Theorem 6.5.1 *are still satisfied if* ω_o *is replaced with* $\omega'_o = c\omega_o$ *for any real number* $c > 0$, *and the reduced ray-optical structures are related by*

$$\hat{N}_{\omega'_o} = c\hat{N}_{\omega_o}. \tag{6.83}$$

In particular, the rays of $\hat{N}_{\omega'_o}$ *coincide with the rays of* $\hat{N}_{\omega_o}$. *If* N *is not only dilation-invariant but also reversible,* (6.83) *carries over to the case* $c < 0$.

Proof. Recall that N is dilation-invariant if and only if $N = cN$ for all real numbers $c > 0$. As we have seen in Sect. 5.4, dilation-invariance implies that, for any $u \in N$ and any local Hamiltonian H of N which is defined around u, the fiber derivative satisfies $\mathbb{F}H(cu) = c\mathbb{F}H(u)$ for all real numbers $c > 0$. If N is not only dilation-invariant but also reversibel, these properties remain valid for $c < 0$. On the basis of these observations the proof of Proposition 6.5.2 is an easy exercise.   □

Proposition 6.5.2 is, of course, in perfect agreement with the basic idea that in a non-dispersive medium the spatial path of a light ray is independent of its frequency.

If we have a reduced ray-optical structure $\hat{N}_{\omega_o}$, constructed by the method of Theorem 6.5.1 from a stationary ray-optical structure, we can integrate each lifted virtual ray of $\hat{N}_{\omega_o}$ over the canonical one-form $\hat{\theta}$ of $T^*\hat{M}$. Quite generally, the integral over the canonical one-form is known as the *action functional* and will play a central role in our discussion of variational principles in Chap. 7 below. In the case at hand, it is helpful to introduce the following definition.

Definition 6.5.3. *Consider the situation of* Theorem 6.5.1 *with* $\omega_o \neq 0$. *For a lifted virtual ray* $\hat{\xi} \colon [s_1, s_2] \longrightarrow \hat{\mathcal{N}}_{\omega_o}$ *of the reduced ray-optical structure* $\hat{\mathcal{N}}_{\omega_o}$,

$$\mathcal{I}\left(\hat{\xi}\right) = \tfrac{1}{\omega_o} \int_{\hat{\xi}} \hat{\theta} = \tfrac{1}{\omega_o} \int_{s_1}^{s_2} \hat{\theta}_{\hat{\xi}(s)} \left(\dot{\hat{\xi}}(s)\right) ds \qquad (6.84)$$

is called the optical path length *of* $\hat{\xi}$. *Here* $\hat{\theta}$ *denotes the canonical one-form on* $T^*\hat{\mathcal{M}}$.

If $\hat{\mathcal{N}}_{\omega_o}$ is everywhere transverse to the Euler vector field $\hat{E}$ on $T^*\hat{\mathcal{M}}$, the integral (6.84) is a strictly monotonous function of the upper bound s_2. In this case the optical path length gives us a distinguished parametrization along each lifted virtual ray of $\hat{\mathcal{N}}_{\omega_o}$. We have already seen that such a distinguished parametrization exists if and only if the Euler vector field is transverse to $\mathcal{N}$, please recall Proposition 5.4.7 and the subsequent discussion.

It is important to realize that the optical path length is "gauge dependent" in the following sense. Under a change of the global timing function the reduced ray-optical structure $\hat{\mathcal{N}}_{\omega_o}$ changes into $\hat{\mathcal{N}}'_{\omega_o}$ according to (6.79). Thereby each lifted virtual ray $\hat{\xi} \colon [s_1, s_2] \longrightarrow \hat{\mathcal{N}}_{\omega_o}$ of $\hat{\mathcal{N}}_{\omega_o}$ changes into a lifted virtual ray $\hat{\xi}' \colon [s_1, s_2] \longrightarrow \hat{\mathcal{N}}'_{\omega_o}$ of $\hat{\mathcal{N}}'_{\omega_o}$ according to (6.80). If we compare the optical path length of $\hat{\xi}'$ with the optical path length of $\hat{\xi}$ we find that they do not coincide but are related by

$$\mathcal{I}\left(\hat{\xi}'\right) = \mathcal{I}\left(\hat{\xi}\right) + \hat{h}\left(\hat{\lambda}(s_2)\right) - \hat{h}\left(\hat{\lambda}(s_1)\right) \qquad (6.85)$$

where $\hat{\lambda} = \tau^*_{\hat{\mathcal{M}}} \circ \hat{\xi} = \tau^*_{\hat{\mathcal{M}}} \circ \hat{\xi}'$.

The following proposition relates the optical path length to the "travel time", measured in terms of the global timing function t. This result is of particular relevance in view of Fermat's principle to be discussed in Chap. 7 below.

Proposition 6.5.3. *Let, in the situation of* Theorem 6.5.1, $\xi \colon [s_1, s_2] \longrightarrow \mathcal{N}$ *be a lifted virtual ray of* $\mathcal{N}$ *along which the conserved momentum* $\theta(\bar{W})$ *takes the value* $-\omega_o \neq 0$. *Then the optical path length of* $\hat{\xi} = \mathrm{red} \circ \xi$ *is given by*

$$\mathcal{I}\left(\hat{\xi}\right) = \tfrac{1}{\omega_o} \int_{s_1}^{s_2} \theta_{\xi(s)} \left(\dot{\xi}(s)\right) ds + t\left(\lambda(s_2)\right) - t\left(\lambda(s_1)\right) . \qquad (6.86)$$

If $\mathcal{N}$ *is dilation-invariant this equation simplifies to*

$$\mathcal{I}\left(\hat{\xi}\right) = t\left(\lambda(s_2)\right) - t\left(\lambda(s_1)\right) . \qquad (6.87)$$

Proof. Since $\hat{\xi} = \mathrm{red} \circ \xi$ and $\theta(\bar{W})$ takes the constant value $-\omega_o$ along ξ,

$$\theta_{\xi(s)} \left(\dot{\xi}(s)\right) = \hat{\theta}_{\hat{\xi}(s)} \left(\dot{\hat{\xi}}(s)\right) - \omega_o \, dt_{\lambda(s)} \left(\dot{\lambda}(s)\right) \qquad (6.88)$$

for $s \in [s_1, s_2]$. To verify this equation we can use a natural chart induced by coordinates $(x^1, \ldots, x^n)$ on $\mathcal{M}$ with $x^n = t$ and $\partial/\partial x^n = W$; then (6.88) is just the trivial identity $p_a \, dx^a = p_\rho \, dx^\rho + p_n \, dx^n$. Integrating (6.88) from s_1 to s_2 yields (6.86). If $\mathcal{N}$ is dilation-invariant, the integral vanishes owing to Proposition 5.4.7.   □

If the reduced ray-optical structure $\hat{\mathcal{N}}_{\omega_o}$ is strongly hyperregular and thus orientable, we know from Proposition 5.2.4 that there is a one-to-one relation between positively oriented virtual rays and positively oriented lifted virtual rays. In that case the optical path length can be viewed as a functional on (positively oriented) virtual rays rather than on lifted virtual rays. Again, this observation is crucial in view of Fermat's principle. As a matter of fact, for any stationary ray-optical structure with relevance to physics the reduced ray-optical structure is indeed strongly hyperregular or at least strongly regular. In the latter case the above-mentioned one-to-one correspondence holds true at least locally. The following proposition gives a useful criterion.

Proposition 6.5.4. *Assume that all the assumptions of* Theorem 6.5.1 *are satisfied and fix a point* $u \in N$ *with* $(\theta(\bar{W}))(u) = -\omega_o$. *Then the reduced ray-optical structure* $\hat{\mathcal{N}}_{\omega_o}$ *is strongly regular at the point* $\hat{u} = \mathrm{red}(u)$ *if and only if the condition*

$$\det \begin{pmatrix} (H^{ab}) & (H^a) & (W^a) \\ (H^b) & 0 & 0 \\ (W^b) & 0 & 0 \end{pmatrix} \neq 0 \tag{6.89}$$

holds at u *in any natural chart. Here we use the same matrix notation as in* (5.15), *with* H *denoting any local Hamiltonian for* N *and* W^a *denoting the components of the vector field* W.

Proof. It is easy to check that (6.89) is independent of which natural chart and which local Hamiltonian has been chosen. We choose a natural chart induced by coordinates $(x^1, \ldots, x^n)$ on $\mathcal{M}$ with $t = x^n$ and $W = \partial/\partial x^n$, and we choose a local Hamiltonian H that is independent of x^n. This is possible owing to Proposition 6.5.1. Then equation (6.78) gives us a local Hamiltonian $\hat{H}$ for $\hat{\mathcal{N}}_{\omega_o}$ around $\hat{u}$. As in the coordinates chosen $W^a = \delta_n^a$, condition (6.89) holds at u if and only if the condition

$$\det \begin{pmatrix} (\hat{H}^{\rho\sigma}) & (\hat{H}^\rho) \\ (\hat{H}^\sigma) & 0 \end{pmatrix} \neq 0 \tag{6.90}$$

holds at $\hat{u}$, where ρ is an index numbering rows and σ is an index numbering columns, both running from 1 to $n - 1$. By Definition 5.2.2, (6.90) holds at $\hat{u}$ if and only if $\hat{\mathcal{N}}_{\omega_o}$ is strongly regular at that point.   □

In many cases of interest condition (6.89) can be checked quickly with the help of the following result from linear algebra.

Proposition 6.5.5. *If the matrix* (H^{ab}) *is invertible,* $H^{ab} H_{bc} = \delta^a_c$, (6.89) *is equivalent to*

$$\det \begin{pmatrix} H_{ab} H^a H^b & H_{cd} H^c W^d \\ H_{ef} H^e W^f & H_{gh} W^g W^h \end{pmatrix} \neq 0. \tag{6.91}$$

Proof. We can assume that (H^a) and (W^a) are linearly independent since otherwise (6.89) and (6.91) are both wrong. With this assumption, (6.89) is satisfied if and only if the image of the $(n-2)$-dimensional vector space $\{(Z_b) \in \mathbb{R}^n \mid H^a Z_a = W^a Z_a = 0\}$ under (H^{ab}) is transverse to the 2-dimensional space spanned by (H^a) and (W^a). This is the case if and only if (H_{ab}) is non-degenerate on the 2-dimensional space spanned by (H^a) and (W^a), i.e., if and only if (6.91) holds true. $\square$

6.6 Stationary ray optics in vacuum and in simple media

In the preceding section we have established the general features of the reduction formalism for stationary ray-optical structures. To illustrate these results by way of example, we shall now carry through the reduction in full detail for stationary vacuum ray-optical structures. To that end we have to assume that we have a g-time-like vector field $W \in \mathcal{G}_{\mathcal{N}}$ where $\mathcal{N} = \mathcal{N}^g$ denotes the vacuum ray-optical structure on $(\mathcal{M}, g)$. According to our results of Sect. 5.3, the condition $W \in \mathcal{G}_{\mathcal{N}}$ means that W is a conformal Killing vector field of the metric g, i.e., that the Lie derivative $L_W g$ is a multiple of g. This implies that W is a Killing vector field of the rescaled metric $e^{-2f} g$,

$$L_W \left(e^{-2f} g \right) = 0 \tag{6.92}$$

where $f = \frac{1}{2} \ln\big(-g(W, W) \big)$. Hence, the one-form

$$\phi = -e^{-2f} g(W, \cdot) \tag{6.93}$$

satisfies

$$\phi(W) = 1 \quad \text{and} \quad L_W \phi = 0. \tag{6.94}$$

Now let us assume that we have a global timing function $t \colon \mathcal{M} \longrightarrow \mathbb{R}$ for W. This gives us a global diffeomorphism $(\pi, t) \colon \mathcal{M} \longrightarrow \hat{\mathcal{M}} \times \mathbb{R}$. The fact that W is a Killing vector field of the rescaled metric $e^{-2f} g$ induces a particular geometrical structure on $\hat{\mathcal{M}}$. To work this out, we use the one-form (6.93) and the differential dt of the global timing function to write the spacetime metric in the form

$$g = e^{2f} \left(e^{-2f} g + \phi \otimes \phi - (\phi - dt + dt) \otimes (\phi - dt + dt) \right) \tag{6.95}$$

which is a trivial identity. Clearly, the symmetric second rank tensor field $e^{-2f} g + \phi \otimes \phi$ satisfies $(e^{-2f} g + \phi \otimes \phi)(W, \cdot) = 0$ and $L_W(e^{-2f} g + \phi \otimes \phi) = 0$, and it is positive definite on the orthocomplement of W. Hence, it must be the pull-back of a (positive definite) Riemannian metric $\hat{g}$ on $\hat{\mathcal{M}}$,

$$e^{-2f} g + \phi \otimes \phi = \pi^* \hat{g} . \tag{6.96}$$

Similarly, the one-form $\phi - dt$ satisfies $(\phi - dt)(W) = 0$ and $L_W(\phi - dt) = 0$. Hence, it must be the pull-back of a one-form $\hat{\phi}$ on $\hat{\mathcal{M}}$,

$$\phi - dt = \pi^* \hat{\phi} . \tag{6.97}$$

With (6.96) and (6.97) inserted into (6.95), the metric g takes the form

$$g = e^{2f} \left(\pi^* \hat{g} - (\pi^* \hat{\phi} + dt) \otimes (\pi^* \hat{\phi} + dt) \right) . \tag{6.98}$$

The conformal factor e^{2f} has no influence on the vacuum light rays. Thus, (6.98) suggests that the metric $\hat{g}$ and the one-form $\hat{\phi}$ are the relevant geometrical objects that determine the reduced ray-optical structures $\hat{\mathcal{N}}_{\omega_o}$. This is indeed the case as we shall see below. But first we want to check if $\hat{g}$ and $\hat{\phi}$ depend on the choice of the global timing function t. If we change t according to (6.73), the metric $\hat{g}$ is obviously unaffected whereas the one-form $\hat{\phi}$ transforms like a gauge potential,

$$\hat{\phi} \longmapsto \hat{\phi}' = \hat{\phi} - d\hat{h} . \tag{6.99}$$

Thus, the two-form

$$\hat{\omega} = d\hat{\phi} \tag{6.100}$$

is independent of which global timing function has been chosen. The geometrical meaning of $\hat{\omega}$ is that it measures the rotation (=twist) of the integral curves of the time-like vector field W. Vanishing of $\hat{\omega}$ characterizes the locally static case, i.e., the equation $\hat{\omega} = 0$ is equivalent to W being locally hypersurface-orthogonal. If $\hat{\mathcal{M}}$ is simply connected, the equation $\hat{\omega} = 0$ is even equivalent to W being globally hypersurface-orthogonal. Let us quickly prove the second statement which implies, of course, the first one. On a simply connected manifold the equation $d\hat{\phi} = 0$ guarantees the existence of a function $\hat{h}$ such that $\hat{\phi} = d\hat{h}$. (We have used this well-known fact already in the proof of Proposition 5.5.2.) Thus, a gauge transformation (6.99) with this function $\hat{h}$ leads to $\hat{\phi}' = 0$. Together with (6.97) this shows that, for $\hat{\mathcal{M}}$ simply connected, the equation $\hat{\omega} = 0$ is equivalent to the existence of a global timing function t' such that $\phi - dt' = 0$ and thus, by (6.93), $g(W, \cdot) = -e^{2f} dt'$. Clearly, the latter equation characterizes the case that W is orthogonal to the hypersurfaces $t' = $ const., i.e., it characterizes the globally static case.

As a preparation for the reduction, we now use the representation (6.98) of the spacetime metric g to write the dispersion relation for vacuum light rays in terms of the spatial metric $\hat{g}$ and the spatial one-form $\hat{\phi}$. If we use local coordinates $(x^1, \ldots, x^n)$ on $\mathcal{M}$ with $x^n = t$ and $\partial/\partial x^n = W$, (6.98) takes the form

$$g_{ab}\, dx^a \otimes dx^b = \tag{6.101}$$

$$e^{2f}\left(\hat{g}_{\rho\sigma}\, dx^\rho \otimes dx^\sigma - (\hat{\phi}_\sigma\, dx^\sigma + dt) \otimes (\hat{\phi}_\rho\, dx^\rho + dt)\right).$$

Here and in the following greek indices are running from 1 to n−1. $\hat{g}_{\rho\sigma}$ and $\hat{\phi}_\sigma$ depend on $(x^1, \ldots, x^{n-1})$ whereas f depends on $(x^1, \ldots, x^n)$. With the covariant metric components g_{ab} given by (6.101) it is an easy exercise to calculate the contravariant metric components g^{ab}. This puts the dispersion relation $g^{ab}\, p_a\, p_b = 0$, by which the vacuum ray-optical structure is determined, into the form

$$\hat{g}^{\rho\sigma}\left(p_\rho - p_n\,\hat{\phi}_\rho\right)\left(p_\sigma - p_n\,\hat{\phi}_\sigma\right) - p_n^2 = 0. \tag{6.102}$$

Here we have introduced the contravariant components $\hat{g}^{\rho\sigma}$ of the metric $\hat{g}$ which are defined by $\hat{g}_{\mu\rho}\,\hat{g}^{\rho\sigma} = \delta^\sigma_\mu$.

We are now ready to construct the reduced ray-optical structures $\hat{\mathcal{N}}_{\omega_o}$ according to Theorem 6.5.1. Let us first check if all the assumptions of this theorem are satisfied in the case at hand. The set $\mathcal{Q}_{\omega_o}$ is non-empty for all real numbers $\omega_o \neq 0$. The transversality condition (i) of Theorem 6.5.1 is satisfied if and only if the one-form dt is nowhere g-light-like whereas the transversality condition (ii) is always satisfied. Thus, we have to assume that the global timing function has been chosen in such a way that the hypersurfaces $t = $ const. are either everywhere space-like or everywhere time-like with respect to g. Then Theorem 6.5.1 gives us a reduced ray-optical structure $\hat{\mathcal{N}}_{\omega_o}$ for all real numbers $\omega_o \neq 0$. Please note that the left-hand side of (6.102) gives us a Hamiltonian H for $\mathcal{N}$ that is independent of the coordinate x^n. As in (6.78) we get a Hamiltonian $\hat{H}$ for the reduced ray-optical structure $\hat{\mathcal{N}}_{\omega_o}$ simply by setting p_n equal to $-\omega_o$,

$$\hat{H}\left(x^1, \ldots, x^{n-1}, p_1, \ldots, p_{n-1}\right) = \tag{6.103}$$

$$\frac{1}{2\,\omega_o}\left(\hat{g}^{\mu\sigma}\,(p_\mu + \omega_o\hat{\phi}_\mu)\,(p_\sigma + \omega_o\,\hat{\phi}_\sigma) - \omega_o^2\right),$$

where the factor $2\omega_o$ was introduced for later convenience. Thus, the dispersion relation of the reduced ray-optical structure reads

$$\hat{g}^{\mu\sigma}\,(p_\mu + \omega_o\hat{\phi}_\mu)\,(p_\sigma + \omega_o\,\hat{\phi}_\sigma) - \omega_o^2 = 0. \tag{6.104}$$

With the Hamiltonian (6.103), Hamilton's equations take the form

$$\dot{x}^\sigma = \frac{1}{\omega_o}\,\hat{g}^{\sigma\rho}\left(p_\rho + \omega_o\,\hat{\phi}_\rho\right), \tag{6.105}$$

$$\dot{p}_\sigma = -\frac{1}{2\omega_o}\frac{\partial \hat{g}^{\rho\mu}}{\partial x^\sigma}\left(p_\rho + \omega_o\,\hat{\phi}_\rho\right)\left(p_\mu + \omega_o\,\hat{\phi}_\mu\right) -$$
$$\hat{g}^{\rho\mu}\frac{\partial \hat{\phi}_\rho}{\partial x^\sigma}\left(p_\mu + \omega_o\,\hat{\phi}_\mu\right). \tag{6.106}$$

Together with (6.104), the equations (6.105) and (6.106) determine the lifted rays of $\hat{\mathcal{N}}_{\omega_o}$ in the special parametrization adapted to $\hat{H}$. (6.105) can be solved for the momentum coordinates; upon inserting the result into (6.104) and (6.106), respectively, we find

$$\hat{g}_{\sigma\mu}\,\dot{x}^\sigma\dot{x}^\mu = 1\,, \tag{6.107}$$

$$\ddot{x}^\mu + \hat{\Gamma}^\mu_{\kappa\lambda}\dot{x}^\kappa\dot{x}^\lambda = \hat{g}^{\mu\rho}\left(\frac{\partial \hat{\phi}_\rho}{\partial x^\sigma} - \frac{\partial \hat{\phi}_\sigma}{\partial x^\rho}\right)\dot{x}^\sigma\,. \tag{6.108}$$

Here we have introduced the Christoffel symbols

$$\hat{\Gamma}^\mu_{\kappa\lambda} = \frac{1}{2}\,\hat{g}^{\mu\rho}\left(\frac{\partial \hat{g}_{\rho\kappa}}{\partial x^\lambda} + \frac{\partial \hat{g}_{\rho\lambda}}{\partial x^\kappa} - \frac{\partial \hat{g}_{\kappa\lambda}}{\partial x^\rho}\right) \tag{6.109}$$

of the metric $\hat{g}$. (6.107) and (6.108) determine the rays of $\hat{\mathcal{N}}_{\omega_o}$ in the parametrization adapted to $\hat{H}$. From (6.107) we read that this is the parametrization by $\hat{g}$-arc length. In the locally static case, $\hat{\omega} = d\hat{\phi} = 0$, the right-hand side of (6.108) vanishes and the rays are exactly the $\hat{g}$-geodesics. If $\hat{\omega}$ does not vanish, the rays deviate from the $\hat{g}$-geodesics in response to the "force term" on the right-hand side of (6.108). This force term has the same formal structure as the *Lorentz force* exerted on a charged particle by a magnetostatic field. In this analogy, the two-form $\hat{\omega}$ corresponds to the magnetic field strength and the one-form $\hat{\phi}$ corresponds to the magnetic potential. This is, of course, only a formal analogy. In the situation at hand $\hat{\omega}$ has nothing to do with a real magnetic field. It is a purely kinematical quantity measuring the rotation (= twist) of the integral curves of W. Physically, the right-hand side of (6.108) can be viewed as a *Coriolis force*.

Since $\hat{\phi}$ enters into (6.108) only in terms of $\hat{\omega} = d\hat{\phi}$, the rays of $\hat{\mathcal{N}}_{\omega_o}$ are gauge invariant although the lifted rays are not. We know already from our discussion following Theorem 6.5.1 that this is a general feature of the reduction formalism. Moreover, as neither (6.107) nor (6.108) involves the parameter ω_o, all the reduced ray-optical structures $\hat{\mathcal{N}}_{\omega_o}$, for $\omega_o \in \mathbb{R} \setminus \{0\}$, give the same rays. This observation exemplifies Proposition 6.5.2 since the vacuum ray-optical structure is dilation-invariant and reversible.

We shall now derive an expression for the optical path length, which was introduced in Definition 6.5.3, in terms of the Riemannian metric $\hat{g}$ and the one-form $\hat{\phi}$. (6.104) and (6.105) determine the lifted virtual rays of $\hat{\mathcal{N}}_{\omega_o}$ in the parametrization adapted to $\hat{H}$. These equations imply

$$p_\sigma\,\dot{x}^\sigma = \omega_o\left(\sqrt{\hat{g}_{\rho\mu}\,\dot{x}^\rho\,\dot{x}^\mu} - \hat{\phi}_\rho\,\dot{x}^\rho\right). \tag{6.110}$$

We can now free ourselves from the particular parametrization. Clearly, an orientation-preserving reparametrization leaves (6.110) unchanged whereas an orientation-reversing reparametrization requires replacing the positive square-root with the negative square-root. Since $\hat{\theta} = p_\sigma \, dx^\sigma$, (6.110) gives us the integrand of the optical path length (6.84). If we switch back to invariant notation, the optical path length of a lifted virtual ray $\hat{\xi} \colon [s_1, s_2] \longrightarrow \hat{\mathcal{N}}_{\omega_o}$ takes the form

$$\mathcal{I}\left(\hat{\xi}\right) = \int_{s_1}^{s_2} \left(\pm \sqrt{\hat{g}(\dot{\hat{\lambda}}, \dot{\hat{\lambda}})} - \hat{\phi}(\dot{\hat{\lambda}}) \right)(s) \, ds \qquad (6.111)$$

where $\hat{\lambda} \colon [s_1, s_2] \longrightarrow \hat{\mathcal{M}}$ is the projection of $\hat{\xi}$ to $\hat{\mathcal{M}}$. As λ is light-like, we can read from (6.98) that

$$\pm \sqrt{\hat{g}(\dot{\hat{\lambda}}, \dot{\hat{\lambda}})} - \hat{\phi}(\dot{\hat{\lambda}}) = dt(\dot{\lambda}) \,. \qquad (6.112)$$

Comparison with (6.93) and (6.97) shows that the positive square-root in (6.112) corresponds to the case that the parametrization of λ is future-oriented with respect to W, i.e., $g(W, \dot{\lambda}) < 0$. Inserting (6.112) into (6.111) demonstrates that, in the case at hand, the optical path length of $\hat{\xi}$ is equal to the travel time with respect to the global timing function used for the reduction. The same result follows from Proposition 6.5.3, using the fact that the vacuum ray-optical structure is dilation-invariant.

(6.111) clearly shows that, up to an orientation-depending sign ambiguity, the optical path length of $\hat{\xi}$ is determined by its projection $\hat{\lambda}$. We have already mentioned that this is true whenever the reduced ray-optical structure $\hat{\mathcal{N}}_{\omega_o}$ is strongly hyperregular. In the case at hand $\hat{\mathcal{N}}_{\omega_o}$ has the additional property that every C^∞ curve in $\hat{\mathcal{M}}$ is a virtual ray. For this reason the optical path length can be viewed as a functional on the set of all C^∞ curves $\hat{\lambda}$ in $\hat{\mathcal{M}}$, given by the right-hand side of (6.111).

(6.111) again exemplifies the gauge dependence of the optical path length. In the globally static case we can choose the global timing function in such a way that $\hat{\phi}$ vanishes. In this distinguished gauge the optical path length coincides with the $\hat{g}$-arc length. In the stationary but non-static case, however, the gauge freedom in the definition of the optical path length cannot be removed.

We have now established all the relevant equations of the reduction formalism for stationary vacuum ray-optical structures. Examples will be given in Chap. 8 below, where the metric $\hat{g}$ and the gauge-dependent one-form $\hat{\phi}$ are calculated for several (conformally) stationary spacetimes $(\mathcal{M}, g)$ with relevance to physics. For examples of this kind we also refer to Abramowicz, Carter and Lasota [3] and to Perlick [109].

For light propagation in matter, the reduction formalism has, in general, rather different features in comparison to the vacuum case. In particular, the reduced ray-optical structure $\hat{\mathcal{N}}_{\omega_o}$ is, in general, not determined by a

Riemannian metric and by a one-form which is unique up to gauge transformations. However, there is a special class of (non-dispersive) media to which our vacuum results immediately carry over, viz., media characterized by a Lorentzian "optical metric". So let us consider a ray-optical structure $\mathcal{N}$ on $(\mathcal{M}, g)$ which is of the kind given in Example 5.1.1, i.e., let us assume that $\mathcal{N} \subset T^*\mathcal{M}$ is the null cone bundle of a Lorentzian metric g_o. If g_o is conformally equivalent to g, then $\mathcal{N}$ is the vacuum ray-optical structure on $(\mathcal{M}, g)$, otherwise $\mathcal{N}$ gives light propagation in a medium. We assume that $\mathcal{N}$ is stationary, i.e., we assume that there is a vector field W which is time-like with respect to g and a conformal Killing field with respect to g_o. If W is time-like with respect to g_o as well, and if we can find a global timing function for W, then we can carry through the reduction procedure in analogy to the vacuum case. Now it is, of course, the optical metric g_o that is decomposed in the form (6.98). Hence, vanishing of the induced one-form $\hat{\phi}$ implies that W is orthogonal to the hypersurfaces $t = $ const. with respect to the optical metric g_o and does not characterize the globally static case. Similarly, it is now the metric g_o with respect to which the hypersurfaces $t = $ const. have to be non-light-like in order to guarantee that the assumptions of Theorem 6.5.1 are satisfied for all $\omega_o \neq 0$. In this situation the rays of the reduced ray-optical structure are, again, determined by equations of the form (6.107) and (6.108), and the optical path length is, again, representable in the form (6.111). Explicit examples of this kind will be given in Chap. 8 below.

One of the most interesting aspects of the reduction formalism for stationary ray-optical structures is that it provides a link between our general relativistic Lorentzian geometry setting of ray optics and the ordinary Euclidean geometry setting of elementary textbook ray optics. Roughly speaking, ray optics in media, as it is treated in elementary optics textbooks, can be viewed as the result of our reduction process applied to an appropriate ray-optical structure on Minkowski space. If $(\mathcal{M}, g)$ is n-dimensional Minkowski space, we can use pseudo-Cartesian coordinates $(x^1, \ldots, x^{n-1}, x^n = t)$ to identify $\mathcal{M}$ with $\mathbb{R}^n$ and to put g into the form

$$g = \delta_{\rho\sigma}\, dx^\rho \otimes dx^\sigma - dt \otimes dt \,. \tag{6.113}$$

Here, as before, the summation convention is used for greek indices running from 1 to $n - 1$. This induces a natural chart $(x^1, \ldots, x^n, p_1, \ldots, p_n)$ globally on $T^*\mathcal{M}$. Up to a minus sign, the momentum coordinate $p_n = \theta(\bar{W})$ gives the frequency with respect to the inertial system $V = W = \partial/\partial t$ for any ray-optical structure $\mathcal{N}$ on $\mathcal{M}$. Now let us consider a ray-optical structure $\mathcal{N}$ on $\mathcal{M}$ such that all the matter functions that enter into the dispersion relation are independent of the time coordinate $x^n = t$. This implies that $W = \partial/\partial t \in \mathcal{G}_\mathcal{N}$, i.e., it implies that $\mathcal{N}$ is stationary. Since $g(W, \cdot) = -dt$, $\mathcal{N}$ is then even globally static and the time coordinate $x^n = t$ gives a distinguished global timing function. For all frequency values $\omega_o \in \mathbb{R}$ for which the assumptions of Theorem 6.5.1 are satisfied, the reduction formalism gives us a reduced ray-optical structure $\hat{\mathcal{N}}_{\omega_o}$ on Euclidean space $\hat{\mathcal{M}} \cong \mathbb{R}^{n-1}$. The physically interesting case is, of course, $n = 4$. Any sort of medium treated in elementary optics textbooks can be modeled in terms of such a one-parameter family of ray-optical structures on $\hat{\mathcal{M}} \cong \mathbb{R}^3$.

Here is a special example of this construction. Let us assume that the ray-optical structure $\mathcal{N}$ on Minkowski space is given by the equation

$$\delta^{\rho\sigma} p_\rho p_\sigma - n(x^1, \ldots, x^{n-1}, -p_n)^2 p_n^2 = 0 \qquad (6.114)$$

where $n \colon \mathbb{R}^{n-1} \times \mathbb{R} \longrightarrow \mathbb{R}^+$ is a C^∞ function. In the terminology of Sect. 6.3, $\mathcal{N}$ is isotropic with respect to the inertial system $V = W = \partial/\partial t$. The function n gives the index of refraction which is assumed to be independent of the time coordinate $t = x^n$ whereas it may depend on the frequency $-p_n$. In this situation the assumptions of Theorem 6.5.1 are satisfied for all $\omega_o \neq 0$. From (6.114) we read that the reduced ray-optical structure $\hat{\mathcal{N}}_{\omega_o}$ is governed by the global Hamiltonian

$$\hat{H}(x^1, \ldots, x^{n-1}, p_1, \ldots, p_{n-1}) =$$
$$\frac{1}{2} \left(\frac{\delta^{\rho\sigma} p_\rho p_\sigma}{n(x^1, \ldots, x^{n-1}, \omega_o)^2} - \omega_o^2 \right) . \qquad (6.115)$$

This implies that the rays of $\hat{\mathcal{N}}_{\omega_o}$ are the geodesics of the conformally flat Riemannian metric

$$\hat{g}_{\rho\sigma} = n(\,\cdot\,, \omega_o)^2 \, \delta_{\rho\sigma} . \qquad (6.116)$$

on $\hat{\mathcal{M}}$ which depends, of course, on the frequency value ω_o. For a lifted virtual ray $\hat{\xi}$ of $\hat{\mathcal{N}}_{\omega_o}$, the optical path length $\mathcal{I}(\hat{\xi})$ equals the $\hat{g}$-arclength of its projection $\hat{\lambda}$ to $\hat{\mathcal{M}}$, where $\hat{g}$ denotes the Riemannian metric given by (6.116). We have thus rediscovered the standard textbook formulae for ray optics in dispersive isotropic media.

7. Variational principles for rays

In this chapter we want to characterize the rays of a ray optical structure in terms of variational principles. In particular, we want to investigate for what kind of ray optical structures some version of Fermat's principle holds true. This question is of interest not only from an abstract theoretical point of view but also in view of applications.

Most elementary optics textbooks, such as, e.g., Born and Wolf [16], give a formulation of Fermat's principle for non-dispersive and isotropic media only. However, generalizations to more complicated media are known, see, e.g., Newcomb [100]. If we allow for dispersive and anisotropic media, Fermat's principle in ordinary optics can be phrased in the following way.

> Fix two points in space and a frequency value ω_o. Consider all possibilities to go, along different spatial paths, from one point to the other at the velocity of light as it is determined, for the frequency ω_o, by the medium considered. Among all these "trial curves", the actual light rays are then the local extrema and saddle-points of a certain functional which is called the optical path length. If the medium is non-dispersive it is not necessary to specify the frequency and the optical path length can be reinterpreted as travel time.

Whenever a variational principle can be viewed as a mathematical reformulation of this statement it is legitimate to consider it as a version of Fermat's principle. In the following we establish several variational principles for the rays of a ray-optical structure, and we discuss their relation to Fermat's principle.

7.1 The principle of stationary action: The general case

In classical mechanics it is usual to define the *action functional* on C^∞ curves $\xi \colon [s_1, s_2] \to T^*\mathcal{M}$ by integration over the canonical one-form θ, i.e.,

$$\mathcal{A}(\xi) = \int_\xi \theta = \int_{s_1}^{s_2} \theta_{\xi(s)}\left(\dot{\xi}(s)\right) ds . \tag{7.1}$$

Actually, the action functional can be defined on curves of a more general differentiability class; for the time being, however, we stick to the C^∞ case.

From standard textbooks on classical mechanics we know that the solutions of Hamilton's equations satisfy a principle of stationary action. Therefore, it should not come as a surprise that the lifted rays of an arbitrary ray-optical structure on $\mathcal{M}$ satisfy a principle of stationary action as well. In comparison to the situation of classical mechanics there are three modifications. First, for an arbitrary ray-optical structure the existence of a Hamiltonian is guaranteed only locally. Second, lifted rays have to satisfy the dispersion relation, i.e., in mechanical terminology they are restricted to the "energy surface" $H = 0$. Third, lifted rays can be reparametrized arbitrarily which is reflected by the arbitrary stretching factor k in the ray equations (5.9). Mimicking standard techniques of classical mechanics, but taking care of these three modifications, we find the following principle of stationary action for lifted rays of an arbitrary ray-optical structure, cf. Figure 7.1.

Theorem 7.1.1. (Principle of stationary action for lifted rays) *Let $\mathcal{N}$ be an arbitrary ray-optical structure on $\mathcal{M}$ and fix a C^∞ immersion $\xi : [s_1, s_2] \longrightarrow \mathcal{N}$. As allowed variations of ξ consider all C^∞ maps $\eta :] - \varepsilon_0, \varepsilon_0[\times [s_1, s_2] \longrightarrow \mathcal{N}$ with $\eta(0, \cdot) = \xi$ for which the tangent vectors to the curves $\eta(\cdot, s_1)$ and $\eta(\cdot, s_2)$ are annihilated by the canonical one-form θ. Then the following holds true.*

(a) *If ξ is a lifted ray of $\mathcal{N}$, then ξ is a stationary point of $\mathcal{A}$,*

$$\tfrac{d}{d\varepsilon}\mathcal{A}\big(\eta(\varepsilon, \cdot)\big)\big|_{\varepsilon=0} = 0 \tag{7.2}$$

for all allowed variations η of ξ.

(b) *Conversely, if (7.2) is true at least for all allowed variations η of ξ with fixed endpoints in $\mathcal{N}$, i.e., with $\eta(\cdot, s_1)$ and $\eta(\cdot, s_2)$ constant, then ξ is a lifted ray.*

Proof. Let η be an allowed variation of ξ and denote the pertaining variational vector field by $Y : [s_1, s_2] \longrightarrow T\mathcal{N}$, i.e.,

$$Y(s) = \eta(\cdot, s)\dot{}(0). \tag{7.3}$$

Then the variational derivative of the action can be calculated in the following way, using standard derivative rules.

$$\tfrac{d}{d\varepsilon}\mathcal{A}\big(\eta(\varepsilon, \cdot)\big)\Big|_{\varepsilon=0} = \int_{s_1}^{s_2} \Big((d\theta)_{\xi(s)}\big(Y(s), \dot{\xi}(s)\big) + \tfrac{d}{ds}\big(\theta_{\xi(s)}(Y(s))\big)\Big)ds =$$

$$- \int_{s_1}^{s_2} \Omega_{\xi(s)}\big(Y(s), \dot{\xi}(s)\big)ds + \theta_{\xi(s_2)}\big(Y(s_2)\big) - \theta_{\xi(s_1)}\big(Y(s_1)\big) =$$

$$\int_{s_1}^{s_2} \Omega_{\xi(s)}\big(\dot{\xi}(s), Y(s)\big)ds. \tag{7.4}$$

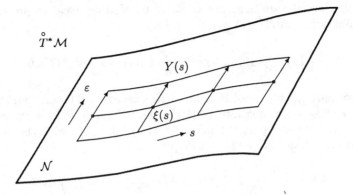

Fig. 7.1. An allowed variation, in the sense of Theorem 7.1.1, must be completely contained in $\mathcal{N}$.

In the last step we have used the fact that $Y(s_1)$ and $Y(s_2)$ are annihilated by θ because η is an allowed variation. Since our allowed variations stay within $\mathcal{N}$, Y must be everywhere tangent to $\mathcal{N}$. Thus, we can read from (7.1) that part (a) of the proposition follows directly from Definition 5.1.2. To prove part (b) we assume that the last integral in (7.1) vanishes for all C^∞ vector fields Y along ξ that are everywhere tangent to $\mathcal{N}$ and vanish at the endpoints. Hence the fundamental lemma of variational calculus implies that ξ must be a lifted ray according to Definition 5.1.2.

Alternative proof: For those readers who feel more comfortable with traditional index notation we give an alternative proof. We assume that ξ can be covered by the domain of a Hamiltonian and of a natural chart, and we write δ for the derivative with respect to ε at $\varepsilon = 0$, as in (5.46), (5.47) and (5.48). Then (7.1) takes the form

$$\delta\mathcal{A} = \delta \int_{s_1}^{s_2} p_a(s)\,\dot{x}^a(s)\,ds =$$

$$\int_{s_1}^{s_2} \delta p_a(s)\,\dot{x}^a(s)\,ds + \int_{s_1}^{s_2} p_a(s)\,\delta\dot{x}^a(s)\,ds =$$

$$\int_{s_1}^{s_2} \delta p_a(s)\,\dot{x}^a(s)\,ds + \tag{7.5}$$

$$p_a(s_2)\,\delta x^a(s_2) - p_a(s_1)\,\delta x^a(s_1) - \int_{s_1}^{s_2} \dot{p}_a(s)\,\delta x^a(s)\,ds =$$

$$\int_{s_1}^{s_2} \delta p_a(s)\,\dot{x}^a(s)\,ds - \int_{s_1}^{s_2} \dot{p}_a(s)\,\delta x^a(s)\,ds\,.$$

Since all allowed variations are confined to $\mathcal{N}$, they have to preserve the constraint equation $H(x(s), p(s)) = 0$, i.e.,

$$\frac{\partial H}{\partial x^a}\big(x(s), p(s)\big)\, \delta x^a(s) \;+\; \frac{\partial H}{\partial p_a}\big(x(s), p(s)\big)\, \delta p_a(s) = 0\,. \qquad (7.6)$$

To prove part (a) we recall that lifted rays satisfy (5.11) and (5.12). Upon inserting these equations into (7.1) the desired result follows immediately with (7.6). To prove part (b) we insert (7.6) into (7.1) with the help of a Lagrange multiplier $k(s)$. This results in

$$\delta \mathcal{A} = \int_{s_1}^{s_2} \delta p_a(s)\, \left(\dot{x}^a(s) - k(s)\, \tfrac{\partial H}{\partial p_a}\big(x(s), p(s)\big) \right)\, ds \;-$$
$$\int_{s_1}^{s_2} \delta x^a(s)\, \big(\dot{p}_a(s) + k(s)\, \tfrac{\partial H}{\partial x^a}\big(x(s), p(s)\big) \big)\, ds\,. \qquad (7.7)$$

By assumption, the right hand side of (7.7) is equal to zero for all smooth δp_a and δx^a that vanish at s_1 and s_2. (The Lagrange multiplier $k(s)$ allows to forget about (7.6).) Hence, the fundamental lemma of variational calculus implies that the ray equations (5.11) and (5.12) have to be satisfied.   □

Theorem 7.1.1 implies, in particular, that ξ is a lifted ray if and only if $\frac{d}{d\varepsilon} \mathcal{A}\big(\eta(\varepsilon, \cdot\,)\big)\big|_{\varepsilon=0} = 0$ for all allowed variations for which the curves $\eta(\,\cdot\,, s_1)$ and $\eta(\,\cdot\,, s_2)$ are vertical, i.e., for variations that keep the endpoints fixed in $\mathcal{M}$.

As $\mathcal{A}(x)$ is obviously invariant under orientation-preserving reparametrizations of ξ, we could equally well allow for variations that change the parameter interval $[s_1, s_2]$. However, such a generalization of Theorem 7.1.1 will not be needed in the following.

We emphasize that it is not justified to use the name "principle of minimal action", rather than "principle of stationary action", for Theorem 7.1.1. In general, a lifted ray can be a local minimum, a local maximum, or a saddle-point of the action functional. This is true even if we restrict to arbitrarily short rays.

The advantage of Theorem 7.1.1 is in the fact that it holds for arbitrary ray-optical structures. In particular, no regularity assumption is needed. Its disadvantage is in the fact that a very big set of allowed variations is used. Apart from the boundary conditions, all curves in $\mathcal{N}$, i.e., all curves for which the momenta satisfy the dispersion relation, are to be considered as "trial curves" $\eta(\varepsilon, \cdot\,)$. This includes curves with arbitrary velocities and not only motions at the velocity of light. For this reason Theorem 7.1.1 cannot be viewed as a version of Fermat's principle as it was stated in the beginning of this chapter. If we restrict the set of trial curves to motions at the velocity of light, in general we will have not enough allowed variations to prove an analogue of part (b) of Theorem 7.1.1. (There is, of course, no problem with part (a).) In Sect. 7.2 we shall see that there are still enough allowed variations if $\mathcal{N}$ is strongly (hyper-)regular.

Theorem 7.1.1 has several interesting consequences. To give just one example, we now show that part (a) of Theorem 7.1.1 leads to the familiar *Huyghens construction* of (lifted) wave surfaces. Here we have to recall the notions of generalized solutions of the eikonal equation and of lifted wave surfaces from Sect. 5.5.

Theorem 7.1.2. (Huyghens construction) *Let $\mathcal{N}$ be a ray-optical structure on $\mathcal{M}$ which is everywhere transverse to the flow of the Euler vector field. Let Φ be the flow of the distinguished characteristic vector field X on $\mathcal{N}$ which is determined by the equation $\theta_{\mathcal{N}}(X) = 1$. Fix a generalized solution $\mathcal{L}$ of the eikonal equation of $\mathcal{N}$ and a lifted wave surface $\bar{S}$ of $\mathcal{L}$. Let $s \in \mathbb{R}$ be such that the set $\Phi_s(\bar{S})$ is non-empty and, thus, again a lifted wave surface of $\mathcal{L}$. Then $\Phi_s(\bar{S})$ can be constructed as the envelope of the surfaces $\Phi_s(\mathcal{N}_q)$ where q ranges over all points in $\mathcal{M}$ that can be represented in the form $q = \tau_{\mathcal{M}}^*(u)$ with some $u \in \bar{S}$.*

Proof. Choose any $u \in \bar{S}$ and let $\xi \colon I \longrightarrow \mathcal{N}$ denote the maximal integral curve of the distinguished characteristic vector field X on $\mathcal{N}$ with $\xi(0) = u$. We have to prove that at the point $\xi(s)$ the $(n-1)$-dimensional surfaces $\Phi_s(\bar{S})$ and $\Phi_s(\mathcal{N}_q)$ are tangent to each other where $q = \tau_{\mathcal{M}}^*(u)$. To that end we consider all C^∞ maps $\eta \colon \,]-\varepsilon_0, \varepsilon_0[\times [0, s] \longrightarrow \mathcal{N}$ with $\eta(0, \cdot) = \xi$ such that the varied curves $\eta(\varepsilon, \cdot)$ are lifted rays with $\eta(\varepsilon, 0) \in \mathcal{N}_q$ and $\eta(\varepsilon, s) \in \Phi_s(\bar{S})$ for all $\varepsilon \in \,]-\varepsilon_0, \varepsilon_0[$. By Theorem 7.1.1 (a), any such η satisfies the condition $\frac{d}{d\varepsilon}\mathcal{A}\big(\eta(\varepsilon, \cdot)\big)\big|_{\varepsilon=0}$. This implies that for any such η the vector $\eta(\cdot, s)\dot{}(0)$ is tangent to $\Phi_s(\mathcal{N}_q)$. Since all elements of $T_{\xi(s)}\big(\Phi_s(\bar{S})\big)$ can be represented in this form, $\Phi_s(\bar{S})$ and $\Phi_s(\mathcal{N}_q)$ must be tangent to each other at $\xi(s)$.   □

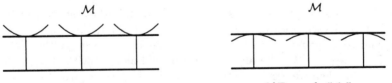

a) Example 5.1.2   b) Example 5.1.5

Fig. 7.2. Theorem 7.1.2 leads to the familiar Huyghens construction in $\mathcal{M}$ according to which wave surfaces are the envelopes of "elementary waves". Please note that transversality of the ray-optical structure to the flow of the Euler vector field is necessary. This is the case, e.g., for Example 5.1.2, where the "elementary waves" are hyperboloids, and for Example 5.1.5, where the "elementary waves" are spheres.

Upon projection to $\mathcal{M}$, Theorem 7.1.2 gives the familiar Huyghens construction according to which the wave surfaces (projections of $\Phi_s(\bar{S})$) are the envelopes of "elementary waves" (projections of $\Phi_s(\mathcal{N}_q)$), see Figure 7.2. Clearly, the projections to $\mathcal{M}$ of $\Phi_s(\bar{S})$ and $\Phi_s(\mathcal{N}_q)$ are smooth submanifolds only if $\Phi_s(\bar{S})$ and $\Phi_s(\mathcal{N}_q)$ are transverse to the fibers.

It is to be emphasized that Theorem 7.1.2 has to presuppose a ray-optical structure $\mathcal{N}$ that is transverse to the flow of the Euler vector field since otherwise a characteristic vector field X with $\theta_{\mathcal{N}}(X) = 1$ does not exist. In particular, Theorem 7.1.2 does not apply to ray-optical structures that give light propagation in a non-dispersive medium on a spacetime. For this reason Theorem 7.1.2 has more relevance for ray-optical structures on space rather than on spacetime.

7.2 The principle of stationary action: The strongly regular case

We have already stressed that Theorem 7.1.1 cannot be viewed as a version of Fermat's principle since the trial curves are not restricted to motions at the velocity of light. What we want to have is a theorem, analogous to Theorem 7.1.1, where only lifted virtual rays are considered as trial curves rather than arbitrary curves in $\mathcal{N}$. (Please recall that lifted virtual rays are defined through Definition 5.2.4.) We are now going to show that such a theorem holds true for strongly regular ray-optical structures. Contrary to Theorem 7.1.1 we restrict to variations with fixed end-points in $\mathcal{M}$.

Theorem 7.2.1. *Let $\mathcal{N}$ be a strongly regular ray-optical structure on $\mathcal{M}$ and fix a lifted virtual ray $\xi \colon [s_1, s_2] \longrightarrow \mathcal{N}$ of $\mathcal{N}$. As allowed variations of ξ we consider all C^∞ maps $\eta \colon\,] - \varepsilon_0, \varepsilon_0\,[\times [s_1, s_2] \longrightarrow \mathcal{N}$ with $\eta(0, \cdot) = \lambda$ for which the curves $\eta(\varepsilon, \cdot)$ are lifted virtual rays for all $\varepsilon \in\,] - \varepsilon_0, \varepsilon_0[$ and the curves $\eta(\cdot, s_1)$ and $\eta(\cdot, s_2)$ are vertical. Then ξ is a lifted ray if and only if $\frac{d}{d\varepsilon}\mathcal{A}\big(\eta(\varepsilon, \cdot)\big)\big|_{\varepsilon=0} = 0$ for all allowed variations η of ξ.*

Proof. Since allowed variations in the sense of this theorem are, in particular, allowed variations in the sense of Theorem 7.1.1, the "only if" part is a trivial consequence of part (a) of Theorem 7.1.1. To prove the "if" part, let $Z \colon [s_1, s_2] \longrightarrow T\mathcal{N}$ be any C^∞ vector field along ξ with $Z(s_1) = 0$ and $Z(s_2) = 0$. Then we can find a C^∞ map $\mu \colon\,] - \varepsilon_0, \varepsilon_0[\times [s_1, s_2] \longrightarrow \mathcal{N}$ with $\mu(0, \cdot) = \xi$, $\mu(\varepsilon, s_1) = \xi(s_1)$ and $\mu(\varepsilon, s_2) = \xi(s_2)$ for all $\varepsilon \in\,] - \varepsilon_0, \varepsilon_0[$ such that Z is the pertaining variational vector field, i.e., $Z(s) = \mu(\cdot, s)\dot{}(0)$. In general the curves $\mu(\varepsilon, \cdot)$ will not be lifted virtual rays, so μ will not be an allowed variation of ξ in the sense of this theorem. Therefore we consider the projection $\kappa = \tau_{\mathcal{M}}^* {\circ} \mu \colon\,]-\varepsilon_0, \varepsilon_0[\times [s_1, s_2] \longrightarrow \mathcal{M}$ of μ to $\mathcal{M}$ which gives a variation of $\lambda = \tau_{\mathcal{M}}^* \circ \xi$ with fixed end-points. Now we have to recall that strong regularity guarantees local solvability of (5.10) and (5.11) for the momenta and for the factor k. By compactness of the interval $[s_1, s_2]$ this guarantees existence and uniqueness of an allowed variation $\eta \colon\,]-\varepsilon_0, \varepsilon_0[\times [s_1, s_2] \longrightarrow \mathcal{N}$

of ξ, for ε_0 sufficiently small, that projects onto κ. We denote the variational vector field of η by Y as in (7.3). Now we have two variations μ and η of ξ both of which project onto κ. Hence, the difference between the pertaining variational vector fields Z and Y must be vertical. Since η is, in particular, an allowed variation in the sense of Theorem 7.1.1, (7.1) holds true. By hypothesis, $0 = \frac{d}{d\varepsilon}\mathcal{A}\big(\eta(\varepsilon, \cdot)\big)\big|_{\varepsilon=0}$. Thus, the last integral in (7.1) has to vanish. Since $Y - Z : [s_1, s_2] \longrightarrow T\mathcal{N}$ is vertical and ξ is a lifted virtual ray, this integral still vanishes if Y is replaced with Z. But Z was an arbitrary C^∞ vector field along ξ tangent to $\mathcal{N}$ that vanishes at both endpoints. Hence, as in the proof of Theorem 7.1.1, the fundamental lemma of variational calculus implies that ξ must be a lifted ray. $\qquad\qquad\square$

If $\mathcal{N}$ is not only strongly regular but even strongly hyperregular, we can choose an orientation for $\mathcal{N}$ and we can construct a one-to-one relation between positively oriented lifted virtual rays and positively oriented virtual rays according to Proposition 5.2.4. In this situation the action functional, which is defined by (7.1) on curves in $T^*\mathcal{M}$, gives a well-defined functional A on the set of all positively oriented virtual rays λ via

$$A(\lambda) = \mathcal{A}(\xi) . \tag{7.8}$$

Here ξ is the unique positively oriented lifted virtual ray that projects onto λ and $\mathcal{A}(\xi)$ is defined by (7.1). Therefore, in the strongly hyperregular case Theorem 7.2.1 can be reformulated as a variational principle for rays, rather than for lifted rays, in the following way.

Theorem 7.2.2. (Principle of stationary action for rays) *Let $\mathcal{N}$ be a ray-optical structure on $\mathcal{M}$ which is strongly hyperregular and thus orientable. Choose an orientation for $\mathcal{N}$ and fix a positively oriented virtual ray $\lambda : [s_1, s_2] \longrightarrow \mathcal{M}$. Consider as allowed variations of λ all C^∞ maps $\kappa : \;] - \varepsilon_0, \varepsilon_0 [\times [s_1, s_2] \longrightarrow \mathcal{M}$ with $\kappa(0, \cdot) = \lambda$ such that $\kappa(\varepsilon, s_1) = \lambda(s_1)$, $\kappa(\varepsilon, s_2) = \lambda(s_2)$, and the curves $\kappa(\varepsilon, \cdot)$ are virtual rays for all $\varepsilon \in \;] - \varepsilon_0, \varepsilon_0 [$. Then λ is a ray if and only if $\frac{d}{d\varepsilon}A\big(\kappa(\varepsilon, \cdot)\big)|_{\varepsilon=0} = 0$ for all allowed variations κ of λ. Here A is defined through (7.8).*

The proof follows immediately from Theorem 7.2.1 and Proposition 5.2.4.

If $\mathcal{M}$ is to be interpreted as space (and not as spacetime), Theorem 7.2.1 and Theorem 7.2.2 can be viewed as versions of Fermat's principle. To put this rigorously, we consider a stationary ray-optical structure and we assume that, for some value $\omega_o \in \mathbb{R}$, all the assumptions of Theorem 6.5.1 are satisfied such that the reduction can be carried through. We can then apply Theorem 7.2.1 to the reduced ray-optical structure $\hat{\mathcal{N}}_{\omega_o}$, provided that $\hat{\mathcal{N}}_{\omega_o}$ is strongly regular. (Criteria for $\hat{\mathcal{N}}_{\omega_o}$ to be strongly regular are given in Propositions 6.5.4 and 6.5.5.) The action functional $\mathcal{A}$ of Theorem 7.2.1 equals the optical path length $\mathcal{I}$ of Definition 6.5.3 up to the constant frequency factor ω_o. If we exclude the pathological case $\omega_o = 0$, varying $\mathcal{A}$ is equivalent to varying $\mathcal{I}$. Thus, Theorem 7.2.1 tells us that, among all lifted virtual rays between two fixed points in space $\hat{\mathcal{M}}$, the lifted rays are characterized by making the optical path length $\mathcal{I}$ stationary. This can be viewed as a version of Fermat's principle. If $\hat{\mathcal{N}}_{\omega_o}$ is even strongly hyperregular, Theorem 7.2.2 gives a more familiar reformulation of this result in terms of rays rather than in terms of lifted rays. In the non-dispersive case the optical path length can be reinterpreted as a travel time, according to Proposition 6.5.3, and the rays of $\hat{\mathcal{N}}_{\omega_o}$ are the same for all positive values of ω_o, according to Proposition 6.5.2.

Viewed in this sense, Theorem 7.2.2 covers virtually all known versions of Fermat's principle for stationary situations. Considering stationary vacuum ray-optical structures, as in the first part of Sect. 6.6, reproduces Fermat's principle for vacuum light propagation on (conformally) stationary Lorentzian manifolds as it is given in many textbooks on general relativity, see, e.g., Landau and Lifshitz [76] or, for the static case, Frankel [43] or Straumann [136]. Considering stationary ray-optical structures on Minkowski space, as in the last part of Sect. 6.6, reproduces all elementary textbook versions of Fermat's principle in ordinary optics.

Theorem 7.2.1 and 7.2.2 also apply to some (necessarily dispersive) ray-optical structures on spacetimes, e.g., to those of Example 5.1.2 giving light propagation in a non-magnetized plasma on a general-relativistic spacetime. In those cases, however, they cannot be interpreted as versions of Fermat's principle since the trial curves have fixed endpoints not only in space but even in spacetime.

7.3 Fermat's principle

Now we want to ask if light rays in an arbitrary general-relativistic medium can be characterized by a version of Fermat's principle. Throughout this section we presuppose a Lorentzian manifold $(\mathcal{M}, g)$ with $\dim(\mathcal{M}) > 2$, as in Chap. 6. Our physical interpretation refers to the case $\dim(\mathcal{M}) = 4$ where $(\mathcal{M}, g)$ can be viewed as a general-relativistic spacetime. For an arbitrary ray-optical structure $\mathcal{N}$ on $\mathcal{M}$, the results of Sect.s 7.1 and 7.2 can then be summarized in the following way.

In any case, the lifted rays of $\mathcal{N}$ are characterized by the variational principle of Theorem 7.1.1. This, however, cannot be viewed as a version of Fermat's principle because the space of trial curves is too big, as outlined above. If $\mathcal{N}$ is strongly regular, which by Corollary 5.4.1 can hold only if $\mathcal{N}$ describes light propagation in a dispersive medium, the lifted rays of $\mathcal{N}$ are characterized by the variational principle of Theorem 7.2.1. This, however, cannot be interpreted as a version of Fermat's principle either because the end-points of the trial curves are fixed in spacetime rather than in space. The results of the preceding sections give a version of Fermat's principle only in the very special case that $\mathcal{N}$ is stationary. More precisely, we have to assume in addition that all the conditions of the reduction theorem (i.e., of Theorem 6.5.1) are satisfied for some $\omega_o \in \mathbb{R} \setminus \{0\}$ and that the reduced ray-optical structure $\hat{\mathcal{N}}_{\omega_o}$ is strongly regular. Then Theorem 7.2.1 applied to $\hat{\mathcal{N}}_{\omega_o}$ gives us a version of Fermat's principle for the lifted rays with frequency constant ω_o in the medium considered. If $\hat{\mathcal{N}}_{\omega_o}$ is even strongly hyperregular, this result can be reformulated as a variational principle for rays, rather than for lifted rays, according to Theorem 7.2.2.

If $\mathcal{N}$ is a non-stationary ray-optical structure on $(\mathcal{M}, g)$, our previous results do not give us a version of Fermat's principle for its rays or lifted rays. Therefore we have to formulate another variational principle which needs some preparation. The first step is to define the space of trial curves.

According to the elementary formulation given at the beginning of Chap. 7 Fermat's principle requires to consider motions "at the velocity of light". In our setting this translates into considering (lifted) virtual rays of $\mathcal{N}$. For convenience we shall restrict to (lifted) virtual rays which are defined on the fixed parameter interval $[0, 1]$.

Then we have to impose boundary conditions by fixing "two points in space". The appropriate way to translate this into a spacetime setting is to fix a point q and a time-like curve γ in spacetime $\mathcal{M}$, and to restrict to virtual rays λ that start at q and terminate on γ. q can be interpreted as "a point in space at a particular time"; γ can be interpreted as "a point in space viewed over some time interval". Allowing the endpoint to float along a time-like curve is necessary since Fermat's principle requires to vary the arrival time. Further physical motivation for considering light rays between a point and a time-like curve will be given in Sect. 8.4 below when we are going to discuss gravitational lensing.

Finally, as we want to include dispersive media, it is necessary to "fix the frequency". This requires to choose an observer field. More precisely, what we shall need is not exactly a time-like vector field on all of $\mathcal{M}$ but rather a time-like vector field along each virtual ray from q to γ. We introduce the following definition, see Figure 7.3.

Definition 7.3.1. *Let γ be a time-like curve in the Lorentzian manifold $(\mathcal{M}, g)$. A generalized observer field for γ is a map W that assigns to each virtual ray λ that terminates on γ a time-like C^∞ vector field W_λ along λ that coincides with $\dot{\gamma}$ at the end-point. I.e., if $\lambda : [0, 1] \longrightarrow \mathcal{M}$ terminates at $\lambda(1) = \gamma(T(\lambda))$, with $T(\lambda)$ denoting some parameter value, then $W_\lambda : [0, 1] \longrightarrow T\mathcal{M}$ satisfies the conditions $W_\lambda(s) \in T_{\lambda(s)}\mathcal{M}$ for all $s \in [0, 1]$ and $W_\lambda(1) = \dot{\gamma}(T(\lambda))$.*

This definition generalizes the notion of observer fields in the following sense. If W is an ordinary observer field, i.e., a time-like C^∞ vector field on $\mathcal{M}$, and if γ is an integral curve of W, then the assignment $\lambda \longmapsto W \circ \lambda$ gives a generalized observer field for γ. Whereas observer fields do not exist on Lorentzian manifolds which are not time-orientable, generalized observer fields always exist for any time-like curve γ. E.g., we may define $W_\lambda(s)$ by parallely transporting the vector $\dot{\gamma}(T(\lambda))$ along λ

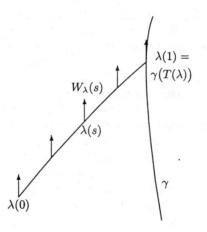

$\lambda(1) = \gamma(T(\lambda))$

$W_\lambda(s)$

$\lambda(s)$

$\lambda(0)$

γ

Fig. 7.3. A generalized observer field for γ, as defined in Definition 7.3.1, assigns to each virtual ray λ that terminates on γ a time-like vector field W_λ along λ.

Upon choosing a generalized observer field for γ we may assign a frequency $\omega(s)$ to each point $\xi(s)$ of a lifted ray by the equation

$$\omega(s) = -(\xi(s))(W_\lambda(s)), \tag{7.9}$$

provided that $\lambda = \tau_{\mathcal{M}}^* \circ \xi$ terminates on γ. As, in general, the frequency does not satisfy a conservation law along rays, the right way to "fix the frequency" is to prescribe a frequency value ω_o for the arrival at γ and to require that the redshift law for lifted rays is everywhere satisfied. (Please recall our discussion of redshift in Sect. 6.2.) Collecting all this together, we are led to defining the space of trial curves $\mathfrak{M}(\mathcal{N}, q, \gamma, W, \omega_o)$ in the following way.

Definition 7.3.2. *Let $\mathcal{N}$ be an arbitrary ray-optical structure on the Lorentzian manifold $(\mathcal{M}, g)$. Fix*

(1) *a point $q \in \mathcal{M}$;*
(2) *a C^∞ embedding $\gamma \colon I \longrightarrow \mathcal{M}$ from a real interval I into $\mathcal{M}$ such that $\dot{\gamma} = W_{\gamma(\cdot)}$;*
(3) *a generalized observer field W for γ;*
(4) *a non-zero real number ω_o.*

Then we define the space of trial curves $\mathfrak{M}(\mathcal{N}, q, \gamma, W, \omega_o)$ as the set of all C^∞ immersions $\xi \colon [0, 1] \longrightarrow T^*\mathcal{M}$ such that

(a) *ξ is a lifted virtual ray of $\mathcal{N}$;*
(b) *$\xi(0) \in T_q^*\mathcal{M}$;*
(c) *there is a $T(\xi) \in I$ such that $\xi(1) \in T_{\gamma(T(\xi))}^*\mathcal{M}$;*
(d) *$(\xi(1))(\dot{\gamma}(T(\xi))) = -\omega_o$;*

(e) ξ *satisfies the redshift law of lifted rays with respect to the generalized observer field* W, *i.e.,* $\Omega_{\xi(s)}\big(\dot{\xi}(s), Q(s)\big) = 0$ *for all* C^∞ *vector fields* $Q \colon [0,1] \longrightarrow T\mathcal{N}$ *along* ξ *with* $T\tau_{\mathcal{M}}^* \circ Q$ *parallel to* $W_{\tau_{\mathcal{M}}^* \circ \xi}$ *along* $\tau_{\mathcal{M}}^* \circ \xi$, *please recall* (6.18).

In terms of a natural chart and a local Hamiltonian, condition (a) requires that the representation $\big(x(s), p(s)\big)$ of ξ has to satisfy (5.10) and (5.11). By condition (e), the equation

$$W^a(s)\,\dot{p}_a(s) = -k(s)\,W^a(s)\,\frac{\partial H}{\partial x^a}\Big(x(s), p(s)\Big) \tag{7.10}$$

has to hold with the same factor $k(s)$ that appears in (5.11); here $W^a(s)$ denotes the components of $W_{\tau_{\mathcal{M}}^* \circ \xi}(s)$.

In the non-dispersive case, i.e., if $\mathcal{N}$ is dilation-invariant, Proposition 6.5.2 gives a natural one-to-one relation between the spaces $\mathfrak{M}(\mathcal{N}, q, \gamma, W, \omega_o)$ and $\mathfrak{M}(\mathcal{N}, q, \gamma, W, c\omega_o)$ for any constant $c > 0$. If $\mathcal{N}$ is not only dilation-invariant but also reversible, this result carries over to the case $c < 0$.

In the stationary case, the distinguished time-like vector field $W \in \mathcal{G}_{\mathcal{N}}$ gives us a generalized observer field $W \colon \lambda \longmapsto W \circ \lambda$ for each integral curve γ of W. (We hope that the reader will not be confused by our using the same symbol W for two mathematical objects which are different but related to each other in an obvious way.) In this special case conditions (d) and (e) of Definition 7.3.2 are equivalent to saying that the momentum $\theta(\bar{W})$ takes the constant value $-\omega_o$ along ξ. If, in addition, all the assumptions of the reduction theorem (i.e., of Theorem 6.5.1) are satisfied, conditions (a), (d) and (e) of Definition 7.3.2 imply that $\hat{\xi} = \mathrm{red} \circ \xi$ is a lifted virtual ray of the reduced ray-optical structure $\hat{\mathcal{N}}_{\omega_o}$. In the non-stationary case, however, there is no reduced ray-optical structure and the trial curves have to be defined in the rather complicated way of Definition 7.3.2.

With the space of trial curves at hand, we are now ready to introduce the functional that is to be extremized.

Definition 7.3.3. *Under the assumptions of* Definition 7.3.2,

$$
\begin{aligned}
\mathcal{F} \colon \mathfrak{M}(\mathcal{N}, q, \gamma, W, \omega_o) &\longrightarrow \mathbb{R} \\
\xi &\longmapsto \mathcal{F}(\xi) = \tfrac{1}{\omega_o}\mathcal{A}(\xi) + \mathcal{T}(\xi)
\end{aligned} \tag{7.11}
$$

is called the generalized optical path length functional. *Here* $\mathcal{A}$ *denotes the action functional* (7.1) *and* $\mathcal{T}$ *denotes the* arrival time functional *defined by condition* (c) *of* Definition 7.3.2.

In the non-dispersive case, i.e., if $\mathcal{N}$ is dilation-invariant, the generalized optical path length reduces to the arrival time, $\mathcal{F}(\xi) = \mathcal{T}(\xi)$, owing to Proposition 5.4.7.

In the stationary case, with W given by the distinguished time-like vector field $W \in \mathcal{G}_{\mathcal{N}}$ and γ an integral curve of W, the reduction theorem (i.e.,

Theorem 6.5.1) can be used to rewrite the generalized optical path length in terms of the reduced ray-optical structure, provided that all the assumtions of the reduction theorem are satisfied. In that case we find that the generalized optical path length is related to the optical path length of Definition 6.5.3 by $\mathcal{F}(\xi) = \mathcal{I}(\text{red} \circ \xi) + \text{const.}$ for all $\xi \in \mathfrak{M}(\mathcal{N}, q, \gamma, W, \omega_o)$, owing to Proposition 6.5.3. This justifies the name "generalized optical path length" for the functional $\mathcal{F}$.

Our goal is, of course, to prove that the stationary points of the functional $\mathcal{F}$ are exactly the lifted rays in $\mathfrak{M}(\mathcal{N}, q, \gamma, W, \omega_o)$. Our previous results suggest that some kind of regularity condition will be necessary to prove this. The crucial point is the following lemma.

Lemma 7.3.1. *Assume that all the assumptions of* Definition 7.3.2 *are satisfied and fix a lifted virtual ray* $\xi \in \mathfrak{M}(\mathcal{N}, q, \gamma, W, \omega_o)$. *Assume that along* ξ *the regularity condition*

$$\det \begin{pmatrix} (H^{ab}) & (H^a) & (W^a) \\ (H^b) & 0 & 0 \\ (W^b) & 0 & 0 \end{pmatrix} \neq 0 \qquad (7.12)$$

holds in any natural chart and with any local Hamiltonian, where $H^a = \partial H / \partial p_a$, $H^{ab} = \partial H / \partial p_a p_b$, *and* W^a *denotes the components of the vector field* $W_{\tau_{\mathcal{M}}^* \circ \xi}$. *Consider as allowed variations of* ξ *the set of all* C^∞ *maps* $\eta \colon] - \varepsilon_0, \varepsilon_0[\times [0,1] \longrightarrow \mathcal{N}$ *with* $\eta(0, \cdot) = \xi$ *and* $\eta(\varepsilon, \cdot) \in \mathfrak{M}(\mathcal{N}, q, \gamma, W, \omega_o)$ *for all* $\varepsilon \in] - \varepsilon_0, \varepsilon_0[$. *Then the following holds true. If* $Z \colon [0,1] \longrightarrow T\mathcal{N}$ *is any* C^∞ *vector field along* ξ *with* $Z(0) = 0$ *and* $Z(1) = 0$, *then there is an allowed variation* η *of* ξ *such that* $T\tau_{\mathcal{M}}^* \circ (Y - Z)$ *is a multiple of* W *along* $\tau_{\mathcal{M}}^* \circ \xi$. *Here* Y *denotes the variational vector field of* η *which is defined by* (7.3).

Proof. Let $Z \colon [0,1] \longrightarrow T\mathcal{N}$ be a C^∞ vector field along ξ with $Z(0) = 0$ and $Z(1) = 0$. Fix a variation of ξ in $\mathcal{N}$ with variational vector field equal to Z that keeps the end-points fixed, i.e., fix a C^∞ map $\mu \colon] - \varepsilon_0, \varepsilon_0[\times [0,1] \longrightarrow \mathcal{N}$ with $\mu(0, \cdot) = \xi$, $\mu(\cdot, s)^{\cdot}(0) = Z(s)$, $\mu(\varepsilon, 0) = \xi(0)$ and $\mu(\varepsilon, 1) = \xi(1)$. In general, μ will not be an allowed variation of ξ. We shall now use this map μ for constructing another variation η of ξ that satisfies all the requirements of the proposition.

In the first step we give the construction of η under the special assumption that ξ can be covered by the domain of a natural chart and of a local Hamiltonian. Moreover, we assume that the natural chart is induced by coordinates $(x^1, \ldots, x^n)$ on $\mathcal{M}$ with $W^a = \delta_n^a$ along the central curve $\tau_{\mathcal{M}}^* \circ \xi$. In that case, our assumption (7.12) allows us to solve the system of equations (5.10), (5.11) and (7.10) for $p_1, \ldots, p_{n-1}, k, \dot{x}^n, \dot{p}_n$ along this central curve. By continuity, the same solvability condition is true for curves which are sufficiently close, i.e., for varied curves with sufficiently small variational parameter ε.

(Here we make use of the fact that our curves are defined on the compact interval $[0, 1]$.) In other words, with $x^1, \ldots, x^{n-1}$ known, (5.10), (5.11) and (7.10) give us algebraic equations for $p_1, \ldots, p_{n-1}$ and first order differential equations for x^n and p_n which have to be satisfied by the varied curves to be constructed. For $\varepsilon \neq 0$, the coordinate representation of the curve $\mu(\varepsilon, \cdot)$ will not satisfy the system of equations (5.10), (5.11) and (7.10). However, we can take the coordinates $x^1(\varepsilon, s), \ldots, x^{n-1}(\varepsilon, s)$ from this representation and determine the quantities $p_1, \ldots, p_{n-1}, k, \dot{x}^n, \dot{p}_n$, locally uniquely, in such a way that (5.10), (5.11) and (7.10) hold. Together with the boundary values $x^n(\varepsilon, 0) = x^n(0, 0)$ and $p_n(\varepsilon, 1) = -\omega_o$ this determines a unique curve $\eta(\varepsilon, \cdot) \colon [0, 1] \longrightarrow \mathcal{N}$ near ξ for all ε sufficiently small. In this way we get an allowed variation η of ξ. Its variational vector field Y satisfies the condition of $T\tau_{\mathcal{M}}^* \circ (Z - Y)$ being parallel to $W_{\tau_{\mathcal{M}}^* \circ \xi}$ since, by construction, the coordinates $x^1, \ldots, x^{n-1}$ coincide along $\eta(\varepsilon, \cdot)$ and $\mu(\varepsilon, \cdot)$.

If ξ cannot be covered by a single chart of the kind considered above, this construction must be supplemented by an appropriate matching procedure. By compactness of the interval $[0, 1]$, we can find finitely many intermediary points, $s_0 = 0 < s_1 < \cdots < s_{m-1} < s_m = 1$, such that, for each $0 \leq i \leq m - 1$, ξ restricted to $[s_i, s_{i+1}]$ can be covered by a chart as considered above. On each interval $[s_i, s_{i+1}]$ we can then construct $\eta(\varepsilon, \cdot) \colon [s_i, s_{i+1}] \longrightarrow \mathcal{N}$ as above, with any choice of initial conditions $x^n(s_i)$ and $p_n(s_i)$. There is a unique choice for these initial conditions such that, by joining these segments together, we get a continuous map $\eta(\varepsilon, \cdot) \colon [0, 1] \longrightarrow \mathcal{N}$ that satisfies the boundary conditions (b) and (d) of Definition 7.3.2. To verify that this map is, indeed, of class C^∞ it suffices to check that the first order coordinate differential equations by which $\eta(\varepsilon, \cdot)$ is piecewise defined have a tensorial transformation behavior. This implies that $\eta(\varepsilon, \cdot)$ at all of its points satisfies an invariant first order differential equation with C^∞ coefficients, so it must be a C^∞ map. Thus, η satisfies all the requirements of the lemma. $\qquad \square$

In the stationary case $W \in \mathcal{G}_\mathcal{N}$, condition (7.12) is equivalent to strong regularity of the reduced ray-optical structure, recall Proposition 6.5.4. In the stationary or non-stationary case, Proposition 6.5.5 gives a useful criterion for (7.12) to hold. This criterion implies, in particular, that for the vacuum ray-optical structure, $\mathcal{N} = \mathcal{N}^g$, (7.12) is automatically satisfied for any time-like W.

We are now ready to prove Fermat's principle.

Theorem 7.3.1. (Fermat's Principle) *Let all the assumptions of Definition 7.3.2 be satisfied and fix a curve $\xi \in \mathfrak{M}(\mathcal{N}, q, \gamma, W, \omega_o)$. Consider as allowed variations of ξ all C^∞ maps $\eta \colon \,] - \varepsilon_0, \varepsilon_0[\, \times [0, 1] \longrightarrow \mathcal{N}$ with $\eta(0, \cdot) = \xi$ and $\eta(\varepsilon, \cdot) \in \mathfrak{M}(\mathcal{N}, q, \gamma, W, \omega_o)$ for all $\varepsilon \in \,] - \varepsilon_0, \varepsilon_0[$. Then the following holds true.*

(a) *For ξ to be a lifted ray it is necessary that $\frac{d}{d\varepsilon}\mathcal{F}\big(\eta(\varepsilon,\,\cdot\,)\big)\big|_{\varepsilon=0}=0$ for all allowed variations η of ξ. Here $\mathcal{F}$ denotes the generalized optical path length functional defined in Definition 7.3.3.*

(b) *If the regularity property (7.12) is satisfied along ξ, this condition is not only necessary but also sufficient.*

Proof. Let η be an allowed variation of ξ and denote the pertaining variational vector field by Y, as in (7.3). Then a calculation analogous to (7.1) yields

$$\frac{d}{d\varepsilon}\mathcal{F}\big(\eta(\varepsilon,\,\cdot\,)\big)\Big|_{\varepsilon=0}=$$
$$\frac{1}{w_o}\frac{d}{d\varepsilon}\mathcal{A}\big(\eta(\varepsilon,\,\cdot\,)\big)\Big|_{\varepsilon=0}+\frac{d}{d\varepsilon}\mathcal{T}\big(\eta(\varepsilon,\,\cdot\,)\big)\Big|_{\varepsilon=0}= \qquad (7.13)$$
$$\frac{1}{w_o}\int_0^1 \Omega_{\xi(s)}\big(\dot{\xi}(s),Y(s)\big)\,ds+\frac{1}{w_o}\theta_{\xi(1)}\big(Y(1)\big)+\frac{d}{d\varepsilon}\mathcal{T}\big(\eta(\varepsilon,\,\cdot\,)\big)\Big|_{\varepsilon=0}.$$

Here we have used the equation $Y(0)=0$ which follows from boundary condition (b) of Definition 7.3.2. Now we consider boundary condition (c) of Definition 7.3.2 which implies that $\tau_{\mathcal{M}}^*\big(\eta(\varepsilon,1)\big)=\gamma\Big(\mathcal{T}\big(\eta(\varepsilon,\,\cdot\,)\big)\Big)$. Differentiation with respect to ε at $\varepsilon=0$ yields $T\tau_{\mathcal{M}}^*\big(Y(1)\big)=W_{\gamma\big(\mathcal{T}(\xi)\big)}\frac{d}{d\varepsilon}\mathcal{T}\big(\eta(\varepsilon,\,\cdot\,)\big)|_{\varepsilon=0}$. Now we apply the covector $\xi(1)$ to this vector equation, and we use boundary condition (d) of Definition 7.3.2. This shows that the last two terms in (7.3) cancel, i.e.,

$$\frac{d}{d\varepsilon}\mathcal{F}\big(\eta(\varepsilon,\,\cdot\,)\big)\Big|_{\varepsilon=0}=\frac{1}{w_o}\int_0^1 \Omega_{\xi(s)}\big(\dot{\xi}(s),Y(s)\big)\,ds \qquad (7.14)$$

for all allowed variations. If ξ is a lifted ray, the integrand vanishes since the curves $\eta(\varepsilon,\,\cdot\,)$ are confined to $\mathcal{N}$. This proves part (a). To prove part (b) we choose an arbitrary C^∞ vector field $Z\colon[0,1]\longrightarrow T\mathcal{N}$ along ξ with $Z(0)=0$ and $Z(1)=0$. By Lemma 7.3.1 we can find an allowed variation η with variational vector field Y such that $T\tau_{\mathcal{M}}^*\circ(Z-Y)$ is a multiple of $W_{\tau_{\mathcal{M}}^*\circ\xi}$ along $\tau_{\mathcal{M}}^*\circ\xi$. By hypothesis, the right-hand side of (7.14) has to vanish. Since ξ satisfies the redshift condition (e) of Definition 7.3.2 this remains true if Y is replaced with Z. As $Z\colon[0,1]\longrightarrow T\mathcal{N}$ was an arbitrary C^∞ vector field along ξ that vanishes at the endpoints, the fundamental lemma of variational calculus implies that ξ must be a lifted ray. □

This theorem may be interpreted as Fermat's principle for light rays in arbitrary media on arbitrary general-relativistic spacetimes. In other words, as far as the medium and the underlying spacetime is concerned, Theorem 7.3.1 is the most general version of Fermat's principle in general relativity. One may think of further generalizations by considering spatially extended light sources and receivers (i.e., replacing the point q and the time-like curve γ by higher-dimensional submanifolds) or by relaxing the C^∞ assumption on the

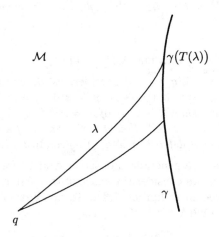

Fig. 7.4. Fermat's principle for vacuum light rays on a Lorentzian spacetime can be phrased in the following way, see Theorem 7.3.2. Among all light-like curves from q to γ the vacuum light rays are the extremals of the arrival time functional T.

trial curves. We shall not be concerned with the first generalization here, but we shall be forced to deal with the second one in Sect. 7.5 on Morse theory below.

It is an important feature of Theorem 7.3.1, somewhat unfamiliar from elementary optics, that the trial curves and the solution curves live in the cotangent bundle rather than in the base manifold. If we want to reformulate this theorem as a variational principle for rays, rather than for lifted rays, we have to assume that the map $\xi \longmapsto \tau^*_{\mathcal{M}} \circ \xi$ is injective on $\mathfrak{M}(\mathcal{N}, q, \gamma, W, \omega_o)$. This property is related to condition (7.12) like hyperregularity is related to regularity. We can then reformulate Theorem 7.3.1 in terms of rays, just as Theorem 7.2.1 could be reformulated as Theorem 7.2.2. If $\mathcal{N}$ is dilation-invariant in addition, we can free ourselves from the necessity to choose a generalized observer field W and a frequency constant ω_o. This results in a considerably simplified version of Fermat's principle. Such a simplification is possible, in particular, for the vacuum ray-optical structure $\mathcal{N} = \mathcal{N}^g$. Roughly speaking, the result is that among all light-like curves between a point and a time-like curve in a Lorentzian manifold the light-like geodesics are exactly the curves of stationary arrival time. The precise formulation reads as follows, cf. Figure 7.4.

Theorem 7.3.2. (Fermat's principle for vacuum light rays) *Let $(\mathcal{M}, g)$ be a Lorentzian manifold. Fix a point $q \in \mathcal{M}$ and a C^∞ embedding $\gamma \colon I \longrightarrow \mathcal{M}$ from a real interval I into $\mathcal{M}$ such that the tangent field of γ is g-time-like everywhere. Let $\mathfrak{L}(q, \gamma)$ denote the set of all virtual rays of the vacuum ray-*

optical structure $\mathcal{N} = \mathcal{N}^g$ (*i.e., all g-light-like* C^∞ *curves*) $\lambda\colon [0,1] \longrightarrow \mathcal{M}$ *with*

(a) $\lambda(0) = q$;
(b) *there is a* $T(\lambda) \in I$ *such that* $\lambda(1) = \gamma(T(\lambda))$.

Fix a virtual ray $\lambda \in \mathfrak{L}(q,\gamma)$ *and consider, as allowed variations of* λ, *all* C^∞ *maps* $\kappa\colon\ ]-\varepsilon_0,\varepsilon_0[\times[0,1] \longrightarrow \mathcal{M}$ *with* $\kappa(0,\,\cdot\,) = \lambda$ *and* $\kappa(\varepsilon,\,\cdot\,) \in \mathfrak{L}(q,\gamma)$ *for all* $\varepsilon \in\ ]-\varepsilon_0,\varepsilon_0[$. *Then* λ *is a ray of* $\mathcal{N} = \mathcal{N}^g$ (*i.e., a geodesic*) *if and only if* $\frac{d}{d\varepsilon}T\big(\kappa(\varepsilon,\,\cdot\,)\big)\big|_{\varepsilon=0} = 0$ *for all allowed variations* κ *of* λ. *Here* T *denotes the arrival time functional defined on* $\mathfrak{L}(q,\gamma)$ *by condition* (b).

Proof. After choosing a generalized observer field W for γ and an arbitrary real number $\omega_o \neq 0$, we consider the space of trial curves $\mathfrak{M}(\mathcal{N}^g,q,\gamma,W,\omega_o)$ as it was introduced in Definition 7.3.2. To our virtual ray $\lambda \in \mathfrak{L}(q,\gamma)$ we assign a map $\xi \longmapsto \mathfrak{M}(\mathcal{N}^g,q,\gamma,W,\omega_o)$ via

$$\xi(s) = \tfrac{1}{k(s)}\, g_{\lambda(s)}\big(\dot{\lambda}(s),\,\cdot\,\big) \tag{7.15}$$

where the function $k\colon [0,1] \longrightarrow \mathbb{R}\setminus\{0\}$ is defined as the solution of the linear differential equation

$$g_{\lambda(s)}\big(W_\lambda(s),\dot{\lambda}(s)\big)\,\dot{k}(s) = g_{\lambda(s)}\big(W_\lambda(s),\nabla_{\dot{\lambda}(s)}\dot{\lambda}\big)\,k(s) \tag{7.16}$$

with

$$k(1) = \tfrac{1}{\omega_o}\, g_{\lambda(1)}\big(W_\lambda(1),\dot{\lambda}(1)\big)\,. \tag{7.17}$$

Please note that the metric function on the left-hand side of (7.16) has no zeros; so k is, indeed, well-defined. (7.15) expresses the fact that ξ is a lifted virtual ray of $\mathcal{N}^g$ that projects onto λ; (7.16) guarantees that ξ satisfies the redshift condition (e) of Definition 7.3.2 whereas (7.17) takes care of part (d) of Definition 7.3.2. Hence, ξ is indeed in $\mathfrak{M}(\mathcal{N}^g,q,\gamma,W,\omega_o)$. The same construction gives a bijective relation between allowed variations κ of λ and allowed variations, in the sense of Theorem 7.3.1, η of ξ. As $\mathcal{N}^g$ is dilation-invariant, the generalized optical path length reduces to the arrival time, $\mathcal{F}\big(\eta(\varepsilon,\,\cdot\,)\big) = \mathcal{T}\big(\eta(\varepsilon,\,\cdot\,)\big) = T\big(\kappa(\varepsilon,\,\cdot\,)\big)$. Moreover, it is easy to check that with the vacuum Hamiltonian $H(x,p) = \tfrac{1}{2}g^{ab}(x)p_ap_b$ the regularity condition (7.12) is satisfied for any time-like W at all points of $\mathcal{N}^g$. Hence, Theorem 7.3.2 is an immediate consequence of Theorem 7.3.1. $\qquad\square$

The idea to formulate Fermat's principle for vacuum light rays in the version of Theorem 7.3.2 is essentially due to Kovner [74] who emphasized the relevance of this result in view of gravitational lensing. We shall comment on applications of Fermat's principle to gravitational lensing in Sect. 8.4 below. The first proof of Theorem 7.3.2 was given in Perlick [108]. Later, Perlick and Piccione [112] have proven a more general version of Theorem 7.3.2 where the point q and the time-like curve γ are replaced with higher-dimensional sub-manifolds. This generalization may be viewed as a vacuum Fermat principle for light sources and receivers which have a spatial extension. We shall not be concerned with this generalization here.

7.4 A Hilbert manifold setting for variational problems

Variational problems can be formulated in two quite different ways. First, there is the traditional formulation we have used so far, where variations are considered in terms of a parameter ε and stationarity of a functional is characterized by vanishing first derivative with respect to ε for all "allowed variations". The advantage of this approach is that it uses nothing but finite dimensional calculus.

The second method of formulating a variational problem is much more sophisticated. It consists in representing the functional to be varied as a differentiable mapping $\mathcal{F}\colon \mathfrak{M} \longrightarrow \mathbb{R}$ defined on some infinite dimensional manifold of maps. Typically, $\mathfrak{M}$ is modeled on a (real) Hilbert or Banach space, e.g. on a Sobolev space. A manifold modeled on a mere Fréchet space might also do in some case or other. However, this is too weak a structure for most applications. The elements of $\mathfrak{M}$ are the "trial maps", i.e., the candidates among which the solutions of the variational problem are to be determined. In this formulation stationarity of the action functional is expressed by vanishing of the Fréchet differential $d\mathcal{F}$, i.e., the solutions to the variational problem are the critical points of $\mathcal{F}$.

Once a variational problem has been cast into a Hilbert manifold setting, several interesting results from global analysis become applicable. This includes, in particular, the body of theorems known as infinite dimensional *Morse theory* which was developed in papers by Palais [104], by Smale [131], and by Palais and Smale [105]. Infinite dimensional Morse theory proved particularly powerful when applied to the geodesic problem on Riemannian manifolds, see, e.g., Palais [104], Schwartz [130] and Klingenberg [72] [73]. Among other things, this approach allows to decide whether a geodesic gives a local minimum, a local maximum, or a saddle-point of the action functional by counting the conjugate points along the geodesic. Moreover, on complete Riemannian manifolds it relates the number of geodesics joining two given points to the topology of the underlying manifold. At least partly, similar results were known already before infinite dimensional Morse theory came into existence. Back in those times it was necessary to use finite dimensional approximation techniques, introduced by Marston Morse in the 1930s, which are detailed in a well-known book by Milnor [96].

Rather than in geodesics on a Riemannian manifold we are interested in rays of a ray-optical structure. To make infinite dimensional Morse theory applicable to this situation we have to cast one of the variational problems treated in the preceding sections into a Hilbert manifold setting. It would be most desirable to work this out for the general Fermat principle given in Theorem 7.3.1. Unfortunately, this would be extremely difficult. For this reason we will be satisfied by establishing a Morse theory for the much simpler variational principle given in Theorem 7.2.2, i.e., for the principle of stationary action for rays of a hyperregular ray-optical structure on $\mathcal{M}$. As outlined above, this can be viewed as a version of Fermat's principle if $\mathcal{M}$ is to be interpreted as space. Setting up a Morse theory for the general Fermat principle of Theorem 7.3.1 will be a challenge for future work. It should be mentioned, however, that for the vacuum version of this variational principle, given in Theorem 7.3.2, a Morse theory has been established by Perlick [110] and, to a fuller extent, by Giannoni, Masiello and Piccione [47] [48].

In Theorem 7.2.2,the trial curves are virtual rays in the sense of Definition 5.2.4 and thus C^∞ curves. The basic idea to get a Hilbert manifold of trial curves is to replace the C^∞ condition with a Sobolev H^r condition. Therefore it will be necessary to recall definition and some basic properties of H^r spaces.

Let $C^r([0,1], \mathbb{R}^N)$ denote the set of all r times continuously differentiable maps from the interval $[0,1]$ to $\mathbb{R}^N$ for any integers $r \geq 0$ and $N \geq 1$. Define

$$< f \mid h >_r = \sum_{i=0}^{r} \int_0^1 f^{(i)}(s) \cdot h^{(i)}(s) \, ds \, , \tag{7.18}$$

$$\|f\|_r = \sqrt{< f \mid f >_r} \tag{7.19}$$

for all $f, h \in C^r([0,1], \mathbb{R}^N)$, where $f^{(i)} : [0,1] \longrightarrow \mathbb{R}^N$ denotes the i-th derivative of f and the dot stands for the standard Euclidean scalar product on $\mathbb{R}^N$. The scalar product (7.18) makes $C^r([0,1], \mathbb{R}^N)$ into a real pre-Hilbert space the completion of which is by definition the *Sobolev space* $H^r([0,1], \mathbb{R}^N)$. Instead of (7.18) some other topologically equivalent scalar product may be used, as is done, e.g., by Palais [104] and by Schwartz [130]. Note that $H^0([0,1], \mathbb{R}^N)$ coincides with the familiar *Lebesgue space* $L^2([0,1], \mathbb{R}^N)$. It is easy to check that $C^\infty([0,1], \mathbb{R}^N)$ is dense in $H^r([0,1], \mathbb{R}^N)$ for all $r \geq 0$.

Integration of the identity

$$f^{(i)}(s_2) = f^{(i)}(s_1) + \int_{s_1}^{s_2} f^{(i+1)}(s) \, ds \tag{7.20}$$

with respect to s_1 from 0 to 1 and application of Schwartz's inequality quickly shows, after renaming the arbitrary element $s_2 \in [0,1]$ into s, that

$$|f^{(i)}(s)| \leq 2 \|f\|_r \, , \quad 0 \leq i \leq r - 1 \tag{7.21}$$

for all $f \in C^r([0,1], \mathbb{R}^N)$ and $r \geq 1$. Hence, if functions $f_m \in C^r([0,1], \mathbb{R}^N)$ form a Cauchy sequence with respect to $\| \cdot \|_r$, then the f_m converge pointwise towards some $f_\infty \in C^{r-1}([0,1], \mathbb{R}^N)$. For this reason $H^r([0,1], \mathbb{R}^N)$ can be, and will be henceforth, identified with a subset of $C^{r-1}([0,1], \mathbb{R}^N)$ for all integers $r \geq 1$. The elements of $H^0([0,1], \mathbb{R}^N) = L^2([0,1], \mathbb{R}^N)$ can, of course, only be identified with equivalence classes of functions $[0,1] \longrightarrow \mathbb{R}^N$.

The notion of an H^r curve in our manifold $\mathcal{M}$ is introduced in the following way. By Whitney's embedding theorem (see, e.g., Golubitsky and Guillemin [49], Proposition 5.9) we can find a C^∞ embedding $j : \mathcal{M} \longrightarrow \mathbb{R}^N$, for some positive integer N. With such an embedding j we define for each integer $r \geq 1$

$$H^r\big([0,1], \mathcal{M}\big) = \{\lambda \colon [0,1] \longrightarrow \mathcal{M} \mid j \circ \lambda \in H^r\big([0,1], \mathbb{R}^N\big)\} \qquad (7.22)$$

where the ring denotes composition of maps as usual. It is easy to see that the set $H^r\big([0,1], \mathcal{M}\big)$ does not depend on the embedding j chosen. Moreover, it is a fundamental result, first stated by Palais and Smale [105], that the map $\lambda \longmapsto j \circ \lambda$ makes the set $H^r\big([0,1], \mathcal{M}\big)$ into a C^∞ submanifold of $H^r\big([0,1], \mathbb{R}^N\big)$ and that the Hilbert manifold structure thereby induced on $H^r\big([0,1], \mathcal{M}\big)$ is equally independent of j. For $r = 1$, the proof can be found in Palais [104] or in Schwartz [130]. Thereupon, the proof for $r > 1$ can be given by induction. Henceforth we use this result and consider $H^r\big([0,1], \mathcal{M}\big)$ as a Hilbert C^∞ manifold in its own right, for all integers $r \geq 1$.

Now we repeat the same construction with $\mathcal{M}$ replaced by $T\mathcal{M}$. This gives, for each integer $r \geq 1$, a Hilbert manifold $H^r\big([0,1], T\mathcal{M}\big)$ that may be viewed as the tangent bundle of $H^r\big([0,1], \mathcal{M}\big)$. More precisely, the tangent space of $H^r\big([0,1], \mathcal{M}\big)$ at a point $\lambda \in H^r\big([0,1], \mathcal{M}\big)$ is given, in the sense of a natural identification, by

$$T_\lambda H^r\big([0,1], \mathcal{M}\big) = \{Z \in H^r\big([0,1], T\mathcal{M}\big) \mid \tau_{\mathcal{M}} \circ Z = \lambda\} \qquad (7.23)$$

for $r \geq 1$. This becomes obvious if the tangent vectors to $H^r\big([0,1], \mathcal{M}\big)$ are expressed with the help of an embedding $j \colon \mathcal{M} \longrightarrow \mathbb{R}^N$ and its tangent map $Tj \colon T\mathcal{M} \longrightarrow T\mathbb{R}^N \cong \mathbb{R}^{2N}$.

We now state two simple lemmas which are readily verified with the help of an embedding $j \colon \mathcal{M} \longrightarrow \mathbb{R}^N$.

Lemma 7.4.1. *For each integer* $r \geq 2$, *a* C^∞ *curve* $\lambda \colon [0,1] \longrightarrow \mathcal{M}$ *is in* $H^r\big([0,1], \mathcal{M}\big)$ *if and only if its tangent field is in* $H^{r-1}\big([0,1], T\mathcal{M}\big)$. *The map*

$$H^r\big([0,1], \mathcal{M}\big) \longrightarrow H^{r-1}\big([0,1], T\mathcal{M}\big), \quad \lambda \longmapsto \dot{\lambda} \qquad (7.24)$$

is a C^∞ *map.*

Lemma 7.4.2. *For each integer* $r \geq 1$ *and each* $s \in [0,1]$ *the evaluation map*

$$\mathrm{ev}_s \colon H^r\big([0,1], \mathcal{M}\big) \longrightarrow \mathcal{M}, \quad \lambda \longmapsto \lambda(s) \qquad (7.25)$$

is a C^∞ *map. Its tangent map at a point* $\lambda \in H^r\big([0,1], \mathcal{M}\big)$ *is given by*

$$T_\lambda \mathrm{ev}_s \colon T_\lambda H^r\big([0,1], \mathcal{M}\big) \longrightarrow T_{\lambda(s)}\mathcal{M}, \quad Y \longmapsto Y(s). \qquad (7.26)$$

Furthermore, we shall need the following important result.

Lemma 7.4.3. *Consider a C^∞ map $\phi\colon \mathcal{M}_1 \longrightarrow \mathcal{M}_2$ between two finite dimensional C^∞ manifolds $\mathcal{M}_1$ and $\mathcal{M}_2$ and fix an integer $r \geq 1$. Then $\varPhi(\lambda) = \phi \circ \lambda$ defines a C^∞ map $\varPhi\colon H^r\big([0,1],\mathcal{M}_1\big) \longrightarrow H^r\big([0,1],\mathcal{M}_2\big)$ and the tangent map $T\varPhi$ is given by $\big((T\varPhi)_\lambda(Z)\big)(s) = (T\phi)_{\lambda(s)}\big(Z(s)\big)$ for all $\lambda \in H^r\big([0,1],\mathcal{M}_1\big)$, $Z \in T_\lambda H^r\big([0,1],\mathcal{M}_1\big)$, and $s \in [0,1]$.*

For $r = 1$, the proof can be found in Palais [104] or in Schwartz [130]. For $r > 1$, the result was first stated in Palais and Smale [105] and can be proven by induction over r.

After these preparations we now turn to the problem we are interested in.

7.5 A Morse theory for strongly hyperregular ray-optical structures

As indicated above, it is our goal to rephrase Theorem 7.2.2 as a variational principle on a Hilbert manifold. To that end we modify our notion of virtual rays in two respects. First, we replace the C^∞ condition on virtual rays by an H^r condition, for an appropriate integer r. Second, we choose a global Hamiltonian and use it for fixing the parametrization of each virtual ray. (Since Theorem 7.2.2 presupposes a strongly hyperregular ray-optical structure, it only applies to situations where the existence of a global Hamiltonian is assured.) We are, thus, led to introduce the following definition.

Definition 7.5.1. *If $\mathcal{N}$ is a ray-optical structure on $\mathcal{M}$ and H a global Hamiltonian for $\mathcal{N}$, we denote by $\mathfrak{V}(H)$ the set of all maps $\lambda\colon [0,1] \longrightarrow \mathcal{M}$ that satisfy the following condition. There is a $\xi \in H^1([0,1],\mathcal{N})$ and a $c \in \mathbb{R}^+$ such that $\tau_\mathcal{M}^* \circ \xi = \lambda$ and $(\tau_\mathcal{M}^* \circ \xi)^\cdot = c\,\mathbb{F}H \circ \xi$.*

Lemma 7.4.1 and Lemma 7.4.3 imply that $\mathfrak{V}(H) \subset H^2([0,1],\mathcal{M})$. It is obvious that each C^∞ curve $\lambda \in \mathfrak{V}(H)$ is a virtual ray in the sense of Definition 5.1.2. Conversely, any such virtual ray which is defined on a compact parameter interval can be made into an element of $\mathfrak{V}(H)$ by a unique reparametrization. It is thus justified to say that Definition 7.5.1 translates our earlier notion of virtual rays into an H^r-setting. In natural coordinates, elements of $\mathfrak{V}(H)$ are represented by exactly those $x \in H^2\big([0,1],\mathbb{R}^n\big)$ for which the system of equations

$$\dot{x}^a(s) = c\,\frac{\partial H}{\partial p_a}\Big(x(s), p(s)\Big)\,, \tag{7.27}$$

$$H\big(x(s), p(s)\big) = 0\,. \tag{7.28}$$

admits a solution $c \in \mathbb{R}^+$, $p \in H^1\big([0,1],\mathbb{R}^n\big)$. In the strongly hyperregular case such a solution must be unique.

We write $\mathfrak{V}(H)$, rather than $\mathfrak{V}(\mathcal{N})$, to indicate that this set depends on the choice of a global Hamiltonian. However, if H and $\tilde{H}$ are two global

Hamiltonians for one and the same ray-optical structure, there is a natural one-to-one relation between $\mathfrak{V}(H)$ and $\mathfrak{V}(\tilde{H})$ given by relating those curves to each other which coincide up to reparametrization.

We are now going to show that, in the strongly hyperregular case, $\mathfrak{V}(H)$ carries a natural Hilbert manifold structure.

Proposition 7.5.1. *Let $\mathcal{N}$ be a ray-optical structure on $\mathcal{M}$. Assume that H is a global Hamiltonian for $\mathcal{N}$ such that the map $\sigma_H \colon \mathcal{N} \times \mathbb{R}^+ \longrightarrow T\mathcal{M}$ defined by (5.16) is a C^∞ diffeomorphism onto its image. (Please recall that, by Definition 5.2.2, such a global Hamiltonian exists if and only if $\mathcal{N}$ is strongly hyperregular.) Then $\mathfrak{V}(H)$ is a C^∞ submanifold of $H^2([0,1], \mathcal{M})$.*

Proof. We denote the image of σ_H by $\mathcal{C}^+$. Since σ_H is a C^∞ diffeomorphism onto its image, its differential has maximal rank. Hence, $\mathcal{C}^+$ is open in $T\mathcal{M}$. As a consequence Lemma 7.4.1 implies that the set

$$H^2([0,1], \mathcal{M}; \mathcal{C}^+) = \{\, \lambda \in H^2([0,1], \mathcal{M}) \mid \dot{\lambda} \in H^1([0,1], \mathcal{C}^+) \,\} \qquad (7.29)$$

is a C^∞ submanifold of $H^2([0,1], \mathcal{M})$. Now we introduce the map

$$\chi \colon H^2([0,1], \mathcal{M}; \mathcal{C}^+) \longrightarrow H^1([0,1], \mathbb{R}^+)$$
$$\lambda \longmapsto \mathrm{pr}_2 \circ \sigma_H^{-1} \circ \dot{\lambda} \qquad (7.30)$$

where $\mathrm{pr}_2 \colon \mathcal{N} \times \mathbb{R}^+ \longrightarrow \mathbb{R}^+$ denotes the projection onto the second factor. It follows immediately from Lemma 7.4.1 and Lemma 7.4.3 that χ is a C^∞ map. This map is defined in such a way that $\mathfrak{V}(H) = \chi^{-1}(\mathbb{R}^+)$ where the set of constant functions from $[0,1]$ to $\mathbb{R}^+$ is identified with $\mathbb{R}^+$. By Lemma 7.4.2, $\mathbb{R}^+$ is a C^∞ submanifold of $H^1([0,1], \mathbb{R}^+)$. Now the statement of the proposition follows if we are able to prove that χ is a submersion. To that end we pick an element $\lambda \in \mathfrak{V}(H) \subset H^2([0,1], \mathcal{M}; \mathcal{C}^+)$. Then the equation $\chi(\lambda) = c$ has to hold with some $c \in \mathbb{R}^+$. By continuous extension of the tangent map

$$T_\lambda \chi \colon T_\lambda H^2([0,1], \mathcal{M}; \mathcal{C}^+) \longrightarrow T_c H^1([0,1], \mathbb{R}^+) \cong H^1([0,1], \mathbb{R}) \qquad (7.31)$$

we get a map

$$\overline{T_\lambda \chi} \colon \overline{T_\lambda H^2([0,1], \mathcal{M}; \mathcal{C}^+)} \longrightarrow H^0([0,1], \mathbb{R}) \,, \qquad (7.32)$$

where $\overline{T_\lambda H^2([0,1], \mathcal{M}; \mathcal{C}^+)}$ denotes the closure of $T_\lambda H^2([0,1], \mathcal{M}; \mathcal{C}^+)$ in $H^1([0,1], \mathcal{M})$. Since $\mathrm{pr}_2 \circ \sigma_H^{-1}$ is homogeneous, any $f \in H^1([0,1], \mathbb{R})$ has to satisfy the equation

$$\overline{T_\lambda \chi}(f\,\dot{\lambda}) = c\,f \,. \qquad (7.33)$$

This proves that the image of $T_\lambda \chi$ is dense in $H^1([0,1], \mathbb{R})$. On the other hand the image of $T_\lambda \chi$ is a closed linear subspace of $H^1([0,1], \mathbb{R})$. Both observations together imply that the image of $T_\lambda\chi$ is all of $H^1([0,1],\mathbb{R})$, i.e., that χ is a submersion at the point λ. As this result holds for all $\lambda \in \mathfrak{V}(H)$, we have proven that $\mathfrak{V}(H) = \chi^{-1}(\mathbb{R}^+)$ is a closed submanifold of $H^2([0,1],\mathcal{M};\mathcal{C}^+)$ and, thus, a submanifold of $H^2([0,1],\mathcal{M})$.   □

If the assumption of strong hyperregularity is satisfied, we can use this result and view $\mathfrak{V}(H)$ as a Hilbert C^∞ manifold in its own right. To impose boundary conditions we need the following proposition.

Proposition 7.5.2. *Under the assumptions of* Proposition 7.5.1, *the map*

$$\Pi: \mathfrak{V}(H) \longrightarrow \mathcal{M} \times \mathcal{M}, \quad \lambda \longmapsto (\lambda(0), \lambda(1)) \tag{7.34}$$

is a C^∞ submersion. Thus,

$$\mathfrak{V}(H;q) = \{\lambda \in \mathfrak{V}(H) \mid \lambda(0) = q\} \tag{7.35}$$

is a closed C^∞ submanifold of $\mathfrak{V}(H)$, and

$$\mathfrak{V}(H;q,q') = \{\lambda \in \mathfrak{V}(H) \mid \lambda(0) = q,\ \lambda(1) = q'\} \tag{7.36}$$

is either empty or a closed C^∞ submanifold of $\mathfrak{V}(H;q)$ for any q and q' in $\mathcal{M}$.

Proof. By Lemma 7.4.2, Π is a C^∞ map. To prove that Π is a submersion, we pick any element $\lambda \in \mathfrak{V}(H)$. Then the equation $\chi(\lambda) = c$ has to hold with some $c \in \mathbb{R}^+$, where the map χ is defined by (7.30). Again by Lemma 7.4.2, the tangent map of Π at the point λ is given by

$$T_\lambda\Pi: T_\lambda\mathfrak{V}(H) \subset T_\lambda H^2([0,1],\mathcal{M}) \longrightarrow T_{\lambda(0)}\mathcal{M} \times T_{\lambda(1)}\mathcal{M},$$

$$Y \longmapsto (Y(0), Y(1)).$$

We consider the continuous extension

$$\overline{T_\lambda\Pi}: \overline{T_\lambda\mathfrak{V}(H)} \longrightarrow T_{\lambda(0)}\mathcal{M} \times T_{\lambda(1)}\mathcal{M}$$

of $T_\lambda\Pi$, where $\overline{T_\lambda\mathfrak{V}(H)}$ denotes the closure of $T_\lambda\mathfrak{V}(H)$ in $T_\lambda H^1([0,1],\mathcal{M})$. Let Y be an arbitrary element of $T_\lambda H^1([0,1],\mathcal{M})$. So, in particular, $Y(0)$ and $Y(1)$ are arbitrary vectors in $T_{\lambda(0)}\mathcal{M}$ and $T_{\lambda(1)}\mathcal{M}$, respectively. We want to find a function $f \in H^1([0,1],\mathbb{R})$ such that

$$f\dot{\lambda} \in T_\lambda\mathfrak{V}(H). \tag{7.37}$$

Since we know from the proof of Proposition 7.5.1 that $\mathfrak{V}(H) = \chi^{-1}(\mathbb{R}^+)$, the function f satisfies (7.37) if and only if

$$\overline{T_\lambda\chi}(Y + f\dot{\lambda}) = \overline{T_\lambda\chi}(Y) + c\dot{f} = \text{const.} \tag{7.38}$$

For any choice of const. (7.38) has a unique solution $f \in H^1([0,1], \mathbb{R})$ with $f(0) = 0$. By integrating (7.38) from 0 to 1 we see that there is a unique choice for const. such that the corresponding solution f satisfies the boundary condition $f(1) = 0$. With this function f we get

$$\overline{T_\lambda \Pi}(Y + f\dot{\lambda}) = (Y(0), Y(1)) \tag{7.39}$$

which proves that $\overline{T_\lambda \Pi}$ is surjective, i.e., that the image of $T_\lambda \Pi$ is dense in $T_{\lambda(0)}\mathcal{M} \times T_{\lambda(1)}\mathcal{M}$. On the other hand, the image of $T_\lambda \Pi$ is a closed subspace of $T_{\lambda(0)}\mathcal{M} \times T_{\lambda(1)}\mathcal{M}$. Thus, Π is a submersion. □

To define the action functional on $\mathfrak{V}(H)$, we have to recall that, by Proposition 5.2.4, for a strongly hyperregular ray-optical structure there is a one-to-one relation between virtual rays and lifted virtual rays. Translated into our H^r-setting, this gives rise to the following proposition.

Proposition 7.5.3. *Under the assumptions of* Proposition 7.5.1, *the map*

$$\Xi : \mathfrak{V}(H) \longrightarrow H^1([0,1], T^*\mathcal{M}), \quad \lambda \longmapsto \mathrm{pr}_1 \circ \sigma_H^{-1} \circ \dot{\lambda} \tag{7.40}$$

is injective and of class C^∞. *Here* $\mathrm{pr}_1 : \mathcal{N} \times \mathbb{R}^+ \longrightarrow \mathcal{N}$ *denotes the projection onto the first factor.*

Proof. Lemma 7.4.1 and Lemma 7.4.3 guarantee that Ξ is a C^∞ map. By Proposition 5.2.4, the restriction of Ξ to $\mathfrak{V}(H) \cap C^\infty([0,1], \mathcal{M})$ is injective. By continuous extension, Ξ must be injective. □

It is not difficult to show that, moreover, Ξ is an immersion. In natural coordinates Ξ is represented by the map $x \longmapsto (x, p)$ given by solving (7.27) and (7.28) for $c \in \mathbb{R}^+$ and $p \in H^1([0,1], \mathbb{R}^n)$. Hence, for every $Z \in T_\lambda \mathfrak{V}(H)$ with coordinate representation $\delta x \in H^2([0,1], \mathbb{R}^n)$, the coordinate representation $(\delta x, \delta p) \in H^1([0,1], \mathbb{R}^{2n})$ of the vector $T\Xi(Z)$ satisfies the system of equations

$$\delta\left(\dot{x}^a - c\frac{\partial H}{\partial p_a}(x, p)\right)(s) = 0, \tag{7.41}$$

$$\delta\left(H(x, p)\right)(s) = 0, \tag{7.42}$$

with some $\delta c \in \mathbb{R}$. As our notation suggests, one should think of δ as of the derivative with respect to a variational parameter ε at the point $\varepsilon = 0$, and one should calculate the left-hand sides of (7.41) and (7.42) with the help of the ordinary chain rule and product rule. This procedure is justified by Lemma 7.4.3. Moreover, δ and $(\cdot)\dot{}$ commute since the derivative map from $H^{r+1}([0,1], \mathbb{R}^N)$ to $H^r([0,1], \mathbb{R}^N)$ is linear.

We are now ready to define the action functional A on $\mathfrak{V}(H)$, i.e., to translate (7.8) into our H^r-setting.

Proposition 7.5.4. *Let all the assumptions of Proposition 7.5.1 be satisfied. Then the action functional* $A: \mathfrak{V}(H) \longrightarrow \mathbb{R}$ *defined by*

$$A(\lambda) = \int_{\Xi(\lambda)} \theta = \int_0^1 \Big(\big(\Xi(\lambda) \big) \big(\dot{\lambda} \big) \Big)(s) \, ds \qquad (7.43)$$

is a C^∞ map. Here Ξ denotes the map introduced in Proposition 7.5.3.

Proof. A is the composition of the following three maps.

$$\mathfrak{V}(H) \rightarrow H^1\big([0,1], T\mathcal{M} \oplus T^*\mathcal{M}\big) \rightarrow H^1\big([0,1], \mathbb{R}\big) \rightarrow \mathbb{R} \qquad (7.44)$$

$$\lambda \;\mapsto\; \big(\dot{\lambda}, \Xi(\lambda) \big) \;\mapsto\; \big(\Xi(\lambda) \big) \big(\dot{\lambda} \big) \;\mapsto\; \int_0^1 \Big(\big(\Xi(\lambda) \big) \big(\dot{\lambda} \big) \Big)(s) \, ds$$

Here $T\mathcal{M} \oplus T^*\mathcal{M}$ denotes the fiber bundle over $\mathcal{M}$ whose fiber at $q \in \mathcal{M}$ is $T_q\mathcal{M} \times T_q^*\mathcal{M}$. The first map in this sequence is a C^∞ map owing to Lemma 7.4.1 and Proposition 7.5.3. The second map is, again, a C^∞ map as can be seen by applying Lemma 7.4.3 to the natural pairing map between vectors and covectors. Finally, the last map in the sequence is obviously linear and, with the help of inequality (7.21), it is easy to check that it is continuous. Thus, A is the composition of three C^∞ maps. $\square$

If H and $\tilde{H}$ are two global Hamiltonians for a ray-optical structure $\mathcal{N}$ both of which satisfy the assumptions of Proposition 7.5.1, Proposition 7.5.4 gives us a C^∞ action functional on $\mathfrak{V}(H)$ and on $\mathfrak{V}(\tilde{H})$. If we identify $\mathfrak{V}(H)$ and $\mathfrak{V}(\tilde{H})$ in the way outlined above, these two action functionals are easily shown to be related in the following way. From Proposition 5.1.3 we know that H and $\tilde{H}$ satisfy, on their common domain of definition, an equation of the form $\tilde{H} = FH$ with a nowhere vanishing function F. If F is positive, the action functionals on $\mathfrak{V}(H)$ and $\mathfrak{V}(\tilde{H})$ coincide. If F is negative, they differ by sign.

A can be represented with the help of natural coordinates in the form

$$A(\lambda) = \int_0^1 p_a(s) \, \dot{x}^a(s) \, ds \,. \qquad (7.45)$$

Here $x \in H^2\big([0,1], \mathbb{R}^n\big)$ denotes the coordinate representation of $\lambda \in \mathfrak{V}(H)$ and $(x, p) \in H^1\big([0,1], \mathbb{R}^{2n}\big)$ denotes the coordinate representation of $\Xi(\lambda)$ in the notation of Proposition 7.5.3. In other words, the function $s \longmapsto p(s)$ is determined by solving the system of equations (7.27) and (7.28) for $c \in \mathbb{R}^+$ and $p \in H^1\big([0,1], \mathbb{R}^n\big)$. We may use the representation (7.45) even if λ cannot be covered by a single chart. We just have to read the integrand as an invariant function that takes the given form locally in any natural chart.

Now we view A as a composition of three maps, as in the proof of Proposition 7.5.4, and we apply Lemma 7.4.3. This shows that the Fréchet differential $(dA)_\lambda: T_\lambda \mathfrak{V}(H) \longrightarrow \mathbb{R}$ can be derived from (7.45) by calculating, in the usual way of differential calculus, the derivative with respect to a variational parameter ε under the integral. Denoting this derivative at $\varepsilon = 0$ by δ we get

$$(dA)_\lambda(Z) = \int_0^1 \left(\delta p_a \, \dot{x}^a + p_a \, \delta \dot{x}^a \right)(s) \, ds = \tag{7.46}$$

$$\int_0^1 \left(\delta p_a \, c \, \frac{\partial H}{\partial p_a}(x,p) - \dot{p}_a \, \delta x^a \right)(s) \, ds + p_a(1) \, \delta x^a(1) - p_a(0) \, \delta x^a(0) =$$

$$\int_0^1 \left(-c \, \delta x^a \, \frac{\partial H}{\partial x^a}(x,p) - \dot{p}_a \, \delta x^a \right)(s) \, ds + p_a(1) \, \delta x^a(1) - p_a(0) \, \delta x^a(0)$$

for $Z \in T_\lambda \mathfrak{V}(H)$. Here we have used (7.27) and (7.42). As in (7.41) and (7.42), $\delta x \in H^2([0,1], \mathbb{R}^n)$ represents Z and $(\delta x, \delta p) \in H^1([0,1], \mathbb{R}^{2n})$ represents $T\Xi(Z)$. It is now easy to prove the desired H^r-reformulation of Theorem 7.2.2.

Theorem 7.5.1. *Consider the situation of* Proposition 7.5.1. *Let* $A_{q,q'}$ *denote the restriction of the action functional* A, *defined in* Proposition 7.5.4, *to the Hilbert manifold* $\mathfrak{V}(H; q, q')$ *of* Proposition 7.5.2. *Then* $\lambda \in \mathfrak{V}(H; q, q')$ *is a ray if and only if the Fréchet differential* $(dA_{q,q'})_\lambda$ *is equal to zero.*

Proof. For $\lambda \in \mathfrak{V}(H; q, q')$, the differential $(dA_{q,q'})_\lambda$ is equal to zero if and only if $(dA)_\lambda(Z) = 0$ for all $Z \in T_\lambda \mathfrak{V}(H)$ with $Z(0) = 0$ and $Z(1) = 0$. By continuous extension, this is the case if and only if the last line in (7.46) vanishes for all $\delta x \in H^1([0,1], \mathbb{R}^n)$ with $\delta x(0) = 0$ and $\delta x(1) = 0$ that represent elements $Z \in \overline{T_\lambda \mathfrak{V}(H)}$, where $\overline{T_\lambda \mathfrak{V}(H)}$ denotes the closure of $T_\lambda \mathfrak{V}(H)$ in $T_\lambda H^1([0,1], \mathcal{M})$. From the proof of Proposition 7.5.2 we know that those δx are of the form $\delta x^a(s) = y^a(s) + f(s) \dot{x}^a(s)$, where y is an arbitrary element of $H^1([0,1], \mathbb{R}^n)$ with $y(0) = 0$ and $y(1) = 0$ and f is some function in $H^1([0,1], \mathbb{R})$ with $f(0) = f(1) = 0$. However, the term proportional to the tangent field gives no contribution to the last integral in (7.46) as is easily verified with the help of (7.27) and (7.28). Hence, the Fréchet differential $(dA_{q,q'})_\lambda$ is zero if and only if the last integral in (7.46) vanishes for all $\delta x \in H^1([0,1], \mathbb{R}^n)$ with $\delta x(0) = 0$ and $\delta x(1) = 0$. Owing to the fundamental lemma of variational calculus, generalized into an H^1-setting by continuous extension, this is the case if and only if

$$\dot{p}_a(s) = -c \, \frac{\partial H}{\partial x^a} \Big(x(s), p(s) \Big), \tag{7.47}$$

i.e., if and only if the H^2 map $s \longmapsto (x(s), p(s))$ satisfies not only (7.27) and (7.28) but all the ray equations. This concludes the proof since, by induction, $s \longmapsto (x(s), p(s))$ must then be an H^r map for all $r \in \mathbb{N}$, i.e., it must be a C^∞ map. $\qquad\square$

With this proposition we have achieved our goal of formulating a variational principle for rays of a strongly hyperregular ray-optical structure in a

Hilbert manifold setting. In a nutshell, the rays from q to q' are the critical points of the functional $A_{q,q'} : \mathfrak{V}(H;q,q') \longrightarrow \mathbb{R}$, i.e., the points where this functional has a local minimum, a local maximum, or a saddle point. These three types of critical points can be distinguished by looking at the second derivative of $A_{q,q'}$, i.e., at the *Hessian* of $A_{q,q'}$ at λ,

$$\mathrm{Hess}_\lambda A_{q,q'} : T_\lambda \mathfrak{V}(H;q,q') \times T_\lambda \mathfrak{V}(H;q,q') \longrightarrow \mathbb{R} \qquad (7.48)$$
$$(\mathcal{X}_\lambda, \mathcal{Y}_\lambda) \longmapsto \mathrm{Hess}_\lambda A_{q,q'}(\mathcal{X}_\lambda, \mathcal{Y}_\lambda) = (\mathcal{X}\mathcal{Y}A_{q,q'})_\lambda.$$

Here $\mathcal{X}$ and $\mathcal{Y}$ denote any C^∞ vector fields (i.e., derivations) on $\mathfrak{V}(H;q,q')$ with values $\mathcal{X}_\lambda$ and $\mathcal{Y}_\lambda$, respectively, at the point λ. If $(dA_{q,q'})_\lambda = 0$, the Hessian is indeed well-defined (i.e., it depends only on the values of $\mathcal{X}$ and $\mathcal{Y}$ at the point λ), and it gives a symmetric continuous bilinear form on the Hilbert space $T_\lambda \mathfrak{V}(H;q,q')$.

Clearly, if the Hessian is non-degenerate, it characterizes the critical point in the following way. Depending on whether the Hessian is positive definite, negative definite, or indefinite, the critical point is a strict local minimum, a strict local maximum, or a saddle point. If the Hessian is degenerate, third or higher order derivatives have to be considered for characterizing the critical point.

In this connection it is helpful to recall the following standard terminology. For a symmetric continuous bilinear form $\Phi: \mathfrak{H} \times \mathfrak{H} \longrightarrow \mathbb{R}$ on a Hilbert space $\mathfrak{H}$, the *index* $\mathrm{ind}(\Phi)$ is, by definition, the maximal dimension of a subspace of $\mathfrak{H}$ on which Φ is negative definite. The *extended index* $\mathrm{ind}_o(\Phi)$ is, by definition, the maximal dimension of a subspace of $\mathfrak{H}$ on which Φ is negative semidefinite. The *nullity* $\mathrm{null}(\Phi)$ is the dimension of the kernel of Φ. Then

$$\mathrm{ind}_o(\Phi) = \mathrm{ind}(\Phi) + \mathrm{null}(\Phi) \qquad (7.49)$$

since, by Hilbert space algebra, the orthocomplement of the kernel of Φ can be decomposed into two orthogonal subspaces on which Φ is positive definite and negative definite, respectively.

For a C^∞ (or, more generally, C^2) function on a Hilbert manifold, the index of the Hessian at a critical point is called the *Morse index* of the critical point. If the Hessian is non-degenerate at all critical points, the function is called a *Morse function*. Clearly, the Hessian at a critical point λ is non-degenerate if and only if the extended Morse index of λ coincides with the Morse index of λ. A strict local minimum has vanishing Morse index (but not necessarily vanishing extended Morse index). Conversely, if the extended Morse index of λ vanishes, λ is a strict local minimum.

In terms of natural coordinates the Hessian of $A_{q,q'}$ at a critical point λ can be calculated from (7.46) with $\delta x(0) = 0$ and $\delta x(1) = 0$ by differentiating with respect to another variational parameter ε'. If we write δ' for this derivative at $\varepsilon' = 0$, we find

$$\text{Hess}_\lambda A_{q,q'}(Z', Z) = -\int_0^1 \delta'\left(\dot{p}_a + c\frac{\partial H}{\partial x^a}(x, p)\right)(s)\, \delta x^a(s)\, ds =$$

$$-\int_0^1 \delta\left(\dot{p}_a + c\frac{\partial H}{\partial x^a}(x, p)\right)(s)\, \delta' x^a(s)\, ds \tag{7.50}$$

for $Z, Z' \in T_\lambda \mathfrak{V}(H; q, q')$. Here $(\delta x, \delta p)$ and $(\delta' x, \delta' p)$ denote the coordinate representations of $T\Xi(Z)$ and $T\Xi(Z')$, respectively. The first equality in (7.50) holds since λ is a ray which implies that (7.47) is satisfied. The second equality in (7.50) holds since the Hessian is symmetric.

We shall now give a criterion for $\text{Hess}_\lambda A_{q,q'}$ to be non-degenerate. To that end we use the notion of Jacobi fields which was introduced, for arbitrary ray-optical structures, in Definition 5.6.2.

Theorem 7.5.2. *Let, in the situation of* Theorem 7.5.1, *$\lambda \in \mathfrak{V}(H; q, q')$ be a ray and $Z \in T_\lambda \mathfrak{V}(H; q, q')$. Then Z is in the kernel of $\text{Hess}_\lambda A_{q,q'}$ if and only if Z is a Jacobi field along λ.*

Proof. As in the proof of Theorem 7.5.1, we work in natural coordinates for notational convenience. The following argument is valid independently of whether or not the considered curve can be covered by a single chart.

Please recall that the coordinate representation $(\delta x, \delta p)$ of $T\Xi(Z)$ has to satisfy (7.41) and (7.42). If Z is a Jacobi field, (5.46), (5.47) and (5.48) must be true. Comparison shows that the function δk must be equal to the constant δc such that (5.48) takes the form

$$\delta\left(\dot{p}_a + c\frac{\partial H}{\partial x^a}(x, p)\right)(s) = 0. \tag{7.51}$$

Now we can read from (7.50) that Z is in the kernel of the Hessian. Conversely, if Z is in the kernel of the Hessian, the last integral in (7.50) vanishes for all $\delta' x$ that represent elements $Z' \in \mathfrak{V}(H; q, q')$. Hence, the last integral in (7.50) vanishes for all $\delta' x \in H^2([0,1], \mathbb{R}^n)$ with $\delta' x(0) = 0$ and $\delta' x(1) = 0$. This follows from the fact that any such $\delta' x$ is the coordinate representation of some $Z' \in T_\lambda \mathfrak{V}(H; q, q')$ up to adding a multiple of the tangent field of λ which drops out from (7.50) anyway. The same trick was used already in the proof of Theorem 7.5.1. Here the situation is even more convenient since λ is a C^∞ curve such that its tangent field is, in particular, an H^2 map and not only an H^1 map. Now the fundamental lemma of variational calculus implies that (7.51) has to hold. To complete the proof that Z is a Jacobi field we still have to verify that Z is a C^∞ map. We know that $(\delta x, \delta p) \in H^1([0,1], \mathbb{R}^{2n})$. By induction, (7.51) and (7.41) imply that $(\delta x, \delta p) \in H^r([0,1], \mathbb{R}^{2n})$ for all $r \in \mathbb{N}$, i.e., that δx and δp are, indeed, C^∞ maps. $\square$

In the terminology of Definition 5.6.3, this proposition implies that $\text{Hess}_\lambda A_{q,q'}$ is degenerate if and only if $q' = \lambda(1)$ is conjugate to $q = \lambda(0)$ along λ, and that the nullity of the Hessian equals the multiplicity of the conjugate point. Please note that in each Jacobi class along a ray $\lambda \in \mathfrak{V}(H)$ there is a unique representative $J \in T_\lambda \mathfrak{V}(H)$.

It is instructive to illustrate these results by specializing to the ray-optical structure of Example 5.1.5 where the rays are the geodesics of a (positive definite) Riemannian metric g_+. For the Hamiltonian given in this example, $\mathfrak{V}(H)$ is the set of all $\lambda \in H^2([0,1], \mathcal{M})$ with $g_+(\dot{\lambda}, \dot{\lambda}) = $ const. and A is the g_+-length functional. In this case $\mathrm{Hess}_\lambda A_{q,q'}$ is (the H^2 version of) the standard index form of Riemannian geometry and the notions of Jacobi fields and of conjugate points are the familiar textbook ones, see e.g. Bishop and Crittenden [15], Chap. 11.

Similarly, specializing to the ray-optical structure of Example 5.1.2 gives the analogous results for time-like geodesics of a Lorentzian metric which should be compared with Beem, Ehrlich and Easley [11], Sect. 10.1. Note that for the Hamiltonian H given in Example 5.1.2 A is the negative Lorentzian length functional on time-like curves. Switching from H to $-H$ yields the positive Lorentzian length functional instead.

Now we want to relate the Morse index of a critical point λ of $A_{q,q'}$ to the number of conjugate points along λ, thereby generalizing the classical Morse index theorem for Riemannian geodesics. Partly as a preparation for that we prove the following criterion for a critical point to be a minimum. This criterion applies to rays that are associated with a classical solution of the eikonal equation. (Please recall Sect. 5.5.) It generalizes a classical theorem of variational calculus, based on the socalled *Weierstrass excess function*, into our setting of ray-optical structures. In the language of traditional variational calculus, rays associated with a classical solution of the eikonal equation are usually characterized as being "embedded in a field of extremals".

Theorem 7.5.3. *Let, in the situation of Theorem 7.5.1, $\lambda_o \in \mathfrak{V}(H; q, q')$ be a ray. Assume that there is a classical solution $S : \mathcal{U} \subset \mathcal{M} \longrightarrow \mathbb{R}$ of the eikonal equation $H \circ dS = 0$ such that the lifted ray $\Xi(\lambda_o)$ is completely contained in $dS(\mathcal{U})$, where Ξ denotes the lifting map of (7.40). Then the following holds true.*

(a) *If, for the Hamiltonian H under consideration, the matrix on the left-hand side of (5.15) is not only non-degenerate but even positive definite along $\Xi(\lambda_o)$ (in one and thus in any natural chart), then λ_o is a strict local minimum of $A_{q,q'}$.*

(b) *If the Hamiltonian $H : \mathcal{W} \longrightarrow \mathbb{R}$ is defined on a domain $\mathcal{W} \subseteq T^*\mathcal{M}$ such that $\mathcal{W} \cap T_q^*\mathcal{M}$ is convex for all $q \in \mathcal{M}$, if the matrix $(\partial^2 H/\partial p_a \partial p_b)$ is positive definite at all points $u \in \mathcal{W}$ (in one and thus in any natural chart), and if the domain $\mathcal{U}$ of S covers all curves $\lambda \in \mathfrak{V}(H; q, q')$, then λ_o is a strict global minimum of $A_{q,q'}$.*

Proof. We use the coordinate representation (7.45) of A and its restriction $A_{q,q'}$. As in the proof of Theorem 7.5.1 and Theorem 7.5.2, coordinates are employed for notational convenience only. The following argument remains valid even if $\mathcal{U}$ cannot be covered by a single chart. Then we find for all $\lambda \in \mathfrak{V}(H; q, q')$ contained in $\mathcal{U}$

$$A_{q,q'}(\lambda) - A_{q,q'}(\lambda_o) = \int_0^1 \left(p_a \dot{x}^a\right)(s)\, ds - \int_0^1 \left(\partial_a S(x_o)\, \dot{x}_o^a\right)(s)\, ds . \quad (7.52)$$

Here (x, p) is the coordinate representation of $\Xi(\lambda)$ whereas $(x_o, \partial S(x_o))$ is the coordinate representation of $\Xi(\lambda_o)$. As $x(0) = x_o(0)$ and $x(1) = x_o(1)$, we can replace x_o by x in the second integral. With (7.27) this puts (7.52) into the form

$$A_{q,q'}(\lambda) - A_{q,q'}(\lambda_o) = \int_0^1 \left((p_a - \partial_a S(x))\, c \frac{\partial H}{\partial p_a}(x, p) \right)(s)\, ds . \quad (7.53)$$

If λ is close to λ_o, $p(s)$ is close to $\partial S(x(s))$. In that case, as H is a C^∞ and thus C^2 function defined on an open neighborhood of $\mathcal{N}$, Taylor's theorem implies

$$H\Big(x(s), \partial S(x(s))\Big) = \quad (7.54)$$

$$H\big(x(s), p(s)\big) + \frac{\partial H}{\partial p_a}\Big(x(s), p(s)\Big)\Big(\partial_a S(x(s)) - p_a(s)\Big) +$$

$$\frac{1}{2}\frac{\partial^2 H}{\partial p_a \partial p_b}\Big(x(s), p'(s)\Big)\Big(\partial_a S(x(s)) - p_a(s)\Big)\Big(\partial_b S(x(s)) - p_b(s)\Big)$$

for some $p'_a(s) = p_a(s) + \theta(s)\Big(\partial_a S(x(s)) - p_a(s)\Big)$ with $0 \le \theta(s) \le 1$. The left-hand side of (7.54) vanishes since S is a classical solution of the eikonal equation. The first term on the right-hand side vanishes since all curves in $\mathfrak{V}(H)$ satisfy (7.28). Hence, inserting (7.54) into (7.53) results in

$$A_{q,q'}(\lambda) - A_{q,q'}(\lambda_o) = \quad (7.55)$$

$$\frac{c}{2} \int_0^1 \left(\frac{\partial^2 H}{\partial p_a \partial p_b}(x, p')\Big(\partial_a S(x) - p_a\Big)\Big(\partial_b S(x) - p_b\Big) \right)(s)\, ds$$

for λ sufficiently close to λ_o. Now the positive definiteness assumption of part (a) implies that the matrix $(\partial^2 H/\partial p_a \partial p_b)(x(s), p(s))$ is positive definite on vertical vectors tangent to $\mathcal{N}$. By continuity, for λ sufficiently close to λ_o, the integrand in (7.55) is strictly positive unless $p_a(s) = \partial_a S(x(s))$. The latter equation holds for all $s \in [0, 1]$ if and only if $\lambda = \lambda_o$ since λ satisfies the same boundary conditions and the same parametrization fixing condition as λ_o. This completes the proof of part (a).

If the convexity assumption of part (b) is satisfied, we can use (7.54) even if λ is not close to λ_o. Thus, under the assumptions of part (b), (7.55) is valid for all $\lambda \in \mathfrak{V}(H; q, q')$, and the integrand is strictly positive unless $p_a(s) = \partial_a S(x(s))$. From the proof of part (a) we know already that the latter equation holds for all $s \in [0, 1]$ if and only if $\lambda = \lambda_o$. $\qquad\square$

From Proposition 5.5.5 we know that for a sufficiently short ray λ_o we can always find a classical solution S of the eikonal equation such that $\Xi(\lambda_o)$ is contained in the image of dS. Hence, under the positive definiteness assumption of Theorem 7.5.3 (a) a sufficiently short ray always gives a strict local minimum of the respective functional $A_{q,q'}$. This positive definiteness assumption is satisfied, e.g., for the Hamiltonian of Example 5.1.5 which gives the geodesics of a (positive definite) Riemannian metric, but also for the Hamiltonian of Example 5.1.2 which gives the time-like geodesics of a Lorentzian metric. Please note that we are always free to change H into $-H$, thereby inverting the sign of the functional A and turning minima into maxima and vice versa.

We are now ready to prove a generalized Morse index theorem.

Theorem 7.5.4. (Morse index theorem) *Let, in the situation of Theorem 7.5.1, $\lambda \in \mathfrak{V}(H; q, q')$ be a ray and assume that, for the Hamiltonian considered, the matrix on the left-hand side of (5.15) is positive definite at all points of $\Xi(\lambda)$, in one and hence in any natural chart. Then the extended Morse index of λ satisfies the equality*

$$\mathrm{ind}_o(\mathrm{Hess}_\lambda A_{q,q'}) = \sum_s n(s) . \qquad (7.56)$$

Here the sum is to be taken over all $s \in {]0,1]}$ such that $\lambda(s)$ is conjugate to $\lambda(0)$ along λ, and $n(s)$ denotes the multiplicity of this conjugate point. (Note that, by Proposition 5.6.3 this sum is finite.)

Proof. Let $\overline{T_\lambda \mathfrak{V}(H; q, q')}$ be the closure of the tangent space $T_\lambda \mathfrak{V}(H; q, q')$ in $T_\lambda H^1([0,1], \mathcal{M})$, and let $\overline{\mathrm{Hess}_\lambda A_{q,q'}}$ be the continuous extension of the Hessian onto this space. To verify that this extension exists, we recall that the Hessian is given, in terms of natural coordinates, by (7.50). (Again, coordinates are used for notational convenience only. The following argument remains true even if λ cannot be covered by a single chart.) If we shift the derivative from $\delta' p_a$ to δx^a by means of a partial integration, we get a manifestly H^1 continuous expression. Thus, the extended Hessian is given by

$$\overline{\mathrm{Hess}_\lambda A_{q,q'}}(Z', Z) = \int_0^1 \left(\delta' p_a \, \delta \dot{x}^a - \delta' \left(c \frac{\partial H}{\partial x^a}(x, p) \right) \delta x^a \right)(s) \, ds . \qquad (7.57)$$

For each $s \in {]0,1]}$ we define a map $\lambda^s : [0,1] \longrightarrow \mathcal{M}$ by $\lambda^s(s') = \lambda(s' s)$. Clearly, λ^s is a critical point of $A_{q,\lambda(s)}$. To ease notation, we write

$$\mathfrak{H}_s = \overline{T_{\lambda^s} \mathfrak{V}(H; q, \lambda(s))} \quad \text{and} \quad \Phi_s = \overline{\mathrm{Hess}_{\lambda^s} A_{q,\lambda(s)}} . \qquad (7.58)$$

If we choose C^∞ vector fields $E_1, \ldots, E_{n-1}$ along λ such that the vectors $E_1(s), \ldots, E_{n-1}(s), \dot{\lambda}(s)$ are linearly independent for each $s \in [0,1]$, then the Hilbert space $\mathfrak{H}_s$ can be identified with the Hilbert space

$$\mathfrak{H} = \{Z \in H^1([0,1], \mathbb{R}^{n-1}) \mid Z(0) = Z(1) = 0\} \qquad (7.59)$$

for each $s \in]0,1]$. Viewed in this sense as a one-parameter family of symmetric bilinear forms $\Phi_s : \mathfrak{H} \times \mathfrak{H} \longrightarrow \mathbb{R}$ on a single Hilbert space, Φ_s depends continuously on s in the weak sense, i.e., the map $s \longrightarrow \Phi_s(Z, Z)$ is continuous for all $Z \in \mathfrak{H}$. This follows from the behavior of the integral (7.57) under parameter transformations. We now introduce the notation

$$i(s) = \text{ind}(\Phi_s) = \text{ind}(\text{Hess}_{\lambda^s} A_{q,\lambda(s)}),$$
$$i_o(s) = \text{ind}_o(\Phi_s) = \text{ind}_o(\text{Hess}_{\lambda^s} A_{q,\lambda(s)}), \qquad (7.60)$$
$$n(s) = \text{null}(\Phi_s) = \text{null}(\text{Hess}_{\lambda^s} A_{q,\lambda(s)}),$$

where in each line the first equality is a definition and the second equality holds since the process of continuous extension leaves index, extended index and nullity unchanged. Note that, by Theorem 7.5.2, $n(s)$ is different from zero if and only if $\lambda(s)$ is conjugate to $\lambda(0)$ along λ and that it gives the multiplicity of this conjugate point. Thus, by Proposition 5.6.3, $n(s)$ is different from zero only at finitely many points $s \in]0,1]$, and at each of those points it takes a finite value, see Figure 7.5. We shall now discuss the behavior of the functions i and $i_o :]0,1] \longrightarrow \mathbb{N}_0^\infty$, where $\mathbb{N}_0^\infty$ denotes the nonnegative integers including infinity. To that end we define, for $0 < s < s' \le 1$, a map $\kappa_{s,s'} : \mathfrak{H}_s \longrightarrow \mathfrak{H}_{s'}$ by

$$\begin{aligned}
(\kappa_{s,s'}(Z))(s'') = Z(2\,s'') \qquad & \text{for } 0 \le s'' \le \frac{1}{2}, \\
(\kappa_{s,s'}(Z))(s'') = 0 \qquad & \text{for } \frac{1}{2} < s'' \le 1.
\end{aligned} \qquad (7.61)$$

Note that this map is, indeed, well defined. (This construction does not work for H^2 curves. Therefore, the extension to H^1 curves was necessary.) Clearly, $\kappa_{s,s'}$ is linear, continuous, and injective (but not surjective, of course). Moreover, with the help of (7.57) we find that $\Phi_{s'}(Z', Z') = \Phi_s(Z, Z)$ for $Z' = \kappa_{s,s'}(Z)$. Hence, if Φ_s is negative definite (or negative semidefinite, respectively) on a certain subspace of $\mathfrak{H}_s$, then this subspace is mapped by $\kappa_{s,s'}$ onto a space of the same dimension on which $\Phi_{s'}$ is negative definite (or negative semidefinite, respectively). This implies that the functions i and i_o are monotonic, i.e.,

$$i(s) \le i(s') \quad \text{and} \quad i_o(s) \le i_o(s') \quad \text{for } 0 < s < s' \le 1. \qquad (7.62)$$

We now fix a parameter value $s \in]0,1]$ and decompose the Hilbert space $\mathfrak{H}_s \cong \mathfrak{H}$ into orthogonal subspaces $\mathfrak{H} = \mathfrak{H}^+ \oplus \mathfrak{H}^- \oplus \mathfrak{H}^o$, where $\mathfrak{H}^o$ is the kernel of Φ_s and Φ_s is positive definite on $\mathfrak{H}^+$ and negative definite on $\mathfrak{H}^-$. Since Φ_s depends continuously on s in the way outlined above, $\Phi_{s+\varepsilon}$ is still positive definite on $\mathfrak{H}^+$ and negative definite on $\mathfrak{H}^-$ for $|\varepsilon|$ sufficiently small, i.e.,

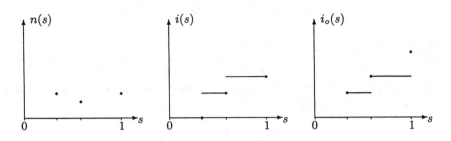

Fig. 7.5. The left-continuous function $i(s)$ and the right-continuous function $i_o(s)$ have jumps at those isolated points where the nullity $n(s)$ is different from zero, see the proof of Theorem 7.5.4.

$$i(s) \leq i(s + \varepsilon) \quad \text{and} \quad i_o(s) \geq i_o(s + \varepsilon) \tag{7.63}$$

for $|\varepsilon|$ sufficiently small. (If $s = 1$, ε must, of course, be negative. Otherwise, (7.63) holds for positive and negative ε.) From (7.62) and (7.63) we can determine the behavior of i and i_o in the following way, see Figure 7.5. On each open interval on which n is equal to zero, i and $i_o = i+n$ coincide. (7.63) shows that $i = i_o$ must be constant on such an interval. Moreover, (7.63) and (7.62) imply that i is left-continuous whereas i_o is right-continuous, i.e., $i(s-0) = i(s)$ and $i_o(s+0) = i_o(s)$. Thus, at each of the finitely many points s where $n(s) \neq 0$, the function i_o jumps by an amount of $i_o(s) - i(s) = n(s)$. This gives the equality

$$i_o(1) = i_o(\varepsilon) + \sum_s n(s) \tag{7.64}$$

where ε must be so small that n vanishes on the interval $]0, \varepsilon]$. From Proposition 5.5.5 we know that a sufficiently short ray can be associated with a classical solution of the eikonal equation. We can thus apply Theorem 7.5.3 (a) to the ray λ^ε. This shows that $i_o(\varepsilon) = i(\varepsilon)$ must be equal to zero. Together with (7.64), this proves the desired result. □

Specialized to the ray-optical structure of Example 5.1.5, where the rays are the geodesics of a Riemannian metric, Theorem 7.5.4 reproduces the classical Morse index theorem for Riemannian geodesics. As a matter of fact, our proof of Theorem 7.5.4 followed the proof of the classical Morse index theorem, as it is given, e.g., in Bishop and Crittenden [15], Chap. 11, as closely as possible.

Specialized to the ray-optical structure of Example 5.1.2, where the rays are the time-like geodesics of a Lorentzian metric, Theorem 7.5.4 reproduces the Morse index theorem for time-like geodesics, cf. Beem, Ehrlich, and Easley [11], Sect. 10.1.

For applications of the Morse index theorem one usually restricts to the case that $\lambda(1)$ is not conjugate to $\lambda(0)$ along λ, i.e., that the Hessian is non-degenerate. In this case (7.56) has the following consequences.

λ is free of conjugate points if and only if λ is a local minimum of $A_{q,q'}$. There is a point $\lambda(s)$ conjugate to $\lambda(0)$ along λ, for some $s \in]0,1[$, if and only if λ is a saddle-point of $A_{q,q'}$. Maxima cannot occur since, by Proposition 5.6.3, the right-hand side of (7.56) is finite.

8. Applications

In this chapter we illustrate our results with examples and indicate some applications to astrophysics and astronomy. In the beginning we show how our formalism can be used to reobtain some standard textbook results. Later we are going to give some more sophisticated applications.

8.1 Doppler effect, aberration, and drag effect in isotropic media

For a light ray passing through a medium, a moving observer will register (a) a different frequency, (b) a different spatial direction and (c) a different velocity in comparison to an observer who is at rest with respect to the medium. It is our goal to calculate the respective formulae for an isotropic medium, thereby determining (a) the Doppler effect, (b) the aberration and (c) the drag effect in such a medium. We perform these calculations on an arbitrary Lorentzian spacetime manifold for a medium in arbitrary motion. However, in essence this is an exercise in special relativity since only algebraic calculations on tangent spaces are involved.

According to Sect. 6.3, light propagation in an isotropic medium on a Lorentzian spacetime manifold $(\mathcal{M}, g)$ is given by a Hamiltonian in terms of an optical metric g_o,

$$H(x,p) = \tfrac{1}{2}\, g_o^{ab}(x)\, p_a\, p_b =$$

$$\tfrac{1}{2} \left(\frac{g^{ab}(x) + U^a(x)\, U^b(x)}{n(x)^2} - U^a(x)\, U^b(x) \right) p_a\, p_b. \tag{8.1}$$

Here the g^{ab} are the contravariant components of the spacetime metric g, the U^a are the components of a vector field U on $\mathcal{M}$ with $g_{ab}\, U^a\, U^b = -1$ that gives the rest system of the medium, and the function $n\colon \mathcal{M} \longrightarrow [1, \infty[$ is the index of refraction. By assuming that n is bounded by 1 and independent of the frequency $-U^a p_a$ we restrict to ray-optical structures that are causal and dilation-invariant. The latter restriction means that the following results apply to non-dispersive media only. At the end of this section we shall briefly comment on the dispersive case.

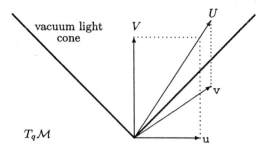

Fig. 8.1. u is the spatial velocity of the medium in the reference frame V and $-v$ is the spatial velocity of the V-observers in the rest system of the medium.

We now consider another vector field V with $g_{ab}\, V^a V^b = -1$. The relative velocity of the observer field V with respect to the observer field U is given by a function $\beta\colon \mathcal{M} \longrightarrow [0,1[$ defined by

$$g_{ab}\, U^a V^b = -\frac{1}{\sqrt{(1-\beta^2)}}\,. \tag{8.2}$$

Here we assume that U and V point into the same half of the g-cone. Then the normalization conditions on U and V imply that, indeed, $g_{ab}\, U^a V^b \le -1$; so β is well-defined. With the help of this function β we introduce vector fields u and v via

$$\mathrm{u}^a = \sqrt{1-\beta^2}\, U^a - V^a \quad \text{and} \quad \mathrm{v}^a = U^a - \sqrt{1-\beta^2}\, V^a\,, \tag{8.3}$$

which obviously satisfy

$$\begin{aligned} g_{ab}\, \mathrm{u}^a V^b &= g_{ab}\, \mathrm{v}^a U^b = 0\,, \\ g_{ab}\, \mathrm{u}^a \mathrm{u}^b &= g_{ab}\, \mathrm{v}^a \mathrm{v}^b = \beta^2\,. \end{aligned} \tag{8.4}$$

u is the spatial velocity vector field of the medium in the reference frame V, whereas $-v$ is the spatial velocity vector field of the V-observers in the rest system of the medium, see Figure 8.1.

At each point of a light ray its momentum can be decomposed, with respect to the observer field V and with respect to the observer field U, into frequency and spatial wave covector according to (6.8) and (6.9),

$$\omega = -p_a V^a \quad \text{and} \quad k_a = p_a - \omega\, g_{ab}\, V^b\,, \tag{8.5}$$

$$\omega^* = -p_a U^a \quad \text{and} \quad k_a^* = p_a - \omega^*\, g_{ab}\, U^b\,. \tag{8.6}$$

Here and in the following quantities in the rest system are marked with an asterisk. In terms of these quantities the dispersion relation $g_o^{ab}\, p_a p_b = 0$ can be written in either of the two following forms.

$$(1 - \beta^2)(g^{ab}\, k_a\, k_b - \omega^2) - (n^2 - 1)(k_a\, u^a - \omega)^2 = 0 \,, \tag{8.7}$$

$$g^{ab}\, k_a^*\, k_b^* - n^2\, \omega^{*2} = 0 \tag{8.8}$$

Moreover, a quick calculation shows that

$$k_a\, u^a = \omega - \sqrt{1 - \beta^2}\, \omega^* \,, \tag{8.9}$$

$$k_a^*\, \mathfrak{v}^a = \sqrt{1 - \beta^2}\, \omega - \omega^* \,. \tag{8.10}$$

The spatial direction of a ray is determined by its ray velocity (6.11) which can be calculated with respect to the observer field V and with respect to the observer field U. With the Hamiltonian (8.1) we find

$$v^a = -\frac{g_o^{ab}\, p_b}{V^c\, g_{cd}\, g_o^{de}\, p_e} - V^a \,, \tag{8.11}$$

$$v^{*a} = -\frac{g_o^{ab}\, p_b}{U^c\, g_{cd}\, g_o^{de}\, p_e} - U^a \,. \tag{8.12}$$

If we use the dispersion relation, a straight-forward calculation puts (8.11) and (8.12) into the forms

$$v^a = \frac{\sqrt{1 - \beta^2}\, g^{ab}\, k_b + (n^2 - 1)\, \omega^*\, u^a}{\sqrt{1 - \beta^2)}\, \omega + (n^2 - 1)\, \omega^*} \,, \tag{8.13}$$

$$v^{*a} = \frac{g^{ab}\, k_b^*}{\omega^*\, n^2} \,. \tag{8.14}$$

This shows that the ray velocity is not parallel to the spatial wave vector unless in the rest system of the medium. (In the vacuum case $n = 1$, every observer field can be viewed as the rest system.) To characterize the spatial direction of the ray we introduce angles θ and θ^* via

$$g_{ab}\, v^a\, u^b = \sqrt{g_{ab}\, v^a\, v^b}\, \sqrt{g_{ab}\, u^a\, u^b}\, \cos\theta \,, \tag{8.15}$$

$$g_{ab}\, v^{*a}\, \mathfrak{v}^b = \sqrt{g_{ab}\, v^{*a}\, v^{*b}}\, \sqrt{g_{ab}\, \mathfrak{v}^a\, \mathfrak{v}^b}\, \cos\theta^* \,. \tag{8.16}$$

We are now ready to derive the desired results.

(a) Doppler effect

After our preparations, the Doppler formula is easily derived by inserting (8.14) into (8.16). With the help of (8.8) and (8.10) this results in

$$\omega = \omega^* \frac{1 + n\,\beta\,\cos\theta^*}{\sqrt{1 - \beta^2}} \,. \tag{8.17}$$

In the vacuum case $n = 1$, (8.17) reduces to the standard Doppler formula which is given in any textbook on special relativity for the case that U and V are inertial systems on Minkowski space. Our argument proves the (rather trivial) fact that, pointwise, the same formula holds for observers in arbitrary motion on an arbitrary Lorentzian spacetime manifold. From (8.17) we read that the transverse Doppler effect ($\theta^* = \pi/2$) is unaffected by n. This reflects the well-known fact that the transverse Doppler effect is caused by time dilation alone. Linearization of the relativistic Doppler formula with respect to β yields the classical Doppler formula. The quadratic corrections to this formula were verified for the first time by Ives and Stilwell [65] in a laboratory experiment with canal rays, cf., e.g., French [44], Sect. 5.7.

The Doppler formula (8.17) should not be confused with the redshift formula (6.23). Contrary to (6.23), (8.17) compares two frequencies at the same point with respect to two different observers. Whenever frequency measurements at two different points are to be compared one should use the redshift formula (6.23). The latter is of paramount importance in cosmology where the influence of a medium is usually considered to be negligible. The redshift formula in a medium has some relevance in view of precision experiments with so-called *microwave links* in our Solar system, see, e.g., Bertotti [13]. In these experiments, microwaves are exchanged between two spacecrafts or between a spacecraft and the Earth, and the emitted and received frequencies are measured with a relative accuracy of 10^{-14} or 10^{-15}. Owing to this high accuracy, the influence of the interplanetary medium (or, for signals grazing the Sun, of the Solar corona) on the frequency shift is very well measurable in experiments of this kind. It is true that such frequency measurements with microwave links are usually called "Doppler measurements"; nonetheless, it is not the Doppler formula (8.17) but rather the redshift formula (6.23) which provides a theoretical basis for those measurements.

As a typical application of the Doppler formula (8.17) to astronomy we consider an inertial system V on Minkowski space and we assume, as an idealization, that our galaxy is at rest with respect to V in the temporal average. We assume that the worldline of the Earth is an integral curve of U. Then, along the worldline of the Earth, the function β defined by (8.2) gives the velocity of the Earth relative to V in units of the vacuum velocity of light. This is mainly determined by the orbital motion of the Solar system around the center of our galaxy, with smaller corrections coming (i) from the peculiar motion of the Solar system, (ii) from the yearly rotation of the Earth around the Sun, and (iii) from the daily rotation of the Earth. This orbital motion takes place with a velocity of about $\beta = 0.00083$ which corresponds to 250 km/s in conventional units. With $\theta^* = \pi$ and $n = 1$ (8.17) yields $\omega^* = 1.00083\,\omega$. Thus, an observer on the Earth sees starlight coming to us ("head-on") from the apex of the Solar motion blueshifted by about 0.083 % in comparison to a fictitious observer at the same place who is at rest with respect to our galaxy. For light coming to us at $\theta^* = \frac{\pi}{2}$, the Doppler effect is purely transverse and yields a tiny redshift of only 0.00003 %. Since these calculations were done with $n = 1$, the influence of our atmosphere was ignored. In the optical regime the atmosphere can be treated as an isotropic non-dispersive medium with $n = 1.0003$. To within the given accuracy, this leaves the above results unchanged. When using the Doppler formula (8.17) in situations like that with $n \neq 1$ it is important to keep in mind that not only the observed frequency ω^* but also the fiducial frequency ω is to be measured in the medium.

(b) Aberration

We now turn to the derivation of the aberration formula by inserting (8.13) into (8.15). After some algebraic manipulations using (8.7), (8.9), and (8.17), we find

$$\cos \theta = \frac{\cos \theta^* + n\,\beta}{\sqrt{(n + \beta\,\cos \theta^*)^2 - (1 - \beta^2)(n^2 - 1)}}. \qquad (8.18)$$

Please note that, by (8.15) and (8.16), θ and θ^* are defined in terms of the ray velocity ($=$ group velocity) and not in terms of the phase velocity. Thus, (8.18) gives the aberration of rays, as it is measured with an ordinary telescope, and not the aberration of wave surfaces, as it is measured with adaptive optics devices. This makes a difference since, as long as $n \neq 1$, the direction of the ray velocity does not coincide with the direction of the phase velocity.

Setting $n = 1$ in (8.18) yields the standard aberration formula for vacuum which is given in any textbook on special relativity for the case that U and V are inertial systems on Minkowski space. With $\beta = 9.92 \cdot 10^{-5}$ (orbital velocity of the Earth around the Sun, in units of the vacuum velocity of light) and $\theta = \frac{\pi}{2}$ (light coming from a star S at the pole of the ecliptic), this vacuum aberration formula yields $\cos \theta^* = -0.000099$. Hence, at the celestial sphere of an observer on the Earth the star S performs a yearly circular motion with radius $20.5''$ around the pole of the ecliptic. By an analogous calculation, a star which is not at the pole of the ecliptic performs a yearly elliptical motion with major semi-axis $20.5''$. This effect was observed already in 1728 by Bradley.

It was found by Airy in the 19th century that the aberrational ellipses are unchanged if they are measured with a telescope filled with water ($n \cong 1.5$) rather than with air ($n \cong 1$), cf., e.g., Preston [121], p. 538. At first sight, this result seems to be at variance with (8.18). However, (8.18) only says that the *relation* between θ and θ^* depends on n. A deeper analysis shows that, if the telescope is filled with water rather than with air, the observed angle θ^* remains unaffected whereas the fiducial angle θ changes. The situation is quite analogous to the Doppler effect. Both the Doppler formula and the aberration formula give a relation between two quantities measured by different observers *at the same place in the same medium*.

(c) Drag effect

Our next goal is to visualize the dependence of the ray velocity on the spatial direction. By (8.11), the dispersion relation $g_o^{ab}\, p_a\, p_b = 0$ implies

$$(g_o)_{ab} \left(v^a + V^a \right) \left(v^b + V^b \right) = 0 . \tag{8.19}$$

Here the

$$(g_o)_{ab} = n^2 g_{ab} + (n^2 - 1) g_{ac} U^c g_{bd} U^d \tag{8.20}$$

are the covariant components of the optical metric, $(g_o)_{ab} \, g_o^{bc} = \delta_a^c$. After some algebraic manipulations (8.19) takes the form

$$n^2 \left(1 - \beta^2 \right) \left(g_{ab} \, v^a \, v^b - 1 \right) + (n^2 - 1) \left(g_{ab} \, u^a \, v^b - 1 \right)^2 = 0 . \tag{8.21}$$

This equation demonstrates that the indicatrix (6.14) of a non-dispersive isotropic medium with respect to an arbitrary observer field V is an ellipsoid, see Figure 8.2. Our causality assumption $n \geq 1$ implies that this ellipsoid is completely contained in the vacuum light sphere, $g_{ab} \, v^a \, v^b \leq 1$. If we pass to the rest system, the indicatrix turns into a sphere,

$$g_{ab} \, v^{*a} \, v^{*b} = \tfrac{1}{n^2} . \tag{8.22}$$

For the sake of completeness we also calculate the figuratrix (6.13) to visualize the dependence of the phase velocity on spatial directions. The definition (6.10) of the phase velocity implies that

$$k_a = \frac{\omega \, w_a}{g^{ab} \, w_a \, w_b} . \tag{8.23}$$

With (8.23), the dispersion relation (8.7) yields

$$\begin{aligned} (1 - \beta^2) \, g^{ab} \, w_a \, w_b \left(1 - g^{ab} \, w_a \, w_b \right) = \\ (n^2 - 1) \left(g^{ab} \, w_a \, w_b - u^a \, w_a \right)^2 . \end{aligned} \tag{8.24}$$

The figuratrix is, thus, a fourth order surface, see Figure 8.2. Our assumption $n \geq 1$ guarantees that $g^{ab} \, w_a \, w_b \leq 1$, i.e., not only the ray velocity but also the phase velocity is bounded by the vacuum veclocity of light. If we pass to the rest system of the medium, the figuratrix turns into a sphere,

$$g^{ab} \, w_a^* \, w_b^* = \tfrac{1}{n^2} . \tag{8.25}$$

Hence, for the rest system figuratrix and indicatrix coincide if we identify tangent space and cotangent space in the usual way with the help of the spacetime metric.

For rays parallel or antiparallel to the relative motion we can use the relations $g_{ab} \, v^a \, u^b = \pm \beta \sqrt{g_{cd} \, v^c \, v^d}$ and $w_a \, u^a = \pm \beta \sqrt{g^{cd} \, w_c \, w_d}$. In this situation (8.21) and (8.24) imply

$$\sqrt{g_{ab} \, v^a \, v^b} = \sqrt{g^{ab} \, w_a \, w_b} = \frac{\frac{1}{n} \mp \beta}{1 \mp \frac{\beta}{n}} . \tag{8.26}$$

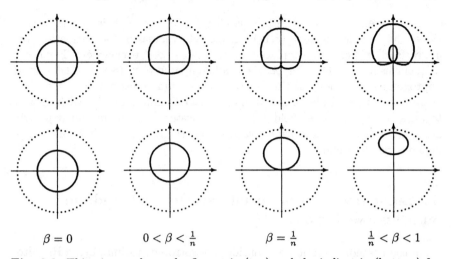

$$\beta = 0 \qquad\qquad 0 < \beta < \tfrac{1}{n} \qquad\qquad \beta = \tfrac{1}{n} \qquad\qquad \tfrac{1}{n} < \beta < 1$$

Fig. 8.2. This picture shows the figuratrix (top) and the indicatrix (bottom) for different values of the observer's velocity β in an isotropic medium with index of refraction $n > 1$. The vertical axis is chosen parallel to the observer's velocity relative to the medium. Analytically, the figuratrix is given by the fourth order equation (8.24) whereas the indicatrix is an ellipsoid given by (8.21). The dashed circle indicates the vacuum light sphere, i.e., figuratrix and indicatrix for $n = 1$. Please note that for $\tfrac{1}{n} < \beta < 1$ the observer's velocity exceeds the velocity of light in the medium whereas it is still limited by the vacuum velocity of light. In each case the intersections with the vertical axis are determined by (8.26).

Note that, by (8.22) and (8.25), $1/n$ is the absolute value of the ray velocity and of the phase velocity in the rest system of the medium. Hence, (8.26) says that, in the relative direction of motion, both the ray velocity and the phase velocity obey the familiar relativistic addition theorem for spatial velocities. For the phase velocity, this result can be tested by interference experiments with light propagating through moving fluids. Such experiments have been performed by Fizeau in 1851 who verified the equation

$$\sqrt{g^{ab}\, w_a\, w_b} = \tfrac{1}{n} \mp \beta \left(1 - \tfrac{1}{n^2}\right) \tag{8.27}$$

which was heuristically suggested already earlier by Fresnel. Obviously, (8.27) follows from (8.26) by neglecting quadratic and higher order terms in β. On the basis of 19th century physics, the factor $(1 - \tfrac{1}{n^2})$ in (8.27) was hard to understand. If light propagates in an ether, and if spatial velocities are to be added in the Newtonian way, then (8.27) seems to suggest that the ether is "partially dragged along" by the medium. If we stick to this outdated terminology, Figure 8.2 illustrates the *drag effect* in an isotropic medium for all spatial directions.

We end this discussion with a quick remark on generalizations to dispersive isotropic media. From Sect. 6.3 we know that then the Hamiltonian (8.1) is still valid, but now n is not only a function of the spacetime point but also of the frequency $\omega^* = -U^a p_a$. It is easy to check that this generalization leaves the Doppler formula (8.17) unchanged, whereas in the aberration formula (8.18) n has to be replaced by $n + \omega^* \, \partial n / \partial \omega^*$ everywhere. For those frequencies for which $\omega^* \, \partial n / \partial \omega^*$ is small compared to n, (8.18) is still a valid approximation. As long as the function $\omega^* \longmapsto n(\omega^*)$ has not been specified, nothing can be said about the form of indicatrix and figuratrix with respect to an arbitrary observer field. In the rest system, indicatrix and figuratrix are spheres as in the non-dispersive case, but the radius of either sphere now depends on the frequency.

8.2 Light rays in a uniformly accelerated medium on Minkowski space

As in the preceding section we consider an isotropic medium, i.e., a Hamiltonian of the form (8.1). This time we specialize to the case that the spacetime metric is the Minkowski metric,

$$g = (dx^1)^2 + (dx^2)^2 + (dx^3)^2 - (dx^4)^2 \,, \qquad (8.28)$$

and we restrict our considerations to the subset

$$\mathcal{M} = \{ (x^1, x^2, x^3, x^4) \in \mathbb{R}^4 \,|\, (x^3)^2 > (x^4)^2 \,\} \qquad (8.29)$$

of Minkowski space. The index of refraction is supposed to be a constant $n \geq 1$ and the medium is supposed to be in uniformly accelerated motion,

$$U = \frac{1}{\sqrt{(x^3)^2 + (x^4)^2}} \left(x^4 \frac{\partial}{\partial x^3} + x^3 \frac{\partial}{\partial x^4} \right) \,, \qquad (8.30)$$

see Figure 8.3. The integral curves of this vector field are known as *Rindler observers*, and $\mathcal{M}$ is known as the *Rindler wedge*, cf. Rindler [123], Sect. 8.6. For the calculation of light rays in this medium it is convenient to introduce new coordinates (x, y, z, t) via

$$x^1 = x \,,$$
$$x^2 = y \,,$$
$$x^3 = z \cosh t \,, \qquad (8.31)$$
$$x^4 = z \sinh t \,.$$

The momenta transform according to

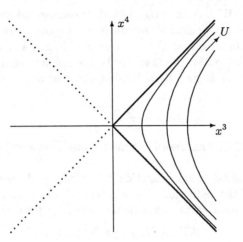

Fig. 8.3. The Rindler observer field U occupies a wedge-shaped region of Minkowski space.

$$p_1 = p_x \,,$$

$$p_2 = p_y \,,$$

$$p_3 = p_z \cosh t - \frac{p_t}{z} \sinh t \,, \tag{8.32}$$

$$p_4 = -p_z \sinh t + \frac{p_t}{z} \cosh t \,.$$

In the new coordinates, the Minkowski metric reads

$$g = dx^2 + dy^2 + dz^2 - z^2 \, dt^2 \,, \tag{8.33}$$

the Rindler wedge is represented as

$$\mathcal{M} = \{\, (x, y, z, t) \in \mathbb{R}^4 \mid z > 0 \,\}, \tag{8.34}$$

and the observer field (8.30) takes the simple form

$$U = \frac{1}{z} \frac{\partial}{\partial t} \,. \tag{8.35}$$

Inserting into (8.1) yields the Hamiltonian

$$H\big(x, y, z, t, p_x, p_y, p_z, p_t\big) = \frac{1}{2} \left(\frac{p_x^2 + p_y^2 + p_z^2}{n^2} - \frac{p_t^2}{z^2} \right). \tag{8.36}$$

As the coordinate t does not appear in the Hamiltonian H, the dispersion relation $H = 0$ determines a stationary ray-optical structure in the sense of

Definition 6.5.1, $W = \partial/\partial t \in \mathcal{G}_{\mathcal{N}}$. (As W is orthogonal to the hypersurfaces $t = $ const., this ray-optical structure is even globally static.) By Proposition 6.2.2, this implies that the function $f = \ln z$ is a redshift potential for the observer field (8.35). In other words, the redshift under which a Rindler observer at $z = z_1$ is seen by a Rindler observer at $z = z_2$ is given by the formula

$$\frac{\omega_2}{\omega_1} = \frac{z_1}{z_2}, \tag{8.37}$$

cf. equation (6.26). This result is true for any (constant) value of the index of refraction n.

With the global timing function t and any real constant $\omega_o \neq 0$, all assumptions of the reduction theorem (i.e., of Theorem 6.5.1) are satisfied. This gives us a reduced ray-optical structure $\hat{\mathcal{N}}_{\omega_o}$ for each $\omega_o \neq 0$ on

$$\hat{\mathcal{M}} = \left\{ (x, y, z) \in \mathbb{R}^3 \mid z > 0 \right\}. \tag{8.38}$$

By (6.78), we find a Hamiltonian $\hat{H}$ for this reduced ray-optical structure simply by setting p_t equal to $-\omega_o$ in (8.36),

$$\hat{H}(x, y, z, p_x, p_y, p_z) = \frac{1}{2n^2}\left(p_x^2 + p_y^2 + p_z^2\right) - \frac{\omega_o^2}{2z^2}. \tag{8.39}$$

Hence the dispersion relation of $\hat{\mathcal{N}}_{\omega_o}$ takes the form

$$\hat{g}^{\mu\nu} p_\mu p_\nu = n^2 \omega_o^2 \tag{8.40}$$

where the $\hat{g}^{\mu\nu}$ are the contravariant components of the Riemannian metric

$$\hat{g} = \frac{dx^2 + dy^2 + dz^2}{z^2}. \tag{8.41}$$

The Riemannian manifold $(\hat{\mathcal{M}}, \hat{g})$ is the socalled *Poincaré half-space* which is dicussed in many textbooks on differential geometry, see, e.g., Thorpe [143], p. 236 and p. 242.

We have thus shown that the rays of $\hat{\mathcal{N}}_{\omega_o}$ coincide with the geodesics of the Poincaré half-space. It is well known and easily verified that the latter are all those half-circles in $\hat{\mathcal{M}}$ that meet the surface $z = 0$ orthogonally, see Figure 8.4. Please note that the rays of $\hat{\mathcal{N}}_{\omega_o}$ are independent of the (constant) index of refraction n. They are, of course, also independent of ω_o which reflects the fact that our medium is non-dispersive.

This calculation exemplifies our findings of Sect. 6.6. There we have seen that the rays of a reduced ray-optical structure are the geodesics of a Riemannian metric $\hat{g}$ whenever the following two properties are satisfied. The stationary ray-optical structure to which the reduction formalism is applied must be given as the null cone of a Lorentzian metric g_o, and the time-like vector field $W \in \mathcal{G}_{\mathcal{N}}$ must be hypersurface-orthogonal with respect to g_o. In particular, the optical path length is then given as the $\hat{g}$-length which implies that Fermat's principle reduces to the geodesic variational problem for the metric $\hat{g}$.

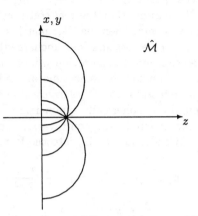

Fig. 8.4. The rays of $\hat{\mathcal{N}}_{\omega_o}$ are the geodesics of the Poincaré half-space which are half-circles.

8.3 Light propagation in a plasma on Kerr spacetime

If we consider the plasma model of Chap. 3, light rays propagating in a non-magnetized plasma on an arbitrary Lorentzian spacetime manifold (M, g) are determined by a Hamiltonian H of the form

$$H(x, p) = \tfrac{1}{2} \left(g^{ab}(x)\, p_a\, p_b + \omega_p(x)^2 \right) . \tag{8.42}$$

Here the g^{ab} are the contravariant components of the spacetime metric g and the spacetime function ω_p is the "plasma frequency" which is determined by the electron density of the plasma according to (3.51). More precisely, we have seen in Chap. 3 that our plasma model gives a dispersion relation with three branches, determined by three Hamiltonians (3.44), (3.45) and (3.46), and that only the third Hamiltonian, which is of the form (8.42), is associated with light rays *passing through* the plasma. If the plasma frequency has no zeros (i.e., if the plasma covers the whole spacetime region under consideration), the ray-optical structure determined by the Hamiltonian (8.42) is of the kind considered in Example 5.1.2.

   In this section we want to discuss the rays of this ray-optical structure for the special case that the underlying Lorentzian manifold (M, g) is the *Kerr* spacetime. In Boyer-Lindquist coordinates $(r, \vartheta, \varphi, t)$, the Kerr metric reads

$$g = \frac{\rho^2}{\Delta}dr^2 + \rho^2 d\vartheta^2 +$$
$$(r^2 + a^2)\sin^2\vartheta\, d\varphi^2 - dt^2 + \frac{2mr}{\rho^2}(a\sin^2\vartheta\, d\varphi - dt)^2 , \tag{8.43}$$

where $\rho^2 = r^2 + a^2 \cos^2 \vartheta$ and $\Delta = r^2 - 2mr + a^2$, see, e.g., Hawking and Ellis [59], Sect. 5.6. We assume that the real constants m and a satisfy the conditions $m > 0$ and $a^2 < m^2$. Then the Kerr metric mathematically models the spacetime region around a rotating (but uncharged) black hole with mass m and angular momentum ma. In the region where r is large enough it also gives a valid approximation for the spacetime around a rotating star. For $a = 0$ the Kerr metric reduces to the *Schwarzschild* metric which models the spacetime region around any spherically symmetric massive body.

From (8.43) we can calculate the contravariant components g^{ab} of the metric. This puts the Hamiltonian (8.42) into the form

$$H(r, \vartheta, \varphi, t, p_r, p_\vartheta, p_\varphi, p_t) = \frac{\Delta p_r^2 + p_\vartheta^2}{2 \rho^2} + \tag{8.44}$$

$$\frac{\rho^2 - 2mr}{2 \Delta \rho^2 \sin^2 \vartheta} \left(p_\varphi - \frac{2mra\sin^2\vartheta}{\rho^2 - 2mr} p_t \right)^2 - \frac{\rho^2 p_t^2}{2\rho^2 - 4mr} + \frac{\omega_p^2}{2}.$$

It should be noted that the Kerr metric is a vacuum solution of Einstein's field equation. Hence, the use of the Hamiltonian (8.44) is physically justified as long as the gravitational field produced by the plasma can be neglected.

In the following we restrict to the region where the vector field $\partial/\partial t$ is time-like, i.e.,

$$r > m + \sqrt{m^2 - a^2 \cos \vartheta}. \tag{8.45}$$

This is the region outside the socalled *ergosphere*. Moreover, we assume that the plasma frequency is independent of t whereas it may depend arbitrarily on r, φ and ϑ. In other words, we assume that the electron density of the plasma is stationary. Please note that, for our plasma model, the *velocity* of the plasma has no influence on the light rays and can therefore be arbitrary. Under these assumptions the vector field $W = \partial/\partial t$ generates a time-like symmetry, i.e., our ray-optical structure is stationary in the sense of Definition 6.5.1, and the coordinate function t is a timing function in the sense of Definition 6.5.2.

We want to carry through the reduction process of Theorem 6.5.1 in order to dicuss the spatial paths of light rays. To that end we have to choose a real number $\omega_o \neq 0$ for the frequency and we have to restrict to the region where the inequality

$$\omega_p^2 < \frac{\rho^2}{\rho^2 - 2mr} \omega_o^2 \tag{8.46}$$

is satisfied. It can be read directly from (8.44) that a ray with $p_t = -\omega_o$ cannot exist outside this region. It is then easy to check that all hypotheses of Theorem 6.5.1 are satisfied, i.e., that we get a reduced ray-optical structure $\hat{\mathcal{N}}_{\omega_o}$ on the 3-manifold $\hat{\mathcal{M}}$ determined by the inequalities (8.45) and (8.46). We get a Hamiltonian for this reduced ray-optical structure simply by replacing the conserved momentum coordinate p_t in (8.44) by the constant $-\omega_o$, please cf. (6.78). As always, we are free to multiply this Hamiltonian with an arbitrary nowhere vanishing function. This shows that

$$\hat{H}\left(r, \vartheta, \varphi, p_r, p_\vartheta, p_\varphi\right) = \qquad (8.47)$$

$$\frac{1}{2\,\omega_o}\left(\frac{\Delta\,p_r^2 + p_\vartheta^2 + \frac{\rho^2 - 2\,m\,r}{\Delta \sin^2\vartheta}\left(p_\varphi + \frac{2\,m\,r\,a\,\omega_o\sin^2\vartheta}{\rho^2 - 2\,m\,r}\right)^2}{\rho^2\left(\frac{\rho^2}{\rho^2 - 2\,m\,r} - \frac{\omega_p^2}{\omega_o^2}\right)} - \omega_o^2\right)$$

is a Hamiltonian for the reduced ray-optical structure $\hat{\mathcal{N}}_{\omega_o}$. This Hamiltonian $\hat{H}$ is of the form (6.103), with the Riemannian metric $\hat{g}$ and the one-form $\hat{\phi}$ given by

$$\hat{g} = \rho^2\left(\frac{\rho^2}{\rho^2 - 2\,m\,r} - \frac{\omega_p^2}{\omega_o^2}\right)\left(\frac{dr^2}{\Delta} + d\vartheta^2 + \frac{\Delta\sin^2\vartheta}{\rho^2 - 2\,m\,r}\,d\varphi^2\right), \qquad (8.48)$$

$$\hat{\phi} = \frac{2\,m\,r\,a\sin^2\vartheta}{\rho^2 - 2\,m\,r}\,d\varphi. \qquad (8.49)$$

The lifted rays of $\hat{\mathcal{N}}_{\omega_o}$ are, thus, determined by (6.104), (6.105), and (6.106), whereas the rays are determined by (6.107) and (6.108). In analogy to (6.111), the optical path length takes the form

$$\mathcal{I} = \pm\int_{s_1}^{s_2}\left(\rho\sqrt{\frac{\rho^2}{\rho^2 - 2\,m\,r} - \frac{\omega_p^2}{\omega_o^2}}\sqrt{\frac{\dot{r}^2}{\Delta} + \dot{\vartheta}^2 + \frac{\Delta\sin^2\vartheta}{\rho^2 - 2\,m\,r}\,\dot{\varphi}^2}\right)(s)\,ds -$$

$$\int_{s_1}^{s_2}\left(\frac{2\,m\,r\,a\sin^2\vartheta}{\rho^2 - 2\,m\,r}\,\dot{\varphi}\right)(s)\,ds. \qquad (8.50)$$

By Fermat's principle, the light rays of frequency ω_o between any two points in $\hat{\mathcal{M}}$ are the extremals of the functional $\mathcal{I}$. In the Schwarschild case $a = 0$, the rays are exactly the $\hat{g}$-geodesics, otherwise they are modified by a kind of Coriolis force. Contrary to the situation considered in (6.111), here the metric $\hat{g}$ depends on the frequency ω_o, thereby reflecting the fact that our plasma is a dispersive medium. By the same token, the optical path length functional (8.3) does not give the travel time with respect to the timing function t, unless in the vacuum case $\omega_p = 0$. Please note that the limit $\omega_o \to \infty$ leads to the same result as setting the function ω_p equal to zero; hence, in the limit of infinite frequency the rays approach the vacuum rays.

In the following we want to use these general results to calculate the total angular deflection of light rays in the equatorial plane $\vartheta = \pi/2$. From now on we assume that the plasma frequency ω_p is a function of r alone, i.e., that the electron density of the plasma is rotationally symmetric. Then the φ-component of (6.105) takes the form

$$\dot{\varphi} = \frac{(r - 2m)\left(\frac{p_\varphi}{\omega_o} + \frac{2ma}{r - 2m}\right)}{r\left(r^2 - 2mr + a^2\right)\left(\frac{r}{r - 2m} - \frac{\omega_p(r)^2}{\omega_o^2}\right)} \tag{8.51}$$

and the φ-component of (6.106) says that p_φ is a constant of motion. On the other hand, (6.107) yields

$$\left(\frac{r}{r - 2m} - \frac{\omega_p(r)^2}{\omega_o^2}\right)\left(\frac{r^2 \dot{r}^2}{r^2 - 2mr + a^2} + \frac{\left(r^2 - 2mr + a^2\right) r \dot{\varphi}^2}{r - 2m}\right) = 1. \tag{8.52}$$

Upon dividing (8.3) by $\dot{\varphi}^2$ and using (8.51) on the right-hand side we find

$$\frac{r^2}{r^2 - 2mr + a^2}\left(\frac{dr}{d\varphi}\right)^2 + \frac{r\left(r^2 - 2mr + a^2\right)}{r - 2m} = \frac{r^2\left(r^2 - 2mr + a^2\right)^2\left(\frac{r}{r - 2m} - \frac{\omega_p(r)^2}{\omega_o^2}\right)}{(r - 2m)^2\left(\frac{p_\varphi}{\omega_o} + \frac{2ma}{r - 2m}\right)^2}. \tag{8.53}$$

For each possible choice of the constant of motion p_φ, this equation determines the orbits of the corresponding light rays. In the following we are only interested in light rays that come in from $r = \infty$, reach a minimum radial coordinate $r = R$, and go out to $r = \infty$ afterwards, i.e., we exclude all light rays that are captured by the central body. Then $dr/d\varphi$ must have a zero at $r = R$ and (8.53) allows to express the constant of motion p_φ in terms of R in the following way.

$$\left(\frac{p_\varphi}{\omega_o} + \frac{2ma}{R - 2m}\right)^2 = \frac{R\left(R^2 - 2mR + a^2\right)}{R - 2m}\left(\frac{R}{R - 2m} - \frac{\omega_p(R)^2}{\omega_o^2}\right). \tag{8.54}$$

With the help of this equation, (8.53) takes the form

$$\pm \frac{\sqrt{r(r - 2m)}}{r^2 - 2mr + a^2}\frac{dr}{d\varphi} = \sqrt{\frac{h(r)^2}{\left(\frac{2ma}{r - 2m} - \frac{2ma}{R - 2m} \pm h(R)^2\right)^2} - 1}, \tag{8.55}$$

where we have introduced the abbreviation

$$h(r) = \sqrt{\frac{r(r^2 - 2mr + a^2)}{r - 2m}\left(\frac{r}{r - 2m} - \frac{\omega_p(r)^2}{\omega_o^2}\right)}. \tag{8.56}$$

Solving (8.55) for $d\varphi$ and integrating over the whole ray results in

$$\Delta\varphi = \tag{8.57}$$

$$\pm 2 \int_R^\infty \frac{\sqrt{r(r-2m)}}{r^2 - 2mr + a^2} \left(\frac{h(r)^2}{\left(\frac{2ma}{r-2m} - \frac{2ma}{R-2m} \pm h(R)^2 \right)^2} - 1 \right)^{-1/2} dr ,$$

where the upper sign is valid for corotating rays ($\dot\varphi > 0$) and the lower sign is valid for counterrotating rays ($\dot\varphi < 0$). The difference between $\Delta\varphi$ and $\pm\pi$ gives the total deflection angle of the ray, see Figure 8.5. If the function $\omega_p(r)$ has been specified, this deflection angle can be calculated to arbitrary accuracy from (8.57), e.g., by numerical integration. The result depends, of course, on the frequency ω_o which is hidden in the function $h(r)$.

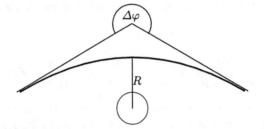

Fig. 8.5. The deviation of $\Delta\varphi$ from $\pm\pi$ gives the light deflection.

In the Schwarzschild case $a = 0$ the formula for the deflection angle simplifies to

$$|\Delta\varphi| = 2 \int_R^\infty \frac{dr}{\sqrt{r(r-2m)}\sqrt{\frac{h(r)^2}{h(R)^2} - 1}} , \tag{8.58}$$

where the function $h(r)$ is now given by

$$h(r) = r \sqrt{\frac{r}{r - 2m} - \frac{\omega_p(r)^2}{\omega_o^2}} . \tag{8.59}$$

This formula can be used, e.g., for calculating the deflection of light rays in the Solar corona. Phenomenological formulae for the electron density $\overset{\circ}{n}(r)$ and, thus, for the plasma frequency

$$\omega_p(r) = \frac{|e|}{\sqrt{m}} \sqrt{\overset{\circ}{n}(r)} \tag{8.60}$$

in the Solar corona can be found in the literature, see, e.g. Zheleznyakov [151]. Actually, the electron density in the Solar corona shows a considerable temporal variation, roughly snychronized with the Solar activity cycle of about 11 years. As an average, one often uses the socalled *Baumbach-Allen formula*

$$\overset{\circ}{n}(r) = \left(1.55\,\frac{r_o^6}{r^6} + 2.99\,\frac{r_o^{16}}{r^{16}}\right)\frac{10^8}{\mathrm{cm}^3}\,,\tag{8.61}$$

where r_o denotes the radius of the Sun. With $\omega_p(r)$ specified by such a phenomenological formula, the integral in (8.58) can be calculated numerically.

In the case $\omega_p(r) = 0$ or, equivalently, for $\omega_o \to \infty$, (8.58) gives the deflection of vacuum light rays in the Schwarzschild metric,

$$|\Delta\varphi| = 2\int_R^\infty \frac{R^2\,dr}{\sqrt{R\,(R-2\,m)\,r^4 - R^4\,r^2 + 2\,m\,R^4\,r}}\,.\tag{8.62}$$

If we linearize this elliptic integral with respect to m/R, we find

$$|\Delta\varphi| = \tag{8.63}$$

$$2\int_R^\infty \frac{R\,dr}{r\sqrt{r^2-R^2}} + \frac{2\,m}{R}\int_R^\infty \frac{R\,(r^3-R^3)\,dr}{r\,\sqrt{r^2-R^2}^3} + O\left(\frac{m^2}{R^2}\right)\,.$$

The two integrals on the right-hand side can be calculated in an elementary fashion with the substitution $u = R/r$. This results in the standard textbook formula

$$|\Delta\varphi| = \pi + \frac{4\,m}{R} + O\left(\frac{m^2}{R^2}\right)\tag{8.64}$$

for vacuum light rays in the Schwarzschild metric, cf., e.g., Wald [146], eq. (6.3.43), or Straumann [136], eq. (3.4.6).

The deflection given by formula (8.64) can be modeled with the help of a logarithmically shaped lens with an index of refraction $n > 1$, see Figure 8.6. For a rotationally symmetric lens with a profile given by the equation

$$x + k\ln y = \mathrm{const.}\,,\tag{8.65}$$

Snell's law implies that rays parallell to the axis are deflected by $\delta = \frac{k}{R}(n-1)$ up to terms quadratic in $\frac{k}{R}$. Comparison with (8.64) shows that, to within linear approximation, this value coincides with the deflection in a Schwarzschild spacetime where $\frac{k}{4}(n-1)$ corresponds to the mass m. (Here it goes without saying that one has to identify the deflection angle δ produced at the surface of the lense with the total deflection angle $\delta = |\Delta\varphi| - \pi$ in the Schwarzschild metric.) Thus, a lens with the appropriate logarithmic shape can be used to approximately visualizing light deflection by a spherically symmetric

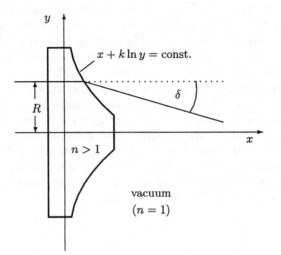

Fig. 8.6. To within certain approximations, the light deflection in a Schwarschild spacetime can be mimicked with the help of a logarithmically shaped lens.

gravitating body. Such plastic lenses have been actually manufactured and are often used in didactic demonstrations. For practical instructions and additional theoretical information we refer, e.g., to Higbie [64] and to Nandor and Helliwell [99].

Similarly to a lens in ordinary optics, a gravitational field can lead to multiple imaging or to the effect that a pointlike light source is seen as an extended object, e.g. as an arc or as a ring. In situations like that we speak of "gravitational lensing". This will be the topic of the next section.

8.4 Gravitational lensing

In the last section we have rediscovered the relevant formulae for light rays being curved by the gravitational field of a massive body. For a light ray not directly influenced by matter, passing a spherically symmetric body of mass m at a minimal radial distance R, the deflection angle is given by formula (8.62) or, to within linear approximation with respect to m/R, by formula (8.64). For a light ray grazing the surface of the Sun, $m \cong 1.5\,km$ and $R \cong 696\,000\,km$, this gives a deflection angle

$$\delta = |\Delta\varphi| - \pi \cong 1.75''. \qquad (8.66)$$

The simple assumption of light particles having a non-vanishing mass, leaving Newtonian physics unchanged otherwise, would lead to only half that value,

as was found by Johann von Soldner already in 1801, see Lenard [79]. It was
the greatest triumph in the history of general relativity when the relativistic
value (8.66) of the deflection angle was confirmed, to within tolerable error
bounds, by observations during a total Sun eclipse in the year 1919. Historical
details on the 1919 expedition, organized by the Royal Astronomical Society
of London and headed by Arthur Eddington, can be found, e.g., in Pais [103],
p. 303. In later years the development of radio telescopes made it possible to
measure the relativistic deflection of rays at any time, not just during a total
Sun eclipse, and with strongly increasing accuracy. Recent measurements,
using very-long-baseline interferometry, have confirmed the relativistic value
to within 0.02 %, see Lebach et al. [78]. Here the influence of the Solar corona
on the deflection of radio rays has to be taken into account. As a matter
of fact, nowadays measurements of this kind are performed chiefly with the
intention to gain information about the Solar corona.

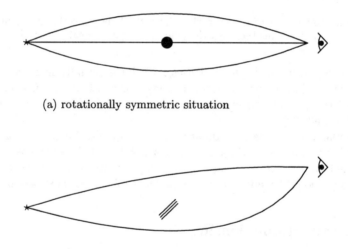

(a) rotationally symmetric situation

(b) non-symmetric situation

Fig. 8.7. In a rotationally symmetric situation, gravitational lensing can lead to
the effect that a pointlike light source is seen as a ring around the deflector. In a
non-symmetric situation, there might be a number of discrete images.

For an observer on the Earth, the deflection of starlight by the gravitational field of the Sun causes only a tiny distortion of the configurations.
However, much more drastic effects are possible if (i) the mass-to-radius ratio

of the deflecting body is bigger than that of the Sun, and/or (ii) the distance between observer and deflecting body is bigger than the distance between Earth and Sun. Then it is even possible that the observer sees more than one image of a light source at his or her celestial sphere. In a rotationally symmetric situation, the observer would see a pointlike light source as a ring around the deflector, in a less symmetric situation there might be a number of isolated images, see Figure 8.7. It has become common to speak of *gravitational lensing* in situations like that. This term was indirectly introduced by Lodge [85] who was the first to discuss the question of whether the effect of the gravitational field of the Sun upon light rays is similar to that of a lens. It should be mentioned that Lodge's discussion cannot be viewed as genuinely general-relativistic since it is based on an ether theory. (Incidentally, it is well known that Sir Oliver Lodge always maintained a skeptical if not rejecting attitude towards general relativity.) Therefore, it is better justified to credit Eddington [34] [35] and Chwolson [28] who independently pointed out the in-principle possibility of gravitational lensing on the basis of general relativity. In particular, Chwolson [28] was the first to mention the ring phenomenon depicted in Figure 8.7 (a). At that time the practical observability of gravitational lensing was a completely open question. In his only publication on this subject, Einstein [39] gave a deeply pessimistic view. (From a scribbled calculation in Einstein's private notebook, discovered only in the 1990s, we know that he had thought about multiple imaging by gravitational fields already in 1912, when the final formalism of general relativity was still to be found.) Zwicky [152] was the first to consider gravitational lensing by extragalactic objects, but his subsequent observations remained without success. It was not before 1979 that the first promising candidate for gravitational lensing was found. In that year Walsh, Carlswell, and Weyman [147] suggested that the double quasar 0957+561 is, actually, only one quasar which is gravitationally lensed by an intervening galaxy. By now, this explanation is accepted by a large majority of astrophysicists, and many other promising candidates for gravitational lensing have been found, including multiple quasars, radio rings, and luminous arcs. For detailed reviews we refer to Schneider, Ehlers, and Falco [128] and to Refsdal and Surdej [122]. In addition, the reader may consult a regularly updated electronic review by Wambsganss [145] and a forthcoming book on mathematical aspects of gravitational lensing by Petters, Levine and Wambsganss [115].

Purely spatial pictures, such as Figure 8.7, are appropriate to illustrate gravitational lensing in stationary situations only. In time-dependent situations (e.g., if the deflector is moving non-stationarily) it is inevitable to switch to a spacetime description. If, in addition, the effect of media on the light rays is to be taken into account, we are led to studying gravitational lensing in terms of ray-optical structures on Lorentzian manifolds, i.e., on general-relativistic spacetimes. In the following we discuss, within such a

differential-geometrical setting, the relevance of Fermat's principle for gravitational lensing. Later we specify to the stationary case.

   To that end we consider the following situation. In a 4-dimensional
Lorentzian manifold $(\mathcal{M}, g)$, to be interpreted as a general-relativistic space-
time, we fix a point $q \in \mathcal{M}$ and a time-like C^∞ embedding $\gamma: I \longrightarrow \mathcal{M}$
from a real interval I into $\mathcal{M}$. We interpret q as an event where an observation takes place, and we interpret γ as the worldline of a light source. The
parametrization of γ could be proper time, $g(\dot{\gamma}, \dot{\gamma}) = -1$, but any other
smooth parametrization would do as well. We interpret the parametrization
of γ as *past-pointing*, as indicated by the arrow in Figure 8.8.

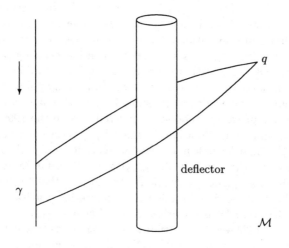

Fig. 8.8. In a gravitational lensing situation there are several light rays from a
light source γ to an observer q.

   We fix a ray-optical structure $\mathcal{N}$ on $\mathcal{M}$, thereby specifying the properties
of the optical medium in which light propagation is to be considered. To
avoid pathologies we assume that $\mathcal{N}$ is causal in the sense of Definition 6.1.1.
Then each light ray, emitted from the light source γ into the future and
received by the observer at q, corresponds to a ray $\lambda: [0, 1] \longrightarrow \mathcal{M}$ of the
ray-optical structure $\mathcal{N}$ with $\lambda(0) = q$ and $\lambda(1) = \gamma(T(\lambda))$, where $T(\lambda)$
denotes some parameter value, with the non-space-like vector $\dot{\lambda}(1)$ pointing
into the same half of the null cone bundle as the time-like vector $\dot{\gamma}(T(\lambda))$,
i.e., $g(\dot{\gamma}(T(\lambda)), \dot{\lambda}(1)) < 0$. If there is more than one such ray, then we are
in a gravitational lensing situation, see Figure 8.8. (Here it goes without

saying that two rays are identified if one is a reparametrization of the other.) There might be a finite or infinite number of denumerable rays, or a whole continuum, e.g., a one-parameter family. In the latter case the observer might see an extended image, such as an arc or a ring, of the pointlike source γ.

(a) The (non-stationary) vacuum case

In the case of vacuum light propagation, $\mathcal{N} = \mathcal{N}^g$, the rays are the light-like geodesics of the spacetime metric g. We can then use Fermat's principle in the version of Theorem 7.3.2 to characterize the rays between q and γ. For the trial curves we have to consider all virtual rays, i.e., all light-like C^∞ curves $\lambda \colon [0,1] \longrightarrow \mathcal{M}$ with $\lambda(0) = q$, $\lambda(1) = \gamma(T(\lambda))$ and $g(\dot{\gamma}(T(\lambda)), \dot{\lambda}(1)) < 0$. By Theorem 7.3.2, such a trial curve is a ray if and only if it makes the arrival time functional T stationary; here the arrival time functional T is defined by the equation $\lambda(1) = \gamma(T(\lambda))$. If there are at least two stationary points λ_1 and λ_2 of the arrival time functional, with λ_2 not just a reparametrization of λ_1, then we are in a gravitational lensing situation.

This version of Fermat's principle has the advantage that it applies to time-dependent gravitational fields. E.g., it can be used to calculate the influence of a gravitational wave sweeping over a gravitational lensing situation. Calculations of this kind have been carried through by Kovner [74] and by Faraoni [42].

If there is a continuous one-parameter family of light rays connecting q and γ, then along any ray of this family the end-point must be conjugate to the initial point in the sense of Definition 5.6.3. For a proof it suffices to observe that a finite portion of the timlike curve γ cannot be contained in the vacuum light cone which is made up by the light-like geodesics issuing from q. In this sense, in a vacuum gravitational lensing situation all parts of an extended image, such as a ring or an arc, show the light source at the same age. This is not necessarily true in a medium.

(b) The (non-stationary) matter case

For light propagation in matter, $\mathcal{N} \neq \mathcal{N}^g$, we have to use Fermat's principle in the more general version of Theorem 7.3.1. If we want to allow for dispersive media, we have to choose a generalized observer field W in the sense of Definition 7.3.1 and we have to choose a frequency constant $\omega_o \in \mathbb{R}$. For the trial curves we have to consider all curves $\xi \in \mathfrak{M}(\mathcal{N}, q, \gamma, W, \omega_o)$, in the sense of Definition 7.3.2, further restricted by the additional assumption $g(\dot{\gamma}(T(\lambda)), \dot{\lambda}(1)) < 0$. By Theorem 7.3.1, such a trial curve ξ is a lifted ray if and only if it makes the generalized optical path length functional $\mathcal{F}$ stationary, provided that the regularity condition (7.12) is satisfied along ξ for one and, thus, for any Hamiltonian H of $\mathcal{N}$. In comparison to the vacuum case, two observations are to be emphasized. First, it is necessary to consider trial curves in $T^*\mathcal{M}$ rather than in $\mathcal{M}$. Second, the variational principle will give us only the light rays for a specific value of the frequency constant ω_o. Please note that ω_o fixes the frequency with which the respective light ray is emitted by γ and that ω_o is given in usual physical units only if γ is parametrized by proper time.

This variational principle can be applied to gravitational lensing in time-dependent gravitational fields and in time-dependent media. As an example, we consider a non-magnetized plasma, i.e., a ray-optical structure $\mathcal{N}$ given by a Hamiltonian of the form (8.42) on an arbitrary Lorentzian spacetime manifold $(\mathcal{M}, g)$ with an arbitrary spacetime function ω_p. The trial curves $\xi \in \mathfrak{M}(\mathcal{N}, q, \gamma, W, \omega_o)$ are characterized, in terms of their representations $(x(s), p(s))$ in a natural chart, by the equations (5.10), (5.11), and (7.10), i.e.,

$$g^{ab}(x(s))\, p_a(s)\, p_b(s) = -\omega_p(x(s))^2 \,, \tag{8.67}$$

$$\dot{x}^a(s) = k(s)\, g^{ab}(x(s))\, p_b(s) \,, \tag{8.68}$$

$$W^a(s)\, \dot{p}_a(s) = \tag{8.69}$$
$$-k(s)\, W^a(s) \left(\frac{1}{2} \frac{\partial g^{cd}(x(s))}{\partial x^a} p_a(s)\, p_b(s) + \omega_p(x(s)) \frac{\partial \omega_p(x(s))}{\partial x^a} \right) \,,$$

supplemented with the boundary conditions that $x^a(0)$ are the fixed coordinates of q, $x^a(1)$ are coordinates of a point on γ, and $W^a(1)\, p_a(1) = -\omega_o$. In addition, we have to restrict to curves with $g_{ab}\, W^a(1)\, \dot{x}^b(1) < 0$. If ω_p has no zeros, (8.67) and (8.68) imply that the projected curves $\lambda = \tau_{\mathcal{M}}^* \circ \xi$ are time-like for every $\xi \in \mathfrak{M}(\mathcal{M}, q, \gamma, W, \omega_o)$. Moreover, it is easy to check that the relation between ξ and λ is one-to-one. The projected trial curves λ are characterized, in terms of their coordinate representations $x(s)$, by the differential equation

$$W^a \left(\frac{\omega_p(x)\, g_{ab}(x)\, \dot{x}^b}{\sqrt{-g_{fh}(x)\, \dot{x}^f\, \dot{x}^h}} \right)^{\cdot} = \tag{8.70}$$
$$-W^a \left(\frac{1}{2} \frac{\partial g^{cd}(x)}{\partial x^a} \frac{\omega_p(x)\, g_{ce}(x)\, \dot{x}^e\, g_{db}(x)\, \dot{x}^b}{\sqrt{-g_{fh}(x)\, \dot{x}^f\, \dot{x}^h}} + \sqrt{-g_{fh}(x)\, \dot{x}^f\, \dot{x}^h}\, \frac{\partial \omega_p(x)}{\partial x^a} \right) \,,$$

supplemented with the boundary conditions that $x^a(0)$ are the coordinates of q, $x(1)$ are coordinates of a point on γ, and

$$\left(W^a\, g_{ab}(x)\, \dot{x}^b \frac{\omega_p(x)}{\sqrt{-g_{fh}(x)\, \dot{x}^f\, \dot{x}^h}} \right)(1) = -\omega_o \,. \tag{8.71}$$

(8.70) and (8.71) fix the pseudo-Euclidean angle between the (projected) trial curve and W. The generalized optical path length, which was introduced in

Definition 7.3.3 as a functional $\xi \longmapsto \mathcal{F}(\xi)$, reduces to a functional on the projected curves, $\lambda \longmapsto F(\lambda)$, given by

$$F(\lambda) = - \int_0^1 \frac{\omega_p(\lambda(s))}{\omega_o} \sqrt{-g\left(\dot\lambda(s), \dot\lambda(s)\right)} \, ds \; + \; T(\lambda) \, . \qquad (8.72)$$

Thus, the light rays emitted on γ with the frequency ω_o are the extremals of the functional (8.72) among all curves λ between q and γ whose coordinate representations $s \longmapsto x(s)$ satisfy (8.70) and (8.71). Please note that F reduces to the arrival time functional T in the limit $\omega_p \to 0$, but that (8.70) and (8.71) cannot be used in this limit since they contain undetermined expressions of the form $0/0$. For this reason, a somewhat inconvenient matching procedure must be used if regions with $\omega_p = 0$ and regions with $\omega_p \neq 0$ are to be treated in a unified setting.

(c) The stationary case

Now we want to consider the situation that $\mathcal{N}$ is a stationary ray-optical structure and that γ is an integral curve of the distinguished time-like vector field $W \in \mathcal{G}_\mathcal{N}$, i.e., that the light source is at rest with respect to this time-like vector field. Moreover, we shall assume that the assumptions of the reduction theorem (i.e., of Theorem 6.5.1) are satisfied. The gravitational lensing situation can then be described in terms of space rather than in terms of spacetime, viz., in terms of the reduced ray-optical structure. If the reduced ray-optical structure is strongly regular (which is true in virtually all situations of physical interest in which the preceding assumptions are valid), the Morse theory developed in Sect. 7.5 can be applied.

We want to illustrate the general features of this approach by way of example. To that end we consider, on a 4-dimensional Lorentzian spacetime manifold $(\mathcal{M}, g)$, a ray-optical structure $\mathcal{N}$ determined by a Hamiltonian of the form (8.42), i.e., a dispersion relation of the form

$$g^{ab}(x)\, p_a\, p_b + \omega_p(x)^2 = 0 \qquad (8.73)$$

which describes light propagation in a non-magnetized plasma. Here g^{ab} are the contravariant components of the spacetime metric and ω_p is the plasma frequency. The spacetime metric is supposed to describe a cosmological model with some local mass concentrations that act as "deflectors"; the light rays are supposed to be influenced by some plasma clouds, situated in regions where the function ω_p is different from zero.

We want to assume that $\mathcal{N}$ is stationary, i.e., that there is a time-like vector field W in the symmetry algebra $\mathcal{G}_\mathcal{N}$. This means that W must be a conformal Killing field of the spacetime metric g,

$$L_W\left(e^{-2f}g\right) = 0 \qquad (8.74)$$

where $f = \frac{1}{2}\ln\left(-g(W, W)\right)$, and that the rescaled plasma density must be constant along each integral curve of W,

$$L_W\left(e^{-f}\omega_p\right) = 0 \,. \tag{8.75}$$

These assumptions are satisfied, e.g., if there is an open subset $\mathcal{D}$ in $\mathcal{M}$, invariant under the flow of W, with the following properties. $(\mathcal{M}, g)$ is a Robertson-Walker spacetime without plasma ($\omega_p = 0$) on $\mathcal{M} \setminus \overline{\mathcal{D}}$, whereas it is a stationary spacetime with a stationary plasma ($L_W g = 0$ and $L_W \omega_p = 0$) on $\mathcal{D}$. $\mathcal{D}$ is to be interpreted as the region where the influence of the deflector mass and of the plasma cloud on the light rays is to be taken into account. Instead of a Robertson-Walker spacetime we could use any other conformally stationary cosmological background on $\mathcal{M} \setminus \overline{\mathcal{D}}$.

To apply the reduction theorem, we have to assume that there is a global timing function $t : \mathcal{M} \longrightarrow \mathbb{R}$ for W which gives us a global diffeomorphism $(\pi, t) : \mathcal{M} \longrightarrow \hat{\mathcal{M}} \times \mathbb{R}$, please recall Figure 6.3. To construct the reduced ray-optical structure according to Theorem 6.5.1, we choose a frequency constant $\omega_o > 0$. From (8.73) we read that rays with $p_a W^a = -\omega_o$ cannot leave the region

$$\mathcal{M}_{\omega_o} = \left\{ q \in \mathcal{M} \mid e^{-2f(q)}\omega_p(q)^2 < \omega_o^2 \right\}. \tag{8.76}$$

If we restrict to this region, all assumptions of Theorem 6.5.1 are satisfied and the reduction can be carried through, giving us a reduced ray-optical structure $\hat{\mathcal{N}}_{\omega_o}$ on the 3-dimensional space $\hat{\mathcal{M}}_{\omega_o} = \mathcal{M}_{\omega_o}/\sim$. In the vacuum case $\omega_p = 0$ we have, of course, $\mathcal{M}_{\omega_o} = \mathcal{M}$ for all $\omega_o > 0$, otherwise it might be necessary to excise some parts from spacetime where the plasma frequency is so large that rays with frequency constant ω_o cannot enter. However, if the function ω_p has spatially compact support we always have $\mathcal{M}_{\omega_o} = \mathcal{M}$ for sufficiently large ω_o.

With the results from Sect. 6.6 it is easy to find a Hamiltonian for the reduced ray-optical structure $\hat{\mathcal{N}}_{\omega_o}$. First we recall that, by (8.74), the space-time metric induces a positive definite metric $\hat{g}$ and a one-form $\hat{\phi}$ on $\hat{\mathcal{M}}$, according to (6.98). The one-form $\hat{\phi}$ vanishes if and only if W is orthogonal to the hypersurfaces $t = \mathrm{const}$. In coordinates with $x^4 = t$ and $\partial/\partial x^4 = W$ the spacetime metric takes the form (6.101). Moreover, (8.75) implies that there is a function $\hat{\omega}_p : \hat{\mathcal{M}} \longrightarrow \mathbb{R}$ such that

$$e^{-f}\omega_p = \pi^*\hat{\omega}_p \,. \tag{8.77}$$

Hence, in coordinates with $x^4 = t$ and $\partial/\partial x^4 = W$ the dispersion relation (8.73) is equivalent to

$$\hat{g}^{\mu\sigma}\left(p_\mu - p_4\hat{\phi}_\mu\right)\left(p_\sigma - p_4\hat{\phi}_\sigma\right) - p_4^2 + \hat{\omega}_p^2 = 0 \tag{8.78}$$

with greek indices running from 1 to 3. According to the general rules found in Sect. 6.5, the left-hand side of (8.78) gives us a Hamiltonian for the reduced ray-optical structure if p_4 is replaced with $-\omega_o$. Since we are always free to multiply the Hamiltonian with a non-zero function, this implies that $\hat{\mathcal{N}}_{\omega_o}$ is generated by the Hamiltonian

$$\hat{H} = \frac{1}{2} \left(\frac{\hat{g}^{\mu\sigma}(p_\mu + \omega_o\hat{\phi}_\mu)(p_\sigma + \omega_o\hat{\phi}_\sigma)}{\omega_o^2 - \hat{\omega}_p^2} - 1 \right). \tag{8.79}$$

To study gravitational lensing we fix two points $\hat{q}$ and $\hat{q}'$ in $\hat{\mathcal{M}}_{\omega_o}$ and we ask how many rays of $\hat{\mathcal{N}}_{\omega_o}$ go from $\hat{q}$ to $\hat{q}'$. It is easy to check that for the Hamiltonian (8.79) the map $\sigma_{\hat{H}} : \hat{\mathcal{N}}_{\omega_o} \times \mathbb{R} \longrightarrow T\hat{\mathcal{M}}_{\omega_o}$ is a global diffeomorphism onto its image, i.e., that $\hat{\mathcal{N}}_{\omega_o}$ is strongly hyperregular according to Definition 5.2.2. Thus, the Morse theory developed in Sect. 7.5 applies. For the Hamiltonian (8.79), the space of trial curves $\mathfrak{V}(\hat{H}, \hat{q}, \hat{q}')$ is equal to the set of all H^2 curves, defined on the interval $[0, 1]$, in $\hat{\mathcal{M}}_{\omega_o}$ from $\hat{q}$ to $\hat{q}'$ with

$$(\omega_o^2 - \hat{\omega}_p^2)\, \hat{g}_{\mu\sigma}\dot{x}^\mu \dot{x}^\sigma = \text{const.} \tag{8.80}$$

and the action functional is given by

$$\hat{A}_{\hat{q},\hat{q}'}(\hat{\lambda}) = \omega_o \int_0^1 \left(\sqrt{\left(1 - \frac{\hat{\omega}_p^2}{\omega_o^2}\right) \hat{g}_{\mu\sigma}\, \dot{x}^\mu\, \dot{x}^\sigma} - \hat{\phi}_\mu \dot{x}^\mu \right) ds \tag{8.81}$$

for each $\hat{\lambda} \in \mathfrak{V}_{\hat{q},\hat{q}'}$ with coordinate representation $x \in H^2([0, 1], \mathbb{R}^3)$. Please note that, up to the factor $\omega_o > 0$, the action functional (8.81) equals the optical path length (6.84) of the lifted ray $\hat{\xi}$ associated with the ray $\hat{\lambda}$. In the vacuum case $\hat{\omega}_p = 0$, the optical path length can be reinterpreted as a travel time according to Proposition 6.5.3.

According to Fermat's principle in the version of Theorem 7.5.1, the light rays from $\hat{q}$ to $\hat{q}'$ are the stationary points of the action functional (8.81) or, equivalently, of the optical path length functional. In the static (i.e., non-rotating) case we can choose the timing function in such a way that $\hat{\phi} = 0$. Then the optical path length functional is equal to the length functional of the frequency-dependent metric

$$\hat{g}_o = \left(1 - \frac{\hat{\omega}_p^2}{\omega_o^2}\right) \hat{g} . \tag{8.82}$$

Please note that in the rotating case the optical path length functional is not invariant under orientation-reversing reparameterizations. Hence, in that case a light ray from $\hat{q}$ to $\hat{q}'$ does not travel along the same path as a light ray from $\hat{q}'$ to $\hat{q}$.

Since, for the Hamiltonian (8.79), the matrix

$$\left(\frac{\partial^2 \hat{H}}{\partial p_\mu \partial p_\sigma} \right) = \frac{(\hat{g}^{\mu\sigma})}{\omega_o^2 - \omega_p^2} \tag{8.83}$$

is positive definite on $\hat{\mathcal{N}}_{\omega_o}$, the Morse index theorem in the version of Theorem 7.5.4 implies that along each ray the extended Morse index is equal to the number of conjugate points counted with multiplicity, see (7.56). In particular, a ray gives a strict local minimum of the optical path length functional if and only if it is free of conjugate points whereas it gives a saddle-point if there is a conjugate point in the interior. Since at each conjugate point neighboring light rays are crossing from one side to the other, an odd Morse index is associated with a side-reversed image in comparison to an even Morse index. This is observable for light sources surrounded by irregular structures, e.g., for quasars with jets or lobes.

For the vacuum rays it is known that the occurence of conjugate points gives rise to multiple imaging situation. Under certain assumptions on the causal and topological structure of spacetime, the converse is also true, i.e., in any multiple imaging situation at least one of the rays must contain a pair of conjugate points. For a general proof of these facts we refer to Perlick [111]. This is an interesting result since, in combination with Einstein's field equation, the existence of conjugate points along a vacuum light ray allows to estimate the matter density along the ray, see Padmanabhan and Subramanian [102]. The above-mentioned Morse index theorem might be useful for generalizing this result to the case of light rays in media, at least for stationary situations and for media which satisfy the positive-definiteness assumption of Theorem 7.5.4.

Finally we want to prove an *odd number theorem*, i.e., we want to show that, under certain reasonable assumptions, a transparent deflector always produces an odd number of images. To that end we generalize a differential-topological argument, first published by McKenzie [95], into our setting of stationary ray-optical structures. For the sake of comparison the reader is refered to Dyer and Roeder [33] who prove an odd number theorem for spherical deflectors, and to Burke [24] and Petters [113] where odd number theorems are given for thin deflectors and weak gravitational fields. An argument very similar to Burke's but under slightly more general assumptions was worked out by Lombardi [86]. A general discussion of odd number theorems can also be found in Schneider, Ehlers, and Falco [128].

The following argument applies to all situations in which the assumptions of the reduction theorem (i.e., of Theorem 6.5.1) are satisfied. As before, we fix two points $\hat{q}$ and $\hat{q}'$ in $\hat{\mathcal{M}}_{\omega_o}$ and we ask how many rays of $\hat{\mathcal{N}}_{\omega_o}$ go from $\hat{q}$ to $\hat{q}'$. We need the following three additional assumptions (see Figure 8.9).

(a) There is an open subset $\hat{\mathcal{B}}$ in $\hat{\mathcal{M}}_{\omega_o}$ with the following properties. $\hat{q} \in \hat{\mathcal{B}}$ and $\hat{\mathcal{B}}$ is contractible to $\hat{q}$, i.e., there is differentiable map $\Phi : [0,1] \times \hat{\mathcal{B}} \longrightarrow \hat{\mathcal{B}}$ with $\Phi(0,\hat{r}) = \hat{r}$ and $\Phi(1,\hat{r}) = \hat{q}$ for all $\hat{r} \in \hat{\mathcal{B}}$. The closure of $\hat{\mathcal{B}}$ is compact in $\hat{\mathcal{M}}_{\omega_o}$. The boundary $\hat{\mathcal{S}} = \partial\hat{\mathcal{B}}$ of $\hat{\mathcal{B}}$ is diffeomorphic to a 2-sphere and $\hat{q}' \in \hat{\mathcal{S}}$.

(b) Every ray of $\hat{\mathcal{N}}_{\omega_o}$ issuing from $\hat{q}$ intersects $\hat{\mathcal{S}}$ if sufficiently extended.

(c) Every vector in $\overset{\circ}{T}_{\hat{q}}\hat{\mathcal{M}}_{\omega_o}$ is the tangent vector of a ray of $\hat{\mathcal{N}}_{\omega_o}$, and this ray is unique up to extension and reparametrization.

In physical terms, conditions (a) and (b) prohibit non-transparent deflectors. Such a non-transparent deflector would to be modeled either as a hole in $\hat{\mathcal{M}}_{\omega_o}$,

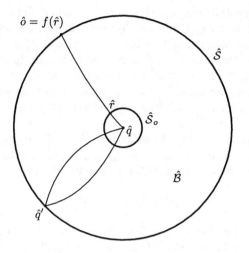

Fig. 8.9. Under the assumptions stated in the text, the rays issuing from $\hat{q}$ define a continuous map from the small sphere $\hat{S}_o$ to the big sphere $\hat{S}$. The degree of this map must be equal to 1 which proves that there is an odd number of rays from $\hat{q}$ to $\hat{q}'$.

thereby violating condition (a), or as a compact region in which some rays are trapped, thereby violating (b). Condition (c), roughly speaking, makes sure that in any spatial direction there is exactly one ray of $\hat{\mathcal{N}}_{\omega_o}$. This condition is satisfied, e.g., if $\mathcal{N}$ is the vacuum ray-optical structure. Please note that condition (c) could not hold if $\hat{\mathcal{N}}_{\omega_o}$ was not strongly regular. In the dispersive case, conditions (a), (b), and (c) have, of course, to be checked for each value of the frequency constant ω_o individually.

Under these assumptions, every ray issuing from $\hat{q}$ intersects an infinitesimally small sphere $\hat{S}_o$ around $\hat{q}$ in exactly one point $\hat{r}$, and it reaches the sphere $\hat{S}$ at some point $f(\hat{r})$. This defines a differentiable map $f\colon \hat{S}_o \longrightarrow \hat{S}$. We now fix a regular value of f, i.e., we fix a point $\hat{o} \in \hat{S}$ such that for all $\hat{r} \in \hat{S}_o$ with $f(\hat{r}) = \hat{o}$ the tangent map $T_{\hat{r}}f\colon T_{\hat{r}}\hat{S}_o \longrightarrow T_{f(\hat{r})}\hat{S}_o$ is a bijection. Please note that, according to the well known *Sard Theorem* (see, e.g., Abraham and Robbin [2], p. 37) almost all points in $\hat{S}$ are regular values of f. Clearly, $\hat{o}$ is a regular value of f if and only if $\hat{o}$ is not conjugate to $\hat{q}$ along any ray in $\hat{\mathcal{B}}$. With a regular value $\hat{o}$ chosen we define the *degree* of f as

$$\deg(f) = \sum_{f(\hat{r})=\hat{o}} \operatorname{sgn}(\hat{r}) \tag{8.84}$$

where $\operatorname{sgn}(\hat{r})$ is equal to $+1$ if the differential $T_{\hat{r}}f$ is orientation preserving and equal to -1 otherwise. Here we refer, of course, to the orientations of the spheres according to which $\hat{q}$ "lies to their inner sides". It is a standard theorem in differential topology that $\deg(f)$ is well-defined, i.e., independent of the choice of $\hat{o}$, see, e.g., Guillemin and Pollack [54] for a detailed discussion. Moreover, our assumption of $\hat{\mathcal{B}}$ being contractible to $\hat{q}$ gives an orientation preserving diffeomorphism from $\hat{\mathcal{S}}$ to $\hat{\mathcal{S}}_o$ and a smooth deformation of f into the identity, i.e., it implies that f is homotopic to the identity map. As it is well known that, for maps between compact manifolds without boundary, the degree is a homotopic invariant, the degree of f must be the degree of the identity, i.e., $\deg(f) = 1$.

We now consider the rays from $\hat{q}$ to $\hat{q}'$. We exclude the exceptional case that $\hat{q}'$ is conjugate to $\hat{q}$ along some ray, i.e, we assume that $\hat{q}'$ is a regular value of f. Then the definition of the degree implies that

$$\deg(f) = n_+ - n_- \tag{8.85}$$

where $n_\pm$ is the number of rays from $\hat{q}$ to $\hat{q}'$ in $\hat{\mathcal{B}}$ such that $\operatorname{sgn}(\hat{r}) = \pm 1$. Here $\hat{r}$ denotes the intersection of the ray with $\mathcal{S}_o$. Clearly, n_+ is the number of rays with an even number of conjugate points and n_- is the number of rays with an odd number of conjugate points. As the degree of f is equal to 1, (8.85) implies that $n_+ + n_- = 1 + 2n_-$, i.e., the number of rays from $\hat{q}$ to $\hat{q}'$ is odd.

For this argument stationarity was, of course, essential since otherwise there is no space $\hat{\mathcal{M}}_{\omega_o}$ in which it could be applied. Even for vacuum rays it is hard to see how a similar degree argument could give an odd number theorem in a spacetime setting, i.e., without assuming stationarity. (This problem was discussed in detail by Gottlieb [51].) For that reason it is important to know that McKenzie [95] was able to give another argument to prove that a transparent deflector produces an odd number of images. This was done for vacuum light rays in a globally hyperbolic spacetime, using Morse theoretical results of Uhlenbeck [144]. Unfortunately, it was necessary for McKenzie to impose some additional assumptions on the spacetime metric the physical meaning of which is obscure. Therefore it seems fair to say that in the non-stationary case a satisfactory odd number theorem is still missing, even for vacuum rays. Infinite dimensional Morse theory, as it was developed for vacuum rays between a point and a time-like curve in a Lorentzian manifold partially by Perlick [110] and, to a fuller extent, by Giannoni, Masiello, and Piccione [47] [48], could be a useful tool. In the non-stationary non-vacuum case, there are not even rudiments of a Morse theory for light rays between a point and a time-like curve. So there is still a lot to be done in the future.

References

1. Abraham R., Marsden J. (1978) Foundations of mechanics. Addison-Wesley, New York
2. Abraham R., Robbin J. (1967) Transversal mappings and flows. Benjamin, New York
3. Abramowicz M., Carter B., Lasota J. P. (1988) Optical reference geometry for stationary and static dynamics. Gen. Rel. Grav. **20**, 1173–1183
4. Anile A.M. (1976) Geometrical optics in general relativity: A study of the higher order corrections. J. Math. Phys. **17**, 576–584
5. Anile A. M., Pantano P. (1977) Geometrical optics in dispersive media. Phys. Lett. **61A**, 215–218
6. Anile A.M., Pantano P. (1979) Foundation of geometrical optics in general relativistic dispersive media. J. Math. Phys. **20**, 177–183
7. Arnold V.I. (1967) Characteristic class entering in quantization conditions. Funct. Anal. Appl. **1**, 1–13
8. Arnold V.I. (1978) Mathematical methods of classical mechanics. Springer, New York
9. Arnold V. I., Gusein-Zade S., Varchenko A. (1985) Singularities of differentiable maps. I. Birkhäuser, Boston
10. Asanov, G. (1985) Finsler geometry, relativity and gauge theories. Reidel, Dordrecht
11. Beem J. K., Ehrlich P.E., Easley K.L. (1996) Global Lorentzian geometry. Marcel Dekker, New York
12. Bel Ll., Martín J. (1994) Fermat's principle in general relativity. Gen. Rel. Grav. **26**, 567–585
13. Bertotti B. (1998) Doppler effect in a moving medium. Gen. Rel. Grav. **30**, 209–226
14. Bičák J., Hadrava P. (1975) General-relativistic radiative transfer theory in refractive and dispersive media. Astron. Astrophys. **44**, 389–399
15. Bishop R.C., Crittenden R.J. (1964) Geometry of manifolds. Academic Press, New York
16. Born M., Wolf E. (1960) Principles of optics. Pergamon, Oxford
17. Bressan A. (1978) Relativistic theories of materials. Springer, Berlin
18. Breuer R.A., Ehlers J. (1980) Propagation of high frequency waves through a magnetized plasma in curved space-time. I. Proc. Roy. Soc. London **A 370**, 389–406
19. Breuer R.A., Ehlers J. (1981) Propagation of high frequency waves through a magnetized plasma in curved space-time. II. Proc. Roy. Soc. London **A 374**, 65–86
20. Brill D. (1972) A simple derivation of the general redshift formula. In Farnsworth D., Fink J., Porter J., Thompson A. (Eds.) Methods of local and

212   References

     global differential geometry in general relativity. Lecture Notes in Physics **14**, Springer, New York, 45–47
21. Brill D. (1973) Observational contacts of general relativity. In Israel W. (Ed.) Relativity, Astrophysics and Cosmology. Proceedings of the Banff Summer School August 1972, Reidel, Dordrecht, 127-152
22. Brillouin L. (1960) Wave propagation and group velocity. Academic Press, New York
23. Bruns H. (1895) Das Eikonal. Hirzel, Leipzig
24. Burke W. A. (1981) Multiple gravitational imaging by distributed masses. Astrophys. J. **244**, L1
25. Carathéodory C. (1937) Geometrische Optik. Springer, Berlin
26. Chazarain J., Piriou A. (1982) Introduction to the theory of linear partial differential equations. North-Holland, Amsterdam
27. Chen K. H. (1961) Asymptotic theory of wave propagation in spatial and temporal dispersive inhomogeneous media. J. Math. Phys. **12**, 743–753
28. Chwolson O. (1924) Über eine mögliche Form fiktiver Doppelsterne. Astronomische Nachrichten **221**, 329
29. Conway A., Synge J. L. (1931) (Eds.) The mathematical papers of William Rowan Hamilton. Cunningham Memoir Nr.XIII, Cambridge Univ. Press, Cambridge
30. Dautcourt G. (1987) Spacetimes admitting a universal redshift function. Astronomische Nachrichten **308**, 293–298
31. Duistermaat J. J. (1974) Oscillatory integrals, Lagrange immersions and unfolding of singularities. Commun. Pure Appl. Math. **27**, 207–281
32. Dwivedi I., Kantowski R. (1972) On the possibility of observing first order corrections to geometrical optics in curved space-time. J. Math. Phys. **13**, 1941–1943
33. Dyer C., Roeder R. (1980) Possible multiple imaging by spherical galaxies. Astrophys. J. **238**, L67–L70
34. Eddington, A. S. (1920) Report on the relativity theory of gravitation. London, Physical Society
35. Eddington, A. S. (1920) Space, time, and gravitation. Cambridge University Press, Cambridge
36. Egorov Yu. V., Shubin M. A. (1992) Partial differential equations. I. Encyclopedia of Mathematical Sciences, vol. 30, Springer, Berlin
37. Ehlers J. (1961) Beiträge zur relativistischen Mechanik kontinuierlicher Medien. Akad. Wiss. Lit. (Mainz), Abh. Math. Kl. **1961(11)**, 791–837
38. Ehlers J. (1967) Zum Übergang von der Wellenoptik zur geometrischen Optik in der allgemeinen Relativitätstheorie. Z. Naturforsch. **22a**, 1328–1332
39. Einstein A. (1936) Lens-like action of a star by the deviation of light in the gravitational field. Science **84**, 506–507
40. Ellis G. F. R. (1971) Relativistic cosmology. In Sachs R. (Ed.) General relativity and cosmology. Enrico Fermi School, Course XLVII, Academic Press, New York
41. Etherington, I. M. H. (1933) On the definition of distance in general relativity. The Philos. Mag. and J. of Science (Ser. 7) **15**, 761–773
42. Faraoni V. (1992) Nonstationary gravitational lenses and the Fermat principle. Astrophys. J. **398**, 425–428
43. Frankel T. (1979) Gravitational curvature. Freeman, San Francisco
44. French A. (1968) Special relativity. MIT Course, Norton, New York
45. Friedrich H., Stewart J. (1983) Characteristic initial data and wavefront singularities. Proc. Roy. Soc. London **A 385**, 345–371
46. Giannoni F., Masiello A. (1996) On a Fermat principle in general relativity: A Morse theory for light rays. Gen. Rel. Grav. **28**, 855–897

47. Giannoni F., Masiello A., Piccione P. (1997) A variational theory for light rays in stably causal Lorentzian manifolds: Regularity and multiplicity results. Commun. Math. Phys. **187**, 375–415

48. Giannoni F., Masiello A., Piccione P. (1998) A Morse theory for light rays on stably causal Lorentzian manifolds. Ann. Inst. H. Poincaré, Physique Theoretique **69**, 359–412

49. Golubitsky M., Guillemin V. (1973) Stable mappings and their singularities. Springer, New York

50. Gordon W. (1923) Zur Lichtfortpflanzung nach der Relativitätstheorie. Annalen der Physik **72**, 421–456

51. Gottlieb D. (1994) A gravitational lens need not produce an odd number of images. J. Math. Phys. **35**, 5507–5510

52. Guckenheimer J.(1973) Catastrophes and partial differential equations. Ann. Inst. Fourier **23**, No. 2, 31–59

53. Guckenheimer J. (1974) Caustics and nondegenerate Hamiltonians. Topology **13**, 127–133

54. Guillemin V., Pollack S. (1974) Differential topology. Prentice-Hall, Eaglewood Cliffs, NJ

55. Guillemin V., Sternberg S. (1977) Geometric asymptotics. Amer. Math. Soc., Providence, Rhode Island

56. Harris S. (1992) Conformally stationary spacetimes. Class. Quantum Grav. **9**, 1823–1827

57. Hasse W., Kriele M., Perlick V. (1996) Caustics of wavefronts in general relativity. Class. Quantum Grav. **13**, 1161–1182; there will be an Erratum to this paper, rectifying the incorrect proof of Theorem 4.4

58. Hasse W., Perlick V. (1988) Geometrical and kinematical characterization of parallax-free world models. J. Math. Phys. **29**, 2064–2068

59. Hawking S., Ellis G. F. R. (1973) The large scale structure of space-time. Cambridge Univ. Press., Cambridge

60. Heintzmann H., Kundt W., Lasota J.P. (1975) Electrodynamics of a charge-separated plasma. Phys. Rev. **A 12**, 204–210

61. Heintzmann H., Schrüfer E. (1977) Lorentz covariant eikonal method in magnetohydrodynamics. I. The dispersion relation. Phys. Lett. **A 60**, 79–80

62. Helfer A. (1994) Conjugate points on spacelike geodesics or pseudo-self-adjoint Morse-Sturm-Liouville systems. Pacific J. Math. **164**, 321–350

63. Herzberger M. (1936) On the characteristic function of Hamilton, the eikonal of Bruns and their use in optics. J. Opt. Soc. Amer. **26**, 177–180

64. Higbie H. (1981) Gravitational lens. Amer. J. Phys. **49**, 652–655

65. Ives H., Stilwell G. (1938) Experimental study of the rate of a moving atomic clock J. Opt. Soc. Amer. **28**, 215–226

66. Jeffrey A., Kawahara T. (1982) Asymptotic methods in nonlinear wave theory. Pitman, Boston

67. Kaufmann A. (1962) Maxwell equations in nonuniformly moving media. Annals of Physics **18**, 264–273

68. Kawaguchi T., and Miron R. (1989) On the generalized Lagarange spaces with the metric $\gamma^{ij}(x) + (1/c^2)y^i y^j$. Tensor **48**, 53–63

69. Kawaguchi T., Miron R. (1989) A Lagrangian model for gravitation and electromagnetism. Tensor **48**, 153–168

70. Keller J. B., Lewis R. M., Seckler, B. D. (1956) Asymptotic solutions of some diffraction problems. Commun. Pure Appl. Math. **9**, 207–265

71. Kermack W. O., McCrea W. H., Whittacker, E. T. (1932) On properties of null geodesics and their application to the theory of radiation. Proc. Roy. Soc. Edinburgh **53**, 31–47

214   References

72. Klingenberg W. (1978) Lectures on closed geodesics. Springer, Berlin
73. Klingenberg W. (1983) Closed geodesics on Riemannian manifolds. Conference Board of the Mathematical Science Regional Conference Series in Mathematics No. 53, Amer. Math. Soc., Providence, Rhode Island
74. Kovner I. (1990) Fermat principle in gravitational fields. Astrophys. J. **351**, 114–120
75. Kravtsov Yu. A. (1968) The geometrical optics approximation in the general case of inhomogeneous and nonstationary media with frequency and spatial dispersion. Sov. Phys. JETP **55**, 1470–1476
76. Landau L. D., Lifshitz E.M. (1959) Course of theoretical physics. II: Theory of fields. Addison–Wesley, Reading, Massachusetts and Pergamon, London
77. Laue M. v. (1920) Theoretisches über neuere optische Beobachtungen zur Relativitätstheorie. Phys. Z. **21**, 659–662
78. Lebach D. E., Corey B. E., Shapiro I. I., Ratner M. I., Webber J. C., Rogers A. E. E., Davis J. L., Herring T. A. (1995) Measurement of the Solar gravitational deflection of radio waves using very-long-baseline interferometry. Phys. Rev. Lett. **75**, 1439–1442 (1995)
79. Lenard P. (1921) Über die Ablenkung eines Lichtstrahls von seiner geradlinigen Bewegung durch die Attraktion eines Weltkörpers, an welchem er nahe vorbeigeht; von J.Soldner, 1801. Annalen der Physik **65** , 593–604
80. Levi-Civita T. (1917) Statica Einsteiniana. Atti della reale accademia dei lincei, Seria quinta, Rendiconti, Classe di scienze fisiche, matematiche e naturali **26**, 458–470
81. Levi-Civita T. (1918) La teoria di Einstein e il principio di Fermat. Nuovo Cimento **16**, 105–114
82. Levi-Civita T. (1924) Fragen der klassischen und relativistischen Mechanik. Vier Vorträge aus dem Jahre 1921, Springer, Berlin
83. Levi-Civita T. (1927) The absolute differential calculus. Blackie, London
84. Lewis R. M. (1965) Asymptotic theory of wave propagation. Arch. Rat. Mech. Anal. **20**, 191–250
85. Lodge O. (1919) Gravitation and light. Nature **104**, 354
86. Lombardi M. (1998) An application of the topological degree to gravitational lenses. Modern Phys. Lett. **A 13**, 83–86
87. Low R. (1998) Stable singularities of wave-fronts in general relativity. J. Math. Phys. **39**, 3332–3335
88. Luneburg R. K. (1948) Propagation of electromagnetic waves. Lecture Notes, New York University
89. Luneburg R. K. (1964) Mathematical theory of optics. Mimeographed notes from 1949, University of California Press, Berkeley
90. Madore J. (1974) Faraday transport in curved space-time. Commun. Math. Phys. **38**, 103–110
91. Marx G. (1954) Das elektromagnetische Feld in bewegten anisotropen Medien. Acta Phys. Hung. **3**, 75–94
92. Mashhoon B. (1987) Wave propagation in a gravitational field. Phys. Lett. **A 122**, 299–304
93. Masiello A. (1994) Variational methods in Lorentzian geometry. Pitman Research Notes in Mathematics Series 309, Longman Scientific & Technical, Essex
94. Maslov V. P. (1972) Théorie des perturbations et méthodes asymptotiques. Dunod, Gauthier-Villars, Paris
95. McKenzie R. (1985) A gravitational lens produces an odd number of images. J. Math. Phys. **26**,1592–1596
96. Milnor J. (1963) Morse theory. Ann. Math. Studies Vol. 51, Princeton University Press, Princeton

97. Miron R., Kawaguchi T. (1991) Relativistic geometrical optics. Int. J. Theor. Phys. **30**, 1521–1543

98. Misner C., Thorne K., Wheeler J. (1973) Gravitation. Freeman, San Francisco

99. Nandor M., Helliwell T. (1996) Fermat's principle and multiple imaging by gravitational lenses. Amer. J. Phys. **64**, 45–49

100. Newcomb W. A. (1983) Generalized Fermat principle. Amer. J. Phys. **51**, 338–340

101. Nityananda R., Samuel J. (1992) Fermat's principle in general relativity. Phys. Rev. **D 45**, 3862–3864

102. Padmanabhan T., Subramanian K. (1988) The focusing equation, caustics and the condition for multiple imaging by thick gravitational lenses. Mon. Not. Roy. Astr. Soc. **233**, 265–284

103. Pais, A. (1982) Subtle is the Lord... Oxford University Press, Oxford

104. Palais R. (1963) Morse theory on Hilbert manifolds. Topology **2**, 299–340

105. Palais R., Smale S. (1964) A generalized Morse theory. Bull. Amer. Math. Soc. **70**, 165–172

106. Pellegrini G. N., Swift A. R. (1995) Maxwell's equations in a rotating medium. Is there a problem? Amer. J. Phys. **63**, 694–705

107. Perlick V. (1990) On redshift and parallaxes in general relativistic kinematical world models. J. Math. Phys. **31**, 1962–1971

108. Perlick V. (1990) On Fermat's principle in general relativity: I. The general case. Class. Quantum Grav. **7**, 1319–1331

109. Perlick V. (1990) On Fermat's principle in general relativity: II. The conformally stationary case. Class. Quantum Grav. **7**, 1849–1867

110. Perlick V. (1995) Infinite dimensional Morse theory and Fermat's principle in general relativity. I. J. Math. Phys. **36**, 6915–6928

111. Perlick V. (1996) Criteria for multiple imaging in Lorentzian manifolds. Class. Quantum Grav. **13**, 529–537

112. Perlick V., Piccione, P. (1998) A general-relativistic Fermat principle for extended light sources and extended receivers. Gen. Rel. Grav. **30**, 1461–1476

113. Petters A. O. (1992) Morse theory and gravitational microlensing. J. Math. Phys. **33**, 1915–1931

114. Petters A. O. (1993) Arnold's singularity theory and gravitational lensing. J. Math. Phys. **34**, 3555–3581

115. Petters A. O., Levine H., Wambsganss J. (1999) Singularity theory and gravitational lensing. To appear with Birkhäuser, Boston

116. Pham Mau Quan (1956) Projections des géodésiques de longeur nulle et rayons électromagnétiques dans un milieu en mouvement permanent. C. R. Acad. Sci. Paris **242**, 875-878

117. Pham Mau Quan (1957) Inductions éléctromagnétique en relativité générale et principe de Fermat. Arch. Rat. Mech. Anal. **1**, 54–80

118. Pham Mau Quan (1958) Sur le principe de Fermat. Enseignement Math.(2) **4**, 41–70

119. Pham Mau Quan (1962) Le principe de Fermat en relativité générale. In Proc. Royaumont Conf. (1959) Les théories relativistes de la gravitation. Editions du Centre Nationale de la Recherche Scientifique, Paris, p. 165

120. Plebański J. (1960) Electromagnetic waves in gravitational fields. Phys. Rev. **118**, 1396–1408

121. Preston T. (1912) Theory of light. Macmillan, London

122. Refsdal S., Surdej J. (1994) Gravitational lenses. Rep. Prog. Phys. **56**, 117–185

123. Rindler W. (1977) Essential relativity. Springer, New York

124. Rund H. (1959) The differential geometry of Finsler spaces. Springer, Berlin

125. Sachs R. K. (1961) Gravitational waves in general relativity. IV. The outgoing radiation condition. Proc. Roy. Soc. London A **264**, 309–338
126. Sachs R. K., Wu H. S. (1977) General relativity for mathematicians. Springer, New York
127. Schmutzer E. (1968) Relativistische Physik. Akademische Verlagsgesellschaft Geest & Portig, Leipzig
128. Schneider P., Ehlers J., Falco E. (1992) Gravitational lenses. Springer, Heidelberg–New York
129. Schrödinger E. (1956) Expanding universes. Cambridge Univ. Press, Cambridge
130. Schwartz J. T. (1969) Nonlinear functional analysis. Notes on Mathematics and its Applications, Gordon & Breach, New York–London–Paris
131. Smale S. (1964) Morse theory and a nonlinear generalization of the Dirichlet problem. Ann. Math. **80**, 382–396
132. Sommerfeld A., Runge I. (1911) Anwendungen der Vektorrechnung auf die Grundlagen der geometrischen Optik. Annalen der Physik **35**, 277–298
133. Stephani H. (1991) Allgemeine Relativitätstheorie. Deutscher Verlag der Wissenschaften, Berlin
134. Stix T. H. (1962) The theory of plasma waves. McGraw-Hill, New York
135. Straubel R. (1902) Über einen allgemeinen Satz der geometrischen Optik und einige Anwendungen. Physikalische Zeitschrift **4**, 114–117
136. Straumann N. (1984) General relativity and relativistic astrophysics. Springer, Berlin
137. Synge J. L. (1925) An alternative treatment of Fermat's principle for a stationary gravitational field. The Philosophical Magazine and Journal of Sciences (6.Ser.) **50**, 913–916
138. Synge J. L. (1937) Geometrical optics. Cambridge University Press, Cambridge
139. Synge J. L. (1937) Hamilton's characteristic function and Bruns's eikonal. J. Opt. Soc. Amer. **27**, 138–144
140. Synge J. L. (1954) Geometrical mechanics and de Broglie waves. Cambridge University Press, Cambridge
141. Synge J. L. (1956) Geometrical optics in moving dispersive media. Comm. Dublin Inst. Adv. Studies, Ser. A , No. 12
142. Synge J. L. (1960) Relativity. The general theory. North-Holland, Amsterdam
143. Thorpe J. A. (1979) Elementary topics in differential geometry. Springer, New York–Heidelberg–Berlin
144. Uhlenbeck K. (1975) A Morse theory for geodesics on a Lorentzian manifold. Topology **14**, 69–90
145. Wambsganss, J. (1998) Gravitational lensing in astronomy. http://www.livingreviews.org/Articles/Volume1/1998-12wamb/
146. Wald R. (1984) General relativity. Chicago University Press, Chicago
147. Walsh D., Carlswell R., Weyman R. (1979) 0957+561 A,B: twin quasistellar objects or gravitational lens? Nature **279**, 381–384
148. Weinstein A. (1971) Symplectic manifolds and their Lagrangian submanifolds. Advances in Mathematics **6**, 329–346
149. Weyl H. (1917) Zur Gravitationstheorie. Annalen der Physik **54**, 117–145
150. Woodhouse N. (1980) Geometric quantization. Clarendon, Oxford
151. Zheleznyakov V. V. (1996) Radiation in astrophysical plasmas. Kluwer Academic Publishers, Dordrecht
152. Zwicky F. (1937) Nebulae as gravitational lenses. Phys. Rev. **51**, 290

Index

aberration   183, 187
action functional   91, 149, 172
adapted coordinates   11
angular diameter distance   130
approximate solution
– of Maxwell equation   18, 36
approximate-plane-wave family   4, 14
arrival time functional   159, 164
asymptotic series solution
– of Maxwell equation   18, 24
asymptotic solution
– of Maxwell equation   4, 17, 36

Baumbach-Allen formula   198
biaxial crystal   31
bicharacteristic curve   32
birefringence   29, 58
branch
– of characteristic variety   27
bundle
– infinitesimal   123, 126

canonical chart   64
canonical equations   36, 64
canonical lift
– of vector field   84
canonical one-form   64
canonical two-form   64
causality
– of ray-optical structure   112, 115,
  123
caustic   5, 100
characteristic curve   32, 73
characteristic determinant   14
characteristic equation   21
characteristic function   62
characteristic matrix   14
characteristic variety   27, 67
characteristic vector field   73
classical solution of eikonal equation
  92
conic Lagrangian submanifold   98

conjugate momentum   26, 64
conjugate point   105, 165, 175, 178,
  203, 207
constant of motion   85, 132
constitutive equations   9
constraints   10, 43
contact manifold   73
corrected luminosity distance   130
cotangent bundle   63

deflection of light   195
dielectricity   3, 5, 9
dilation invariance   87, 116, 138
dimensional reduction   86, 135
dispersion   44, 87, 116
dispersion relation   44
– for linear medium   27
– for plasma   54
Doppler effect   183, 185
double refraction   29, 58
drag effect   183, 187

eikonal   15, 62
eikonal equation   4, 15, 92
– classical solution of   92
– for linear medium   19, 21, 24
– for plasma   53
– generalized solution of   95
– partial   27, 54
eikonal function   15
eikonal surface   16, 93
energy density   9, 39
energy flux   39
equivalence of Hamiltonians   29, 75
Euler equation   47
Euler vector field   88
evolution equations   43
– in linear medium   10, 12
– in plasma   50
excess function   176
excitation
– electric   8

– electromagnetic 8
– magnetic 8
expansion
– of light bundle 127
extraordinary ray 36
extraordinary wave 31

Fermat principle 62
– and gravitational lensing 203
– for arbitrary ray-optical structure
  156, 161
– for Kerr metric 195
– for Rindler model 192
– for Schwazschild metric 195
– for stationary ray-optical structure
  155, 165
– for vacuum ray-optical structure
  163
– in ordinary optics 149
fiber derivative 65
field strength
– electric 8
– electromagnetic 8
– magnetic 8
figuratrix 116, 188, 189
Finsler metric 28
Fizeau experiment 189
Fourier synthesis 17, 45
frequency 16, 114, 136, 185
Fresnel formula 189

gauge transformation 137
generalized observer field 157
generalized optical path length 159
generalized solution of eikonal equation
  95
generalized wave surface 99
geodesic 63
geometric optics approximation
– of Maxwell fields 24
gravitational lensing 199
group velocity 113, 114, 188, 189

H^r space 166
Hamilton equations 4, 36, 64
– vertical part 65
Hamilton-Jacobi equation 4, 27, 92
Hamiltonian 36, 64
– equivalence of 29, 75
– for isotropic ray-optical structure
  121
– for linear medium 27
– for plasma 53
– for ray-optical structure 68, 74

– for stationary ray-optical structure
  132
– hyperregular 65
– partial 27, 53
– regular 65
– transformation of 29, 75, 76
Hamiltonian vector field 64
Hessian 174
high frequency limit 16, 19, 52
Hilbert manifold 165
homogeneous background limit 51, 52
homogeneous Lagrangian submanifold
  98
Huyghens construction 153
hyperregularity
– of Hamiltonian 65
– of ray-optical structure 77

index
– Morse 174, 178, 207
– of refraction 30, 122, 124, 183, 190
indicatrix 116, 188, 189
inertial system 16
initial value problem 11
isotropic medium 10, 30
isotropic ray-optical structure 121
isotropic submanifold 93
isotropy subgroup 86

Jacobi class 101, 176
– lifted 101
Jacobi field 101, 124, 175
– lifted 101
JWKB method 4

Kerr spacetime 193

Lagrange bracket 93
Lagrangian submanifold 5, 93
– conic 98
– homogeneous 98
Landau gauge 49
lensing 199
Levi-Civita connection 6
Levi-Civita tensor 7
lifted Jacobi class 101
lifted Jacobi field 101
lifted ray 73
light bundle 123
light cone
– for ray-optical structure 79
light-like 63
Lorentz force 47

Lorentzian metric   63

Malus theorem   128
Maxwell equations   3, 7
- in linear medium   3, 4, 7
- in plasma   3, 46
medium
- accelerated   190
- biaxial   31
- dielectric   3, 7, 9, 69
- dispersive   5, 44, 87, 112, 116, 147,
  189
- isotropic   3, 10, 30, 62, 121, 147, 183,
  190
- linear   3, 7, 9, 69
- non-linear   45
- permeable   3, 9, 69
- uniaxial   30, 69
metric
- Lorentzian   63
- optical   30, 31, 36, 69
- spacetime   69
microwave links   186
Minkowski energy-momentum tensor
  39
momentum
- conjugate   26, 64
- of symmetry   86
Morse index   174, 178
Morse index theorem   178
- and gravitational lensing   207
Morse theory   165, 168
- and gravitational lensing   205
multiplicity
- of conjugate point   105, 175, 178
- of solution of eikonal equation   32

natural chart   64

observer field   113
- generalized   157
odd number theorem   208
optical metric   30, 31, 36, 69, 124, 146,
  183
optical path length   139
- as travel time   139, 145, 149
- generalized   159, 205
ordinary ray   36
ordinary wave   31
orientation
- for ray-optical structure   76, 81

partial eikonal equation   27
partial Hamiltonian   27

partial transport vector field   34
permeability   3, 5, 9
permeable   7
phase surface   16, 93
phase velocity   113, 114, 188, 189
plasma   3, 5, 46, 70, 205
- on Kerr spacetime   193
plasma frequency   55, 70, 72, 193, 205
plasma oscillation   55, 72
Poincaré half-space   193
polarization condition
- for linear medium   22
- for plasma   55, 57
Poynting vector   39
principal determinant   14
principal matrix   14
principle of stationary action   150, 154

ray   31, 32, 34, 36, 54, 73
- and energy flux   41
- extraordinary   36
- in vacuum   36, 111
- lifted   73
- ordinary   36
- oriented   81
- virtual   81
ray method   4, 17
ray optical structure
- symmetry of   82
ray velocity   114, 188, 189
ray-optical structure   67
- causal   112, 115, 123
- dilation invariant   87, 116, 123
- examples   69
- hyperregular   77
- isotropic   121, 123
- Lorentz invariant   120, 123
- on Lorentzian manifold   111
- orientable   75
- regular   77
- reversible   87
- stationary   131, 155
- strongly hyperregular   78, 168
- strongly regular   78, 140
- vacuum   111
reciprocity theorem   129
redshift   116, 131, 186
- of vacuum rays   118
redshift potential   119, 132
- for Rindler observers   192
reduction   86
- and gravitational lensing   206
- of Kerr spacetime   195

– of Rindler model   192
– of stationary ray-optical structure
    134
– of stationary vacuum ray-optical
    structure   141
reduction theorem   135
reference system   8
regularity
– of Hamiltonian   65
– of ray-optical structure   76
reversibility   87
Rindler model   190
Robertson-Walker spacetime   206
rotation
– of light bundle   127

Sachs bein   124
Schwarzschild metric   194
shear
– of light bundle   127
sine condition   131
Sobolev space   166
space-like   63
stationarity
– of ray-optical structure   131
strong hyperregularity   78
strong regularity   78
structure group   86
superposition   17
symmetry algebra   85
symmetry group   83
symmetry of ray optical structure   82
symplectic geometry   61
symplectic manifold   64

tangent bundle   63
time orientability   111
time-like   63
timing function   133
transport equation
– for linear medium   19, 22, 33, 35
– for plasma   57
transport vector field
– for linear medium   32, 34
– for plasma   54
– partial   34, 54
travel time   145

uniaxial crystal   30, 69

vacuum ray   36, 111
vacuum ray-optical structure   111
variational principle   149
variational vector field   150
vector of normal slowness   115
velocity coordinates   64
vertical part of Hamilton equations
    65
virtual ray   81

wave
– extraordinary   31
– ordinary   31
wave amplitude   19
wave covector   17, 114
wave surface   4, 16, 19, 92, 99, 153
– generalized   99
– lifted   99, 153
Weierstrass excess function   176

Druck: Strauss Offsetdruck, Mörlenbach
Verarbeitung: Schäffer, Grünstadt

Lecture Notes in Physics

For information about Vols. 1–504
please contact your bookseller or Springer-Verlag

Vol. 505: D. Benest, C. Froeschlé (Eds.), Impacts on Earth. XVII, 223 pages. 1998.

Vol. 506: D. Breitschwerdt, M. J. Freyberg, J. Trümper (Eds.), The Local Bubble and Beyond. Proceedings, 1997. XXVIII, 603 pages. 1998.

Vol. 507: J. C. Vial, K. Bocchialini, P. Boumier (Eds.), Space Solar Physics. Proceedings, 1997. XIII, 296 pages. 1998.

Vol. 508: H. Meyer-Ortmanns, A. Klümper (Eds.), Field Theoretical Tools for Polymer and Particle Physics. XVI, 258 pages. 1998.

Vol. 509: J. Wess, V. P. Akulov (Eds.), Supersymmetry and Quantum Field Theory. Proceedings, 1997. XV, 405 pages. 1998.

Vol. 510: J. Navarro, A. Polls (Eds.), Microscopic Quantum Many-Body Theories and Their Applications. Proceedings, 1997. XIII, 379 pages. 1998.

Vol. 511: S. Benkadda, G. M. Zaslavsky (Eds.), Chaos, Kinetics and Nonlinear Dynamics in Fluids and Plasmas. Proceedings, 1997. VIII, 438 pages. 1998.

Vol. 512: H. Gausterer, C. Lang (Eds.), Computing Particle Properties. Proceedings, 1997. VII, 335 pages. 1998.

Vol. 513: A. Bernstein, D. Drechsel, T. Walcher (Eds.), Chiral Dynamics: Theory and Experiment. Proceedings, 1997. IX, 394 pages. 1998.

Vol. 514: F. W. Hehl, C. Kiefer, R. J. K. Metzler, Black Holes: Theory and Observation. Proceedings, 1997. XV, 519 pages. 1998.

Vol. 515: C.-H. Bruneau (Ed.), Sixteenth International Conference on Numerical Methods in Fluid Dynamics. Proceedings. XV, 568 pages. 1998.

Vol. 516: J. Cleymans, H. B. Geyer, F. G. Scholtz (Eds.), Hadrons in Dense Matter and Hadrosynthesis. Proceedings, 1998. XII, 253 pages. 1999.

Vol. 517: Ph. Blanchard, A. Jadczyk (Eds.), Quantum Future. Proceedings, 1997. X, 244 pages. 1999.

Vol. 518: P. G. L. Leach, S. E. Bouquet, J.-L. Rouet, E. Fijalkow (Eds.), Dynamical Systems, Plasmas and Gravitation. Proceedings, 1997. XII, 397 pages. 1999.

Vol. 519: R. Kutner, A. Pękalski, K. Sznajd-Weron (Eds.), Anomalous Diffusion. From Basics to Applications. Proceedings, 1998. XVIII, 378 pages. 1999.

Vol. 520: J. A. van Paradijs, J. A. M. Bleeker (Eds.), X-Ray Spectroscopy in Astrophysics. EADN School X. Proceedings, 1997. XV, 530 pages. 1999.

Vol. 521: L. Mathelitsch, W. Plessas (Eds.), Broken Symmetries. Proceedings, 1998. VII, 299 pages. 1999.

Vol. 522: J. W. Clark, T. Lindenau, M. L. Ristig (Eds.), Scientific Applications of Neural Nets. Proceedings, 1998. XIII, 288 pages. 1999.

Vol. 523: B. Wolf, O. Stahl, A. W. Fullerton (Eds.), Variable and Non-spherical Stellar Winds in Luminous Hot Stars. Proceedings, 1998. XX, 424 pages. 1999.

Vol. 524: J. Wess, E. A. Ivanov (Eds.), Supersymmetries and Quantum Symmetries. Proceedings, 1997. XX, 442 pages. 1999.

Vol. 525: A. Ceresole, C. Kounnas, D. Lüst, S. Theisen (Eds.), Quantum Aspects of Gauge Theories, Supersymmetry and Unification. Proceedings, 1998. X, 511 pages. 1999.

Vol. 526: H.-P. Breuer, F. Petruccione (Eds.), Open Systems and Measurement in Relativistic Quantum Theory. Proceedings, 1998. VIII, 240 pages. 1999.

Vol. 527: D. Reguera, J. M. G. Vilar, J. M. Rubí (Eds.), Statistical Mechanics of Biocomplexity. Proceedings, 1998. XI, 318 pages. 1999.

Vol. 528: I. Peschel, X. Wang, M. Kaulke, K. Hallberg (Eds.), Density-Matrix Renormalization. Proceedings, 1998. XVI, 355 pages. 1999.

Vol. 529: S. Biringen, H. Örs, A. Tezel, J.H. Ferziger (Eds.), Industrial and Environmental Applications of Direct and Large-Eddy Simulation. Proceedings, 1998. XVI, 301 pages. 1999.

Vol. 530: H.-J. Röser, K. Meisenheimer (Eds.), The Radio Galaxy Messier 87. Proceedings, 1997. XIII, 342 pages. 1999.

Vol. 531: H. Benisty, J.-M. Gérard, R. Houdré, J. Rarity, C. Weisbuch (Eds.), Confined Photon Systems. Proceedings, 1998. X, 496 pages. 1999.

Vol. 532: S. C. Müller, J. Parisi, W. Zimmermann (Eds.), Transport and Structure. Their Competitive Roles in Biophysics and Chemistry. XII, 400 pages. 1999.

Vol. 533: K. Hutter, Y. Wang, H. Beer (Eds.), Advances in Cold-Region Thermal Engineering and Sciences. Proceedings, 1999. XIV, 608 pages. 1999.

Vol. 534: F. Moreno, F. González (Eds.), Light Scattering from Microstructures. Proceedings, 1998. XII, 300 pages. 2000

Vol. 536: T. Passot, P.-L. Sulem (Eds.), Nonlinear MHD Waves and Turbulence. Proceedings, 1998. X, 385 pages. 1999.

Vol. 537: S. Cotsakis, G. W. Gibbons (Eds.), Mathematical and Quantum Aspects of Relativity and Cosmology. Proceedings, 1998. XII, 251 pages. 1999.

Vol. 538: Ph. Blanchard, D. Giulini, E. Joos, C. Kiefer, I.-O. Stamatescu (Eds.), Decoherence: Theoretical, Experimental, and Conceptual Problems. Proceedings, 1998. XII, 345 pages. 2000.

Vol. 539: A. Borowiec, W. Cegła, B. Jancewicz, W. Karwowski (Eds.), Theoretical Physics. Fin de Siècle. Proceedings, 1998. XX, 319 pages. 2000.

Vol. 540: B. G. Schmidt (Ed.), Einstein's Field Equations and Their Physical Implications. Selected Essays. 1999. XIII, 429 pages. 2000

Vol. 541: J. Kowalski-Glikman (Ed.), Towards Quantum Gravity. Proceedings, 1999. XII, 376 pages. 2000.

Monographs

For information about Vols. 1–17
please contact your bookseller or Springer-Verlag

Vol. m 18: H. Carmichael, An Open Systems Approach to Quantum Optics. X, 179 pages. 1993.

Vol. m 19: S. D. Bogan, M. K. Hinders, Interface Effects in Elastic Wave Scattering. XII, 182 pages. 1994.

Vol. m 20: E. Abdalla, M. C. B. Abdalla, D. Dalmazi, A. Zadra, 2D-Gravity in Non-Critical Strings. IX, 319 pages. 1994.

Vol. m 21: G. P. Berman, E. N. Bulgakov, D. D. Holm, Crossover-Time in Quantum Boson and Spin Systems. XI, 268 pages. 1994.

Vol. m 22: M.-O. Hongler, Chaotic and Stochastic Behaviour in Automatic Production Lines. V, 85 pages. 1994.

Vol. m 23: V. S. Viswanath, G. Müller, The Recursion Method. X, 259 pages. 1994.

Vol. m 24: A. Ern, V. Giovangigli, Multicomponent Transport Algorithms. XIV, 427 pages. 1994.

Vol. m 25: A. V. Bogdanov, G. V. Dubrovskiy, M. P. Krutikov, D. V. Kulginov, V. M. Strelchenya, Interaction of Gases with Surfaces. XIV, 132 pages. 1995.

Vol. m 26: M. Dineykhan, G. V. Efimov, G. Ganbold, S. N. Nedelko, Oscillator Representation in Quantum Physics. IX, 279 pages. 1995.

Vol. m 27: J. T. Ottesen, Infinite Dimensional Groups and Algebras in Quantum Physics. IX, 218 pages. 1995.

Vol. m 28: O. Piguet, S. P. Sorella, Algebraic Renormalization. IX, 134 pages. 1995.

Vol. m 29: C. Bendjaballah, Introduction to Photon Communication. VII, 193 pages. 1995.

Vol. m 30: A. J. Greer, W. J. Kossler, Low Magnetic Fields in Anisotropic Superconductors. VII, 161 pages. 1995.

Vol. m 31 (Corr. Second Printing): P. Busch, M. Grabowski, P.J. Lahti, Operational Quantum Physics. XII, 230 pages. 1997.

Vol. m 32: L. de Broglie, Diverses questions de mécanique et de thermodynamique classiques et relativistes. XII, 198 pages. 1995.

Vol. m 33: R. Alkofer, H. Reinhardt, Chiral Quark Dynamics. VIII, 115 pages. 1995.

Vol. m 34: R. Jost, Das Märchen vom Elfenbeinernen Turm. VIII, 286 pages. 1995.

Vol. m 35: E. Elizalde, Ten Physical Applications of Spectral Zeta Functions. XIV, 224 pages. 1995.

Vol. m 36: G. Dunne, Self-Dual Chern-Simons Theories. X, 217 pages. 1995.

Vol. m 37: S. Childress, A.D. Gilbert, Stretch, Twist, Fold: The Fast Dynamo. XI, 406 pages. 1995.

Vol. m 38: J. González, M. A. Martín-Delgado, G. Sierra, A. H. Vozmediano, Quantum Electron Liquids and High-Tc Superconductivity. X, 299 pages. 1995.

Vol. m 39: L. Pittner, Algebraic Foundations of Non-Com-mutative Differential Geometry and Quantum Groups. XII, 469 pages. 1996.

Vol. m 40: H.-J. Borchers, Translation Group and Particle Representations in Quantum Field Theory. VII, 131 pages. 1996.

Vol. m 41: B. K. Chakrabarti, A. Dutta, P. Sen, Quantum Ising Phases and Transitions in Transverse Ising Models. X, 204 pages. 1996.

Vol. m 42: P. Bouwknegt, J. McCarthy, K. Pilch, The W3 Algebra. Modules, Semi-infinite Cohomology and BV Algebras. XI, 204 pages. 1996.

Vol. m 43: M. Schottenloher, A Mathematical Introduction to Conformal Field Theory. VIII, 142 pages. 1997.

Vol. m 44: A. Bach, Indistinguishable Classical Particles. VIII, 157 pages. 1997.

Vol. m 45: M. Ferrari, V. T. Granik, A. Imam, J. C. Nadeau (Eds.), Advances in Doublet Mechanics. XVI, 214 pages. 1997.

Vol. m 46: M. Camenzind, Les noyaux actifs de galaxies. XVIII, 218 pages. 1997.

Vol. m 47: L. M. Zubov, Nonlinear Theory of Dislocations and Disclinations in Elastic Body. VI, 205 pages. 1997.

Vol. m 48: P. Kopietz, Bosonization of Interacting Fermions in Arbitrary Dimensions. XII, 259 pages. 1997.

Vol. m 49: M. Zak, J. B. Zbilut, R. E. Meyers, From Instability to Intelligence. Complexity and Predictability in Nonlinear Dynamics. XIV, 552 pages. 1997.

Vol. m 50: J. Ambjørn, M. Carfora, A. Marzuoli, The Geometry of Dynamical Triangulations. VI, 197 pages. 1997.

Vol. m 51: G. Landi, An Introduction to Noncommutative Spaces and Their Geometries. XI, 200 pages. 1997.

Vol. m 52: M. Hénon, Generating Families in the Restricted Three-Body Problem. XI, 278 pages. 1997.

Vol. m 53: M. Gad-el-Hak, A. Pollard, J.-P. Bonnet (Eds.), Flow Control. Fundamentals and Practices. XII, 527 pages. 1998.

Vol. m 54: Y. Suzuki, K. Varga, Stochastic Variational Approach to Quantum-Mechanical Few-Body Problems. XIV, 324 pages. 1998.

Vol. m 55: F. Busse, S. C. Müller, Evolution of Spontaneous Structures in Dissipative Continuous Systems. X, 559 pages. 1998.

Vol. m 56: R. Haussmann, Self-consistent Quantum Field Theory and Bosonization for Strongly Correlated Electron Systems. VIII, 173 pages. 1999.

Vol. m 57: G. Cicogna, G. Gaeta, Symmetry and Perturbation Theory in Nonlinear Dynamics. XI, 208 pages. 1999.

Vol. m 58: J. Daillant, A. Gibaud (Eds.), X-Ray and Neutron Reflectivity: Principles and Applications. XVIII, 331 pages. 1999.

Vol. m 59: M. Kriele, Spacetime. Foundations of General Relativity and Differential Geometry. XV, 432 pages. 1999.

Vol. m 60: J. T. Londergan, J. P. Carini, D. P. Murdock, Binding and Scattering in Two-Dimensional Systems. Applications to Quantum Wires, Waveguides and Photonic Crystals. X, 222 pages. 1999.

Vol. m 61: V. Perlick, Ray Optics, Fermat's Principle, and Applications to General Relativity. X, 220 pages. 2000.